# ACTIVATED SLUDGE PROCESS
# DESIGN AND CONTROL

WATER QUALITY MANAGEMENT LIBRARY
**VOLUME 1**
SECOND EDITION

# ACTIVATED SLUDGE PROCESS DESIGN AND CONTROL
## THEORY AND PRACTICE

EDITED BY

# W. Wesley Eckenfelder, D.Sc., P.E.
# Petr Grau, D.Sc.B.

LIBRARY EDITORS
W. W. ECKENFELDER, D.Sc., P.E.   J. F. MALINA, JR., Ph.D., P.E., D.E.E.   J. W. PATTERSON, Ph.D.

**CRC Press**
Taylor & Francis Group
Boca Raton London New York

CRC Press is an imprint of the
Taylor & Francis Group, an **informa** business

## Water Quality Management Library—Volume 1

First published 1998 by Technomic Publishing Company, Inc.

Published 2018 by CRC Press
Taylor & Francis Group
6000 Broken Sound Parkway NW, Suite 300
Boca Raton, FL 33487-2742

ISBN 13: 978-0-367-44778-6 (pbk)
ISBN 13: 978-1-56676-643-2 (hbk)
ISBN 13: 978-1-56676-660-9 (set)

**Visit the Taylor & Francis Web site at**
**http://www.taylorandfrancis.com**

**and the CRC Press Web site at**
**http://www.crcpress.com**

Main entry under title:
   Water Quality Management Library—Volume 1/Activated Sludge Process Design and
      Control: Theory and Practice, Second Edition

Library of Congress Catalog Card No. 98-85168

# Table of Contents

# Foreword

IN 1992 the United States National Committee of IAWQ (International Association on Water Quality) organized eight specialty courses offered in conjunction with the 1992 IAWQ Biennial Conference in Washington, D.C. Designed for the practicing engineer, the specialty courses covered critical topics in environmental quality management water pollution control, wastewater treatment, toxicity reduction, and residuals management. These courses were compiled in an eight-volume series as the Water Quality Management Library. Experts from the United States and many countries contributed their expertise and experience to the preparation of these state-of-the-art texts.

The success of this series prompted the editors to expand the series to include volumes on such timely topics as water reuse, non-point source control and aeration and oxygen transfer. Additional volumes are presently being considered. In addition to the new topics, in order to keep pace with this rapidly developing field, many of the original volumes have been updated to reflect current advances in the field. In addition to providing an up-to-date technical reference for the practicing engineer and scientist as in the first series, these volumes will provide a text for continuing education courses and workshops.

The Water Quality Management Library should provide a unique reference source for professional and education libraries.

W. WESLEY ECKENFELDER
JOSEPH F. MALINA, JR.
JAMES W. PATTERSON

# List of Contributors

PAUL F. COOPER, BTech, MSc, C Eng, MI Chem E, WRc Swindon, England

A. L. DOWNING, D.SC., FI.CHEM.E., Herts, England

W. WESLEY ECKENFELDER, D.SC., P.E., Eckenfelder, Inc., Nashville, Tennessee, U.S.A.

MARK HSU, PH.D., P.E., D.E.E., Greeley & Hansen, Chicago, Illinois, U.S.A.

JACK L. MUSTERMAN, PH.D., P.E., Musterman & Associates, Nashville, Tennessee, U.S.A.

JIRI WANNER, PH.D., Prague Institute of Chemical Technology, Prague, Czechoslovakia

THOMAS E. WILSON, PH.D., P.E., D.E.E., Greeley & Hansen, Chicago, Illinois, U.S.A.

# CHAPTER 1

# Process Theory: Biochemistry, Microbiology, Kinetics, and Activated Sludge Quality Control

## ACTIVATED SLUDGE PROCESS

THE activated sludge process has been used in practice already for more than eighty years. Its theory has been developing steadily, rather slowly, and in small steps.

Initially, only simple design criteria were developed and used by traditional sanitary engineers. Some of the criteria, such as those expressing specific loading rates—for instance activated sludge tank volumetric loading rate $B_V$ and sludge loading rate $B_X$ (or $F/M$)—are still quite widely used in practice to characterize process intensity.

The second phase of development can be described as the formal application of chemical reaction type kinetics. Kinetic equations, initially of the first, later of the second, and finally of a non-integer order, were used to describe rates of various phenomena in activated sludge systems [1,2,3].

The third phase of development can be characterized as the application of chemical and subsequently biological reactor engineering principles, techniques, and methods. Their use is accompanied by large matrixes of equations, stoichiometric constants, and other values. Such extensive mathematical expressions can be efficiently handled, manipulated, and solved only by computers. Consequently, the present phase of development of theory can be characterized as mathematical modelling [4,5].

The activated sludge process is a rather unique biotechnological process not having many similarities with the other processes utilized in practice

Jiri Wanner, Prague Institute of Chemical Technology, Prague, Czech Republic.

1

and frequently called fermentations. Typical features of the activated sludge process are:

(1) Multifarious substrate in terms of chemical composition and variety of particle size

(2) Multispecies biological culture, desirably growing in aggregates (flocs)

(3) Widely fluctuating flows, temperatures, and changes in the influent wastewater concentration and composition

(4) Ability to metabolize a vast number of organic compounds and to oxidize/reduce/polymerize etc. compounds containing nitrogen, phosphorus, sulphur, and others

(5) A variety of reactor configurations used, e. g., completely stirred tanks, plug—flow, sequencing batch, oxic, anoxic, and anaerobic selectors, etc.

The activated sludge process does exist in a large number of modifications and variations. In the last decade or so the most important development in practice can be observed in industrial wastewater treatment, nutrient removal (N and P), and bulking control technologies. This is also reflected by extension of appropriate theories.

**PROCESS THEORY**

**TERMINOLOGY AND CLASSIFICATION OF MICROBIAL SYSTEMS**

In modern activated sludge systems the following processes are important from the point of view of activated sludge population dynamics:

- organic carbon removal
- reactions and importance of inorganic carbon
- reactions of different forms of nitrogen
- processes of biological removal of phosphorus
- metabolism of sulphur compounds.

These processes form rather complex biochemical systems. To describe the systems properly, and to avoid misunderstanding, it is necessary to use the same terms for describing the same processes. Unfortunately, an international unified terminology does not exist. Sometimes significant differences can be found between the terminology used by microbiologists and by chemical engineers. Therefore the meaning of the terms used below should be clarified. The following definitions are based on terminology which is used in papers and reports of:

(1) IAWQ Task Group on Mathematical Modelling for Design and Operation of Biological Wastewater Treatment
(2) IAWQ Specialist Group on Nutrient Removal Processes from Wastewater
(3) IAWQ Specialist Group on Activated Sludge Population Dynamics

For the description of parameters in biological wastewater treatment the notation recommended by IAWQ will be used [6].

### Substrate

Substrate is defined as the source of energy for living cells. Principally, there are three kinds of substrates in aquatic systems:

- light—source of energy for phototrophic microorganisms
- inorganic compounds—The energy is generated from the oxidation of reduced forms of such elements as N, S, Fe and Mn. The microbes obtaining energy in this way are called chemolithotrophs (chemoautotrophs).
- organic compounds—The energy is produced by biochemical oxidation of organic carbon to carbon dioxide. The microorganisms performing this reaction are called chemoorganotrophs.

Both the inorganic and the organic compounds serve as electron donors in biochemical oxidation-reduction reactions.

### Carbon Source

Besides energy the living cells require a source of carbon for the synthesis of new biomass. Carbon can be metabolized in form of:

- inorganic carbon: Inorganic carbon (dissolved $CO_2$, carbonate, bicarbonate) can be converted to organic cell materials by microbes called autotrophs. If the autotrophs use light (solar energy) as the substrate, they are known as photoautotrophs. In the case of using inorganic compounds as the substrate, the microbes are called chemolithotrophs (chemoautotrophs).
- organic carbon: Most aquatic microorganisms utilize organic carbon for the synthesis of new cells. Thus for the chemoorganotrophs the organic carbon represents both the substrate and the source of carbon for synthesis purposes. These microbes are often termed (chemo)heterotrophs in the literature.

The use of organic carbon as substrate and carbon source is energetically more advantageous for the microorganisms than the chemolithotrophy. Thus in complex environment, such as wastewaters, where both organic and inorganic substrates are present, the organotrophs will generally grow faster and easier than the lithotrophs with a rather complicated mechanism of obtaining energy and forming new biomass. This is an important selective factor governing the composition of open mixed cultures like activated sludge.

From the point of view of a microbial cell, the substrates and the carbon sources can be divided into two major groups:

- external substrates and carbon sources: These materials are present in the medium in which the cells are cultivated. Before they can be utilized by the cells, they should be transported from the environment to the cells' interior. This transport requires energy and is carried out by transfer enzymes. The extracellular materials are either originally present in the cultivation medium (e.g., organic pollution in wastewaters) or synthesized by cells themselves (e.g., organic polymers). The metabolism when the cells use external substrates and carbon sources is termed *exogenous*.
- internal substrates and carbon sources: A typical situation of activated sludge process is when the external substrates are exhausted or present only in limiting concentrations. In such a case the cells do not cease their activities but switch the metabolism from exogenous to *endogenous*. During the endogenous metabolism the materials accumulated and/or stored in the cells are metabolized. When these intracellular substrates are exhausted, the so-called autooxidation of cellular protoplasm may occur.

In connection with the terms of *substrate* and *carbon source* the phenomenon of *mixotrophy* should be mentioned. The mixotrophic microorganisms can obtain energy by both organotrophic oxidation of organic compounds and chemolithotrophic oxidation of reduced inorganic compounds (especially sulphur compounds). The phenomenon of mixotrophy represents another selective factor in mixed cultures.

### Nutrients

In general microbiology the term nutrients means all chemical elements which are utilized as building materials for cell synthesis. In wastewater treatment terminology only two elements are referred to as nutrients, viz., nitrogen and phosphorus. The reason is that those two elements are considered to be the limiting nutrients for the growth of algae in eutrophic surface waters.

From the viewpoint of bacterial growth, elements as nitrogen, phosphorus and sulphur are termed macronutrients because of their significant content in microbial biomass (in activated sludge 6–8% N and 2% P related to dry matter). The elements such as Fe, Ca, Mg, K, Mo, Zn and Co can be classified as micronutrients. The mass fraction of these elements in biomass is negligible but they play an important role in cells metabolism as constituents of many enzymes. In cultivation conditions other than oxic (the definition see below), oxygen (chemically bound) can be considered a nutrient only. For assessing the amount of nutrients required in different systems, rather complicated approaches have to be used [7].

### Electron Acceptors and Cultivation Conditions

The energy for maintaining cells activities is generated from substrates by means of biochemical oxidation. The process of biooxidation means a transfer of electrons from oxidized substrates (donors) to the compounds termed acceptors which are reduced in this process.

The modern terminology of cultivation conditions is based on different electron acceptors which participate in individual biochemical reactions.

#### *Oxic Conditions*

In oxic conditions electrons from both organic and inorganic substrates are transferred to dissolved molecular oxygen which is reduced and bound in a molecule of water. The most important biochemical processes in oxic conditions are as follows:

- oxic oxidation of organic compounds (organotrophic metabolism)
- chemolithotrophic oxidation of ammonia and nitrite to nitrate (nitrification) and of reduced sulphur compounds to sulphate
- synthesis of intracellular polyphosphate polymers (this process is not connected with an electron transport between donors and acceptors but it requires energy released from organotrophic metabolism of intracellular organic compounds).

#### *Anoxic Conditions*

In anoxic conditions the electron acceptor oxygen is replaced by nitrogen at the oxidation state +5 (nitrate) or less frequently +3 (nitrite). By accepting five or three electrons, nitrogen is reduced to oxidation stage zero (molecular nitrogen, $N_2$).

Two principal reactions are possible under anoxic conditions:

- anoxic organotrophic oxidation of organic compounds (when

nitrate nitrogen as electron acceptor is reduced to nitrogen gas, the process is termed *denitrification*)
- anoxic chemolithotrophic oxidation of sulphide and elementary sulphur to sulphate

The synthesis of intracellular polyphosphates in anoxic conditions seems to be possible as the intracellular storage products can also be oxidized anoxically [44,45]. An experimental proof of the so-called anoxic phosphate uptake was given by Sorm et al. [46,47].

### Anaerobic Conditions

Anaerobic conditions are characterized by the absence of both molecular oxygen and nitrate/nitrite nitrogen. The electrons in biochemical oxidation-reduction reactions can be transferred from organic substrates to:

(1) Organic compounds (processes of fermentation, i.e., acido- and aceto-genesis; methanogenesis is not considered in the activated sludge process)

(2) Sulphate sulphur: In this dissimilatory sulphate reduction the sulphur is reduced by chemoorganotrophic microbes to elementary sulphur and to sulphide. The chemolithotrophic sulphate reduction is also possible but not very important in wastewater treatment.

The energy fixed in intracellular polyphosphates can be released under anaerobic conditions and used for the synthesis of organic intracellular storage materials from the products of anaerobic fermentation. In this reaction the oxidation stage of organic carbon remains principally the same, so that no electron acceptors are needed.

The above defined individual cultivation conditions can be distinguished easily and exactly by measuring the oxidation-reduction potential (ORP). The values of ORP in oxic conditions are positive (higher than +50 mV as standard ORP) while the ORP values in anaerobic conditions are negative (less than −50 mV). The ORP values around zero are typical of anoxic conditions.

### Medium

A cultivation medium represents the real environment in which cells grow. The medium contains substrate, carbon source, nutrients, electron acceptors and other components. Besides its chemical composition, the medium is also characterized by parameters like temperature, pH, ORP, osmotic pressure, etc. In biological wastewater treatment a mixture of treated wastewater and mixed culture of activated sludge can be referred to as the cultivation medium. This medium is often called *mixed liquor*.

## ACTIVATED SLUDGE MICROORGANISMS

Activated sludge should be understood as an artificial living ecosystem which is under continuous influence of abiotic and biotic factors. Because of the necessity to reach rather low effluent concentrations of organic compounds (carbon and energy source) and inorganic nutrients, the activated sludge is cultivated under limiting conditions. This fact leads to a strong competition between individual groups of microorganisms and only the best adapted microbes win in this competition. As the influencing factors are not constant in wastewater treatment plants, the winners of the competition, dominating the activated sludge population, can change. Thus the microbial composition of activated sludges is not constant but reflects all the effects to which the activated sludge system is exposed.

Another characteristic feature of the mixed culture called activated sludge is that individual microbial cells are not separated in the cultivation medium but they grow in aggregates. The ability of activated sludge microorganisms to flocculate is the most decisive property of activated sludges which enables us to use this mixed culture in large scale installations. The flocculated aggregates (flocs of activated sludge) exhibit technologically acceptable sedimentation rates and the gravity sedimentation is the only economically possible way of separating the biomass from treated wastewaters in full-scale wastewater treatment plants. On the other hand, the ability to flocculate can be considered as a primary selective pressure in the mixed culture of activated sludge. The microorganisms which can clump or form flocs or at least which can be fixed into the flocs have the following selective advantage over freely growing cells:

- The microorganisms in flocs are retained in the activated sludge system while the dispersed, freely growing cells are washed out.
- The growth in flocs protects most microbial cells against predators.

From the microbiological point of view the activated sludge microorganisms can be divided into two major groups:

(1) Decomposers, which are responsible for biochemical degradation of polluting substances in wastewaters: This group is represented chiefly by bacteria, fungi and colourless cyanophyta. Osmotrophic protozoa can also ingest soluble organic substances but at the low concentrations of these materials in wastewater they cannot efficiently compete with bacteria.

(2) Consumers, which utilize bacterial and other microbial cells as substrates: This group belongs to the so-called activated sludge microfauna and consists of phagotrophic protozoa and microscopic metazoa.

About 95 percent of the microbial population of activated sludges is formed by decomposers, especially by bacteria. This indicates that the role of microfauna in the removal of organic pollution and nutrients is only marginal.

### Oxic Organotrophic Microorganisms

The genera *Bacillus, Pseudomonas, Alcaligenes, Moraxella* and *Flavobacterium* are reported to be able to degrade complex organic substrates by exo- and endoenzymes. Bacteria specialized to specific substrates can be concentrated in activated sludges after proper adaptation of the mixed culture to a given wastewater. Such specialized bacteria are, for example, *Proteus* spp. (proteinaceous materials) or *Achromobacter* spp. (lipides, acids and alcohols).

Besides organotrophic bacteria, microscopic fungi (micromycetes) and colourless cyanophyta (cyanobacteria) are involved in oxic oxidation of organic substrates, especially of saccharidic and polysaccharidic compounds.

The organotrophic bacteria are both floc-forming and filamentous. *Zoogloea* spp. are considered typical floc-formers of activated sludges. About 30 different kinds of bacteria and cyanobacteria can grow in filaments in activated sludges (see below).

### Fermentative Bacteria

A fermentative conversion of organic compounds to volatile fatty acids is extremely important for the EBPR (Enhanced Biological Phosphate Removal) mechanism. In this connection, the presence of *Aeromonas punctata* and of genera *Pasteurella* and *Alcaligenes* is stressed in the literature (*inter alia* [48]).

### Anoxic Organotrophic Microorganisms (Denitrifiers)

The ability to use nitrate nitrogen as the final electron acceptor in biochemical reactions seems to be quite widespread among the activated sludge microorganisms. At least forty species of aquatic microorganisms can denitrify. The genera *Achromobacter, Alcaligenes, Arthrobacter, Bacillus, Flavobacterium, Hypomicrobium, Moraxella* and *Pseudomonas* are typical organotrophic denitrifiers of activated sludges [49–51]. Grabinska-Loniewska [52] estimates that 82–97 percent of microorganisms in activated sludges from systems with an anoxic zone are able to denitrify. Activated sludge fungi perform only the nitrate respiration, i.e., the first step of denitrification. The ability of filamentous microorganisms to utilize nitrate nitrogen as electron acceptor is crucial from the point of view of metabolic selection. In general, common filamentous microorganisms perform only the first step of denitrification, i.e., the reduction of nitrate to nitrite [22].

### Nitrifiers

The nitrifying bacteria are originally soil microbes. In aquatic environment the following genera of nitrifiers are reported in the literature [50,51,53]:

- *Nitrosomonas, Nitrosococcus, Nitrosospira* and *Nitrosocystis* for the oxidation of ammonia
- *Nitrobacter, Nitrospina* and *Nitrococcus* for the final oxidation of nitrite to nitrate.

In activated sludge process the chemolithotrophic bacteria *Nitrosomonas* and *Nitrobacter* are considered the main nitrifiers. The heterotrophic nitrification can be attributed mainly to micromycetes which are not important in activated sludge systems.

### Polyphosphate Accumulating Microorganisms

The ability to remove phosphate from wastewater by the EBPR mechanism is generally attributed to the genus *Acinetobacter* (*Acinetobacter calcoaceticus* var. lwoffi) which was identified in the isolates from EBPR activated sludges by means of fluorescent antibody techniques [54]. The other microorganisms which may contain polyphosphate granules and thus contribute to the EBPR belong to genera *Aeromonas, Arthrobacter, Klebsiella, Moraxella* and *Pseudomonas*. In the literature on activated sludge process the polyphosphate accumulating bacteria are generally referred to as poly-P bacteria. Some of poly-P bacteria are seemingly able also to denitrify, i.e., to take up phosphate not only under oxic but also under anoxic conditions [45–47]. The process of simultaneous denitrification and phosphorus removal is generally called anoxic phosphate uptake.

### Sulphur Bacteria

From numerous sulphur bacteria the colourless filamentous bacteria *Beggiatoa* and especially *Thiothrix* are important in activated sludge process because they can cause bulking problems. *Thiothrix* seems to be a mixotrophic organism.

### Microfauna of Activated Sludges

The microfauna of activated sludges consists of the following groups of protozoa and metazoa:

(1) Protozoa

- flagellates
- rhizopods
- ciliates: free swimming, crawling (grazing), attached (stalked)

(2) Metazoa

- nematodes
- rotifers
- higher microfauna

Flagellates are the smallest protozoa found in activated sludges. Flagellates and rhizopods are mostly osmotrophic feeders, which means that they utilize soluble substrates like bacteria. Because the specific surface of protozoan cells is lower than that of bacterial cells, the efficiency of gaining substrate is less for osmotrophic protozoa than for bacteria. Thus the osmotrophic flagellates and rhizopods are typical only for start-up periods of activated sludge process prior to the stabilization of bacterial population. Phagotrophic flagellates using particulate substrates (for instance bacterial cells) follow the osmotrophic protozoa but are soon replaced by better organized ciliates in the development of activated sludge. Thus ciliates are characteristic protozoa of activated sludges cultivated in steady state conditions. Free-swimming ciliates are connected with large numbers of dispersed bacteria in the bulk liquid. On the other hand, grazing and attached ciliates requiring less energy because of lower motility can be found in well flocculating activated sludges with limited numbers of dispersed bacteria. The attached ciliates can grow alone or in large colonies which can even form whole flocs. Metazoa occupy similar niche as attached ciliates, so that they are also found mostly in well established activated sludges with good flocculation.

Protozological literature attribute several important functions to microfauna in biological wastewater treatment, namely:

- increased flocculation of bacteria
- removal of dispersed bacteria by adsorbing them onto protozoan metabolites and by predation
- increase of the F/M ratio by reducing the number of bacteria as a result of predation
- direct uptake of substrates.

## BASIC METABOLIC PROCESSES IN ACTIVATED SLUDGE

### Organic Carbon Removal

The metabolism of organic carbon is the most important process in biological wastewater treatment. The removal of organic compounds from wastewater after its contact with activated sludge can be divided into several steps:

- enmeshment of particles into the structure of activated sludge flocs
- entrapment and adsorption of colloidal matters
- sorption of high molecular soluble organic compounds
- accumulation of single organic compounds with small molecules (readily biodegradable substrates) in bacterial cells [8]

The above processes are very fast and most of organic substances are removed from the bulk liquid shortly after the contact of wastewater with the activated sludge. Eikelboom [55] introduced the term *biosorption* to describe this simultaneous processes of initial, rapid organic substance removal. The extent of the substrate removed by biosorption depends on the ratio of the mass of substrate in contact with the mass of activated sludge. Eikelboom [55] named this ratio *floc loading:*

$$\text{floc loading} = \frac{(COD_i - COD_e) \cdot Qi}{Xr \cdot Qr} \qquad (1.1)$$

where

$COD_i$ = influent COD
$COD_e$ = effluent COD
$Q_i$ = influent flow
$Q_r$ = flow of return activated sludge
$X_r$ = concentration of return activated sludge

The biosorption also depends on the microbial composition of the particular activated sludge. The extent of biosorption is greater for activated sludges with good settling properties, i.e., activated sludges with minor fraction of filamentous population. This different capability of biosorption represents a selective factor which is used for practical control of filamentous bulking.

Most of organic substances sequestered by the activated sludge immediately after its contact with wastewaters (biosorption) are in a form which is not available for intracellular metabolism. Only a small fraction (15 to

30 mg/l) of filterable solids can be found in the size range between 0.1 and 1.0 µm and most of solids are colloidal and particulate in sewage and municipal wastewater [9]. The molecules of sorbed organic compounds are too large, so that they cannot permeate through the cell membranes. From the chemical point of view these substrates which remain sorbed in the flocs of activated sludge and do not penetrate directly into the individual cells are organic polymers. Polysaccharides, lipides and proteins are the most common high molecular weight compounds in wastewaters.

These polymers have to be degraded to structures with only few monomers in the chain or directly to monomers before the enzymatic transport through the cell membrane. In biochemistry this process is termed hydrolysis and is performed by specialized enzymes called hydrolases. The rate of hydrolysis depends on cultivation conditions. It has been found that the rate of hydrolysis under anoxic and anaerobic conditions is smaller than under oxic cultivation conditions [12]. The degradation of organic polymers to single monomers forms the first part of organic carbon metabolism depicted in Figure 1.1.

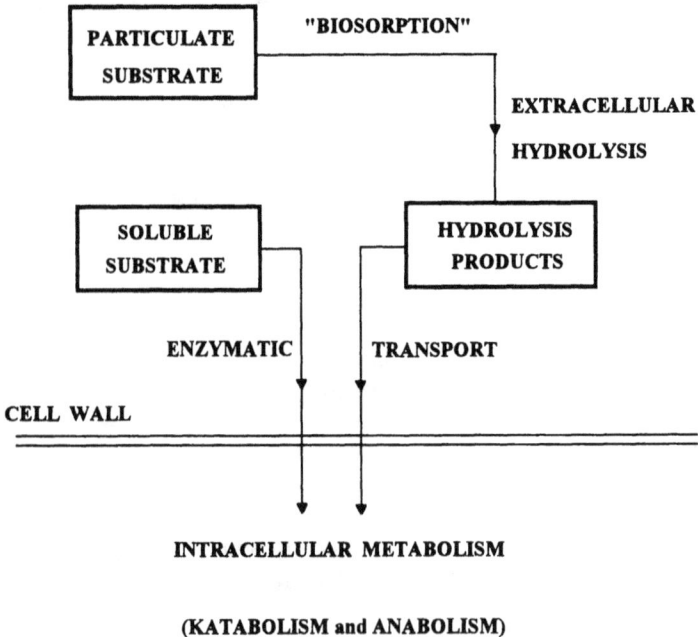

**Figure 1.1** Schematic description of organic pollution removal from wastewaters by activated sludge microorganisms.

After the extracellular hydrolysis of organic polymers, the fragments of polymers and single molecules are transferred to the cells where they are metabolized by cells' internal enzymatic apparatus. The metabolism consists of two simultaneous processes:

- katabolism, which generates energy from substrates for covering all energetic needs of cells
- anabolism (assimilation) leading to a synthesis of new biomass

The katabolism can be divided into three phases. The first phase finished the extracellular breakdown of complex molecules into simple low molecular weight compounds (monomers). The aim of the second phase of katabolism is to transform the numerous products of hydrolysis into few compounds entering the third, energy producing phase.

The final part of the katabolism in which the energy released from chemical bounds in the previous reactions is gained in a utilizable form is the respiration chain. The respiration chain is a sequence of enzymatic steps transferring the electrons and hydrogen protons to their final acceptors. Because of the transfer of electrons, this chain is also termed electron transport system (ETS).

Energy generated in the ETS is not dissipated but conserved for a future use in the cell in a form of adenosinetriphosphate (ATP). ATP is the most important and universal energy transferring compound in living cells.

The ability of the pair ATP + ADT to fix and release energy interconnects the mechanisms of energy production (katabolism) and energy consumption (anabolism). The energy in a form of ATP cannot be stored in the cells and should be converted to some energy "cans" like polysaccharides (both intra- and extracellular), lipides and possibly proteins. In general, the ability of microbial cells to produce such "energy cans" is rather limited in comparison with animal cells. However, if some microbial species exhibits this ability, it would represent a great metabolic advantage in the competition with other microorganisms in mixed cultures.

Anabolic processes of biosynthesis follow principally the same metabolic pathways but in the reverse direction. The energy required for biosynthesis is derived from the katabolism and transferred by means of pyridine nucleotide $NADP^+$ (nicotinamideadeninedinucleotidephosphate) produced in the third phase of katabolism.

From the point of view of microbial selection in mixed cultures, the anabolic pathways leading to synthesis of storage product are of crucial importance. The organic storage products can store chemical energy from ATP for a long period of time and the energy stored inside the cells can even be transferred from one cultivation condition (reactor) to other one(s). The reason is that the solubility of organic storage products in water is very low, so that they do not exhibit any significant osmotic pressure. The consumption of energy for retaining these materials inside the cells is thus negligible. Typical storage products are polysaccharides but contrary to plant cells, in which starch is the storage product, the microbial cells produce either glycogen (polysaccharide with the structure of molecule similar to amylopectin) or extracellular heteropolysaccharides. The polysaccharides are generally synthesized from glucose which is formed from products of the glyoxylic acid cycle. However, the organic storage product most frequently found in microbial cells is poly-β-hydroxybutyric acid (PHB). This polymer is synthesized by both aerobic and anaerobic organotrophic bacteria and by other microbes like cyanobacteria. A simplified scheme of the synthesis of PHB from acetic acid (anabolism) and its degradation (katabolism) is shown in Figure 1.2.

### Nitrogen Metabolism

#### Degradation of Organic Nitrogenous Compounds

Proteinaceous organic substances are metabolized during the metabolism of organic compounds. The biodegradable organic nitrogen from the amino group is converted to ammonia nitrogen by hydrolytic reactions, i.e., the oxidation stage of nitrogen does not change in this process. As ammonia results from the degradation of organic nitrogenous compounds, the process is also called ammonification.

#### Nitrification

Nitrification is a typical example of chemolithotrophic process when the microbes utilize energy generated from the oxidation of inorganic compounds. During this oxidation the oxidation stage of nitrogen gradually changes from $-3$ to $+5$ as indicated in Figure 1.3. The first step is the oxidation of ammonium nitrogen to hydroxyl amine catalyzed by monooxygenase. This reaction is endergonic and the energy is produced only in the following steps, i.e., in the oxidation of hydroxyl amine and of nitrite. In the final step, the electrons are transferred from nitrite nitrogen to molecular oxygen via cytochrom $a_1$. The process of nitrification can formally be described by the following chemical equations:

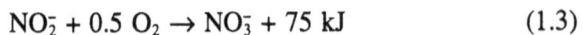

$$NH_4^+ + 1.5\ O_2 \rightarrow NO_2^- + H_2O + 2\ H^+ + 250\ kJ \qquad (1.2)$$

$$NO_2^- + 0.5\ O_2 \rightarrow NO_3^- + 75\ kJ \qquad (1.3)$$

ACETIC ACID

ACETYL CoA

ACETOACETYL CoA

ACETOACETIC ACID

BETA-HYDROXYBUTYRYL CoA

BETA-HYDROXYBUTYRIC ACID

POLY-BETA-HYDROXYBUTYRATE

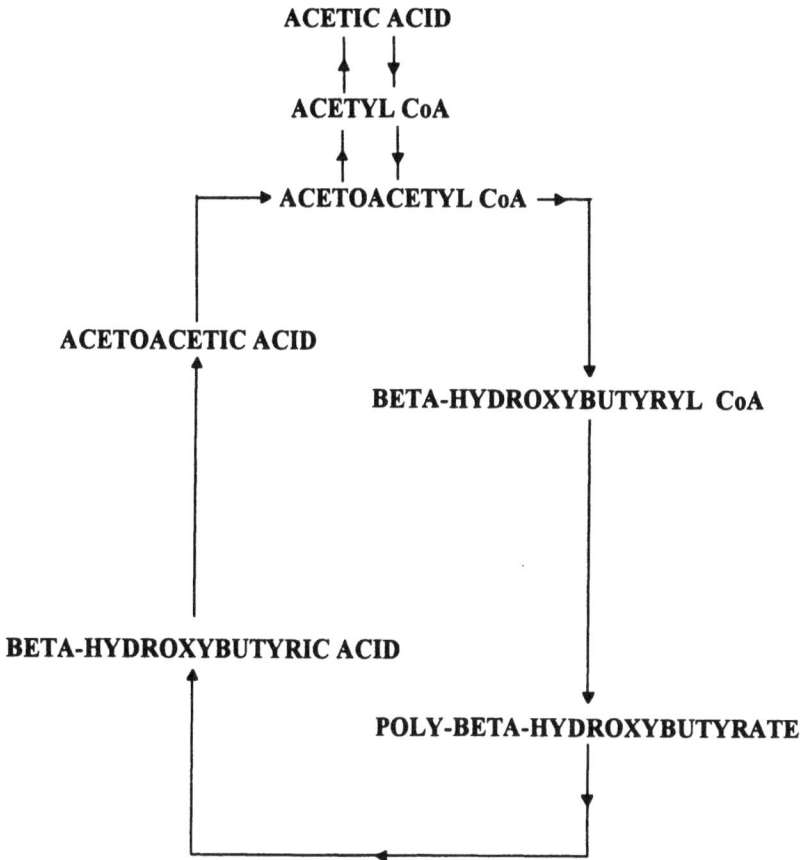

**Figure 1.2** Simplified scheme of PHB synthesis and degradation [48].

Stoichiometrically the oxygen requirement for the first reaction is 3.43 g/g ($O_2$/N) and for the oxidation of nitrite 1.14 g/g ($O_2$/N). The total specific consumption of oxygen for ammonia oxidation is 4.57 g/g ($O_2$/N). If we take into account that in simultaneous anabolic processes some electrons are transferred to carbon dioxide, which is thus fixed and converted to a new biomass, the specific consumption of oxygen for ammonia oxidation should be less than 4.57 g/g ($O_2$/N). The value of 4.2 g/g ($O_2$/N) is most commonly found in the literature [51].

The biochemical principles of chemolithotrophic nitrification in aquatic environment as well as the responsible microorganisms are well understood and described today [50,51]. For our purposes it can be summarized that:

- Nitrification is performed in two distinct processes by different kinds of chemolithotrophic microorganisms.

**OXIDATION**
   **STAGE**

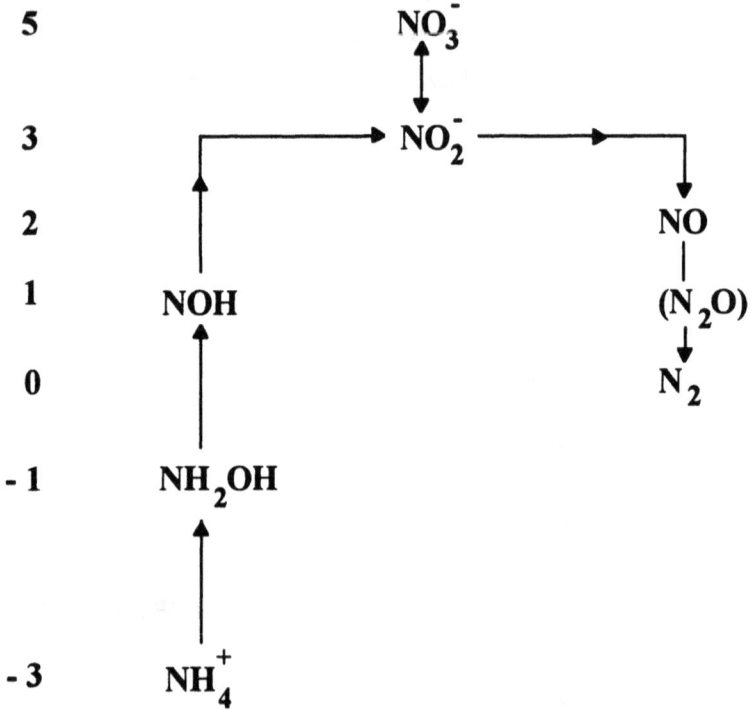

| STAGE | | |
|---|---|---|
| 5 | | $NO_3^-$ |
| 3 | | $NO_2^-$ |
| 2 | | NO |
| 1 | NOH | $(N_2O)$ |
| 0 | | $N_2$ |
| - 1 | $NH_2OH$ | |
| - 3 | $NH_4^+$ | |

**Figure 1.3** The changes in the nitrogen oxidation stage during nitrification and denitrification.

- Most energy is produced and most of oxygen is consumed in the first process.
- Chemolithotrophic nitrifying microorganisms are slow growing and susceptible to various inhibitory effects including the self-inhibition by substrate (un-ionized forms of nitrogen, i.e., $NH_3$ and $HNO_2$).
- Under normal conditions the rate of the oxidation of nitrite is higher than that of ammonia; thus nitrite does not accumulate in nitrifying activated sludge systems.
- Protons released during the oxidation of ammonia affect the buffer capacity; if the alkalinity is not high enough (at least 1.5–2 mmol/l at the end of nitrification), a significant decrease in pH may occur.
- The specific consumption of oxygen for nitrification is rather high, which should be reflected in the design of aeration systems

(dissolved oxygen concentration in the range of 1.5–2.0 mg/l is recommended).

- Although microbiological literature refers to nitrifying microorganisms as strict "aerobes," the process of nitrification can also be stabilized in activated sludges which are exposed for several hours to anoxic or anaerobic conditions; the enzymatic apparatus of nitrifying bacteria is probably more complex than it was thought in the past.

### Denitrification

As shown in Figure 1.3, the oxidation stage of nitrogen can also be biochemically reduced. When the reduced nitrogen is incorporated into newly synthesized biomass, the process is termed *assimilative nitrate reduction*. When nitrate nitrogen is reduced to elementary nitrogen and serves as an electron acceptor in the ETS, the process is known as *denitrification*. The process of denitrification can formally be described by the following equation:

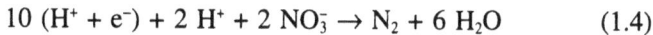

$$10 \ (H^+ + e^-) + 2 \ H^+ + 2 \ NO_3^- \rightarrow N_2 + 6 \ H_2O \qquad (1.4)$$

The donors of the activated hydrogen $(H^+ + e^-)$ in Equation (1.4) are predominantly external readily biodegradable substrates or intracellular accumulated substrates or organic storage products (PHB). When all these substrates are exhausted, slow endogenous anoxic respiration (denitrification) with $(H^+ + e^-)$ derived from cellular materials is also possible. As the rate of hydrolysis under anoxic conditions is slow, the availability of particulate substrates for denitrification is rather limited.

Denitrification can be considered as an equivalent alternative to oxic respiration from both metabolic and thermodynamic point of view. From the stoichiometry of oxic and anoxic respiration, it can be calculated that 1g $NO_3^-$–N equals 2.86g $O_2$ in oxidation-reduction reactions. However, the actual consumption of readily biodegradable substrates for a full denitrification of 1g $NO_3^-$–N expressed in COD units is estimated to be about 8.

From Equation (1.4) another important feature of denitrification can be seen. While by nitrification the protons are generated, during denitrification the protons are consumed. Thus in activated sludge systems with nitrification and denitrification the alkalinity can partially be replenished in anoxic conditions [15].

The process of denitrification is well understood from the point of view of its mechanism and kinetics [50,51,53]. Nevertheless, there is still some uncertainty concerning the inhibitory effect of one denitrification intermetabolite, nitric oxide NO, on the denitrifying microorganisms [56].

### Phosphorus Metabolism

Phosphorus in a form of orthophosphate originates in municipal waste-waters from the degradation of phosphorus containing organic substances and from the hydrolysis of polyphosphates commonly used in commercial detergents. In conventional activated sludge systems, phosphorus from wastewaters is utilized only for the synthesis of new biomass components. The phosphorus content in activated sludges from the conventional systems averages two percent in dry biomass. When the activated sludge is alternately exposed to anaerobic and oxic conditions, phosphorus is taken up by the cells in excess to synthesis purposes and the content of P in dry biomass may reach more than 10 percent. This phenomenon is called enhanced biological phosphorus removal (EBPR). The mechanism of EBPR represents an efficient way of transferring energy between anaerobic and oxic cultivation conditions. The transfer of energy by means of organic storage products (OSP) and polyphosphates (PP) is schematically depicted in Figure 1.4.

The phosphorus taken up in excess is stored in the cells in the form of polyphosphates counterbalanced with ions of $Ca^{2+}$, $Mg^{2+}$ and $K^+$. The

**CULTIVATION CONDITIONS**

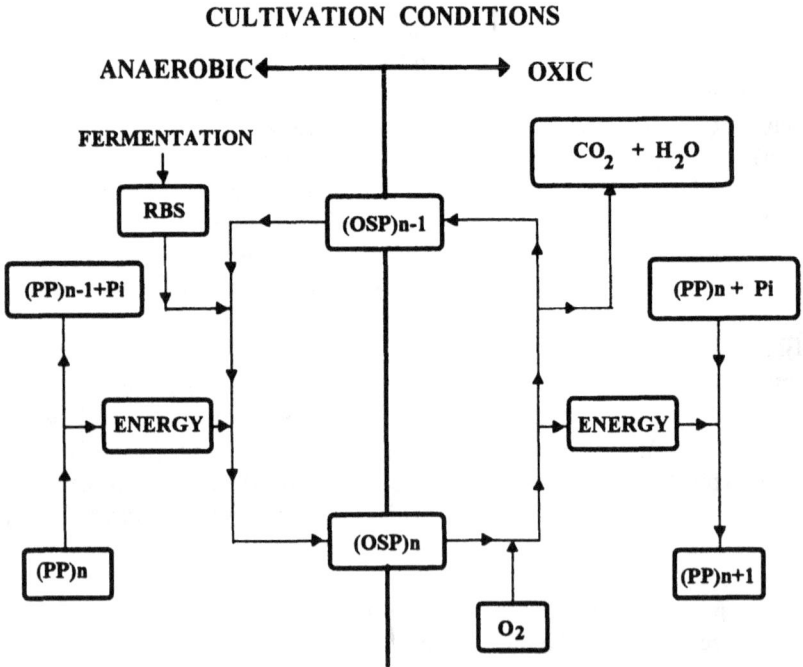

Figure 1.4 Transport of energy between anaerobic and oxic conditions.

polyphosphate together with lipidic and proteinaceous materials forms intracellular granules called volutine. The primary purpose of the stored polyphosphate in most bacteria is that it serves as a phosphorus source for periods of phosphorus starvation. In some bacterial strains polyphosphate acts also as an energy source similarly to the organic storage products (PHB, glycogen). The role of polyphosphate as intracellular energy storage material is decisive for the mechanism of the EBPR as well as for the metabolic selection of bacterial species in activated sludge systems with anaerobic zone.

Sulphur Metabolism

In aquatic environment sulphur can be present in the following stable forms:

- reduced inorganic sulphur (oxidation stage −2, sulphidic sulphur)
- organically bound sulphur (oxidation stage −2, mostly in amino acids)
- elemental sulphur $S^0$
- oxidized inorganic sulphur (oxidation stage +6, sulphate sulphur)

Accordingly, the metabolism of sulphur consists of three basic processes which are to a large extent analogous to the reactions of nitrogen metabolism:

- mineralization of organic sulphur compounds (analogy: ammonification)
- oxidation of reduced inorganic sulphur compounds (analogy: nitrification)
- reduction of sulphate sulphur (analogy: denitrification)

The above mentioned phenomenon of mixotrophy may result in bulking by *Thiothrix* [13]. The heterotrophic sulphate reduction in anaerobic zones of nutrient removal systems may also inhibit the mechanism of enhanced biological phosphate removal [13,14].

KINETICS OF BACTERIAL GROWTH

Besides the ability of different microbial species to utilize organic carbon under various cultivation conditions (metabolic selection) is the composition of mixed bacterial consortia like activated sludge governed by the rate of growth and substrate utilization (kinetic selection). In continuous cultivation systems the final biomass concentration is given by its increase by growth in balance with its losses by decay, wash-out and controlled biomass removal.

## Growth

Cell growth is understood as a synthesis of the biomass indicated by increased concentration of materials such as ATP. Increase in weight is not a good measure of growth since it also includes formation of storage materials. There may be an increase of weight without real growth. In practice, the real growth and the increase in weight are rarely differentiated due to technical problems and the cost of special chemical analyses. Concentration of biomass is expressed as sludge volatile fraction concentration (MLVSS). Such practice caused many problems and uncertainties.

The primary concern is growth rate limitation. In a simple Monod equation, the influence of a single limiting substrate can be described as follows:

$$\mu = \mu_{max} \left[ S_n/(K_S + S_n) \right] \qquad (1.5)$$

where

$\mu$ = the specific growth rate
$\mu_{max}$ = the maximum specific growth rate
$K_S$ = the saturation rate constant
$S_n$ = the rate-limiting substrate concentration

Specific growth rate $\mu$ is defined as

$$\mu = dX/X \, dt \qquad (1.6)$$

where

$X$ = concentration of biomass
$t$ = chronological (running) time

Equation (1.5) presumes that all other substrates (nutrients) are in excess. Actually, multiple limitation is more frequent than was supposed in the past. A multiple limitation was discovered first in the phototrophs [10]. It seems to be equally important in chemoautotrophs (e.g., nitrifiers) and in heterotrophs.

Such limitation can be described as follows:

$$\mu = \mu_{max} \left[ S_1/(K_{S1} + S_1) \right] \left[ S_2/(K_{S2} + S_2) \right] \ldots \left[ S_n/(K_{Sn} + S_n) \right] \qquad (1.7)$$

where the indexes 1 to $n$ refer to the individual nutrients.

According to Equation (1.7), any single nutrient can be growth-rate limiting and so can all nutrients multiplicatively.

Growing biomass utilizes all needed nutrients present in the medium. In a closed system (batch cultivation), concentrations of nutrients decrease as they are utilized by the culture. While the concentration of nutrients decreases, the growth rate slows according to Equation (1.7). Finally, when one of the nutrients is depleted, the growth ceases. That nutrient is limiting the biomass crop. It can be imagined that two or perhaps more nutrients would be depleted at the same time. Such a situation is, however, improbable. A probable situation is that only one of the nutrients limits the production of biomass.

Proportions between substrate utilized and biomass produced are as follows:

$$dX/dt = -Y_{obs}(dS/dt) \qquad (1.8)$$

or

$$X_t - X_0 = Y_{obs}(S_0 - S_t) \qquad (1.9)$$

where

$S$ = substrate concentration
$X$ = biomass concentration
$Y_{obs}$ = the observed yield coefficient
$S_0 - S_t$ = substrate utilized
$X_t - X_0$ = biomass produced

The indexes 0 and $t$ denote initial and actual values in time $t$, respectively.

If the biomass crop is nutrient-limited, then the smaller amount of biomass increment $\Delta X$ results in a smaller amount of substrate utilized, regardless of the reaction time. Additional substrate is utilized for maintenance energy only. In other words, under nutrient-limiting situations, even in reactors with sufficiently long detention times, the biomass concentration is limited (due to the limited biomass crop caused by the deficient nutrient), and consequently the substrate removal efficiency is decreased. In nutrient-limited activated sludge plants the effluent BOD concentration can be several times higher than in nutrient-balanced systems.

### Decay

Specific biomass decay rate $b$ is defined similarly as the specific growth rate $\mu$.

$$b = -dX/Xdt \qquad (1.10)$$

The net growth rate is consequently

$$\mu_{net} = \mu - b \tag{1.11}$$

When exogenous substrate is available, $\mu$ is generally greater than $b$ [except at very low concentration of substrate, when $\mu$ is, according to Equation (1.5) low]. Consequently, $\mu_{net}$ is a positive value and biomass concentration increases. When exogenous substrate is not available, $\mu = 0$ and $\mu_{net}$ has a negative value; biomass concentration decreases.

The traditional term *decay* actually includes a number of phenomena related to the endogenous phase. A cell that does not have access to the exogenous substrate first utilizes endogenous accumulated/stored substrate if available. After its depletion, other cell materials are utilized which are not absolutely necessary to restore metabolic and growth function when substrate is available again. Finally, the cell loses its vital functions, dies, and usually ruptures. Remaining cell materials are thus available to other living cells in the culture. This concept is reflected in the term *maintenance energy*, which has been used recently instead of the term *decay rate*.

### Growth and Decay

By definition

$$-dS/dt = \mu X/Y \tag{1.12}$$

$$dX/dt = (\mu - b)X \tag{1.13}$$

Combining Equations (1.12) and (1.13) yields

$$-dX/dS = Y_{obs} = [(\mu - b)/\mu]Y \tag{1.14}$$

where

$Y_{obs}$ = the observed coefficient of biomass production
$Y$ = the theoretical biomass yield coefficient

### Kinetic and Stoichiometric Characteristics

These terms come from chemical reactions terminology where they are considered constants. In biological processes they represent lump parameters that rarely, if ever, describe a single reaction. Due to this and other phenomena, they cannot be considered constants but only relatively stable coefficients or characteristics. They also vary in time even at strictly controlled (pseudo) steady state laboratory cultivations.

Kinetic behaviour of each bacterial species and each substrate (a reaction pair) can be described by a set of four coefficients, namely $\mu_{max}$, $K_s$, $b$, and $Y$. These four characteristics introduce minimum required information for a rather simplistic description of a reaction pair.

The true yield coefficient $Y$ does not vary much for various reaction pairs. In applications to municipal wastewater treatment its value is around 0.65 kg/kg (BOD, MLVSS). For single substrates the value of $Y$ can be estimated from the free energy of substrate oxidation. All substrates produce approximately the same amount of biomass per unit of utilized free energy of substrate oxidation, about 24 g/MJ. Similarly, the decay coefficient $b$ is a fairly constant value, $b = 0.15 \times 1.047^{T-20}$.

When applying these values to engineering calculations, care has to be taken to consider the influent wastewater suspended solids, the degradable, nondegradable, and mineral fractions of activated sludge, etc. Appropriate detailed equations and computing procedures are available in the literature [7].

In contrast, the value of $\mu_{max}$ and $K_S$ vary quite a lot from one reaction pair to another. It can be postulated that those two are the values in which the various reaction pairs really differ. Such differences offer a valuable tool to selection of desired species in activated sludge biomass. For instance, oxic selectors were developed and theoretically justified by recognizing various values of $\mu_{max}$ and $K_S$ at filamentous and non-filamentous microorganisms [11].

## ACTIVATED SLUDGE SEPARATION PROBLEMS

### SETTLING PROPERTIES OF ACTIVATED SLUDGE

Activated sludge should exhibit certain features which are of primary importance to the operation activated sludge system. Well settling activated sludge:

- settles fast, with zone settling velocities 3 m/h and more
- does not occupy an excessive volume after settling and thickening in secondary clarifier
- after sedimentation leaves a clear supernatant ($X = 15$ mg/l and less)
- does not rise within at least 2–3 h period after sedimentation

Only when the activated sludge exhibits these properties secondary clarifiers may play their two basic roles, viz.

- to separate activated sludge from wastewater
- to thicken the separated activated sludge so that the water content in excess and return activated sludge can be minimized

Maximum separation efficiency is dictated by the need to protect receiving waters against additional pollution from secondary effluents. The suspension escaping from secondary settling tanks is not formed by inert particles but by flocs (agglomerates) of living microorganisms. These microbes respire and consume oxygen dissolved in receiving waters. The biomass not separated in secondary settling tanks increases not only residual $BOD_5$ but also the concentration of nitrogen and phosphorus in secondary effluent. Thus the poor activated sludge separation may deteriorate the overall performance of activated sludge systems.

The second, thickening function of secondary settling tanks is crucial for the proper functioning of the activated sludge system itself. Good thickening properties of activated sludge enable us to keep required biomass concentration (and sludge age) in aeration basin and to reduce hydraulic loading of sludge handling facility.

In wastewater treatment practice, six major problems related to microbial biomass quality can be distinguished [19,62], as listed here.

### Dispersed Growth

The activated sludge microorganisms are dispersed freely in the cultivation medium as individual cells or small clumps with a diameter of up 10–20 $\mu$m. The sedimentation rate of individual cells or bacterial clumps is too low for gravity sedimentation; no zone settling occurs in secondary settling tanks. This has two impacts on activated sludge process:

(1) The separation efficiency of secondary settling tank is very low; final effluent is turbid.

(2) Because of poor separation efficiency, a significant amount of biomass escapes from the system. Therefore only low values of sludge age $\Theta_x$ can be maintained in the system. The system with dispersed growth resembles more a chemostat than a continuous cultivation system with biomass recycle.

The poor bioflocculation is caused by a low production of extracellular biopolymers (glycocalyx) creating a matrix of firm activated sludge flocs. One of the typical reasons for the dispersed growth is a very high organic loading of biomass when bacteria are not forced to produce glycocalyx. Another reason may be toxicity of treated wastewater.

## Unsettleable Microflocs

The outer symptom of this case of separation problems looks very similar to the dispersed growth at the first sight—the final effluent from secondary clarifier is not clear and contains many microparticles of escaping biomass. However, the nature of the problem is different, as can be seen under microscope. The unsettleable particles are of larger dimension (about 50–100 μm) than in the case of dispersed growth, the particles are roughly spherical and compact. These microflocs result from the disintegration of initially firm and sound flocs.

During the settling test the activated sludge is quickly divided into two parts. The larger flocs settle rapidly and when the sludge volume index is calculated on the basis of the volume of these larger flocs, its value is quite low (around 50 ml/g). But the supernatant in the cylinder is turbid and a substantial fraction of total biomass remains in these unsettleable particles.

The reasons for the disintegration of activated sludge flocs are:

- insufficient production of glycocalyx or its consumption by bacteria inside the flocs due to a low organic loading of biomass (typical of high sludge age systems, extended aeration)
- absolute absence of filamentous microorganisms which form a ''backbone'' of most flocs larger than 80–100 μm [18,19]
- disintegration by shearing effects, for instance by inappropriate mechanical aerators [16]

## Rising Sludge

When this problem is the case, the water surface in secondary clarifier is covered by patches (or in worse cases completely) of floating activated sludge. When the phenomenon is observed in a glass cylinder, two phases can be distinguished:

- First, the activated sludge settles rapidly and a rather compact bottom layer of settled sludge and a clear supernatant are formed.
- After a certain period of time (at elevated temperatures even less than 30 minutes, which may raise difficulties in the SVI test), a part or a whole volume of the settled and thickened sludge starts to float and move up to the water surface.

The floating material is full of gas bubbles. The nature of this phenomenon is endogenous denitrification which takes place in the settled and thickened layer of activated sludge. Thanks to a high biomass concentration, the dissolved oxygen from the previous aeration is quickly depleted and anoxic conditions can be thus established provided nitrification occurs in the system. The bubbles of nitrogen liberated during this endogenous denitrification act as sludge "carrier."

## Viscous Bulking

The rather broad term *viscous bulking* describes the consequences but not the reasons. The activated sludge contains an excessive amount of extracellular biopolymers which impart a slimeous, jelly-like consistency to the sludge. As the biopolymers are hydrophilic colloids, the activated sludge becomes highly water retentive. Such a "hydrous" activated sludge exhibits low settling and compaction velocities. The biopolymers are also natural surface active agents. When the viscous activated sludge is intensively aerated, a strong foaming may appear.

The production of biopolymers is characteristic of most floc forming microorganisms but under normal conditions (no toxic compounds, nutrient balanced growth) the amount of generated biopolymers is just enough for formation of firm flocs. On the other hand, zoogloeal bacteria produce always large amounts of biopolymers because the individual cells of *Zoogloea* are fixed in slimeous colonies. Thus some authors use the term "zoogloeal bulking" for describing the settling problems caused by an excessive presence of highly hydrated zoogloeal colonies in activated sludge.

## Filamentous Bulking

Filamentous bulking is a typical problem of poor activated sludge compaction which results in:

- low return and waste activated sludge concentrations
- difficulties to maintain the required activated sludge concentration in reaction basins
- poor sludge dewaterability
- hydraulic overloading of sludge handling facilities

Filamentous microorganisms interfere with the sedimentation and compaction of activated sludge flocs in two ways [19]:

(1) Some kinds of filamentous microorganisms grow preferably in flocs interiors. They produce flocs with a very diffuse open structure. These open flocs provide a lot of space for water inside the flocs, so that although the aggregation of individual flocs is not mechanically hindered by filaments protruding from the flocs, still too much water remains "captured" in the settled sludge.

(2) The second way how filaments can deteriorate the sedimentation and compaction of activated sludge flocs is much more common. Most of filamentous microorganisms observed in activated sludges protrude from rather compact and firm flocs into the bulk liquid. The filaments, which in low numbers form a backbone of firm flocs, in large number are able to prevent mechanically the compaction of individual flocs.

### Foaming Caused by Filamentous Microorganisms

The biological foaming by "foam-forming" filamentous microorganisms is a complex of physico-chemical and biochemical processes leading to the stabilization of a three-phase system: air–water–microbial cells. The stabilization of biological foams results from the following features of foam-forming filaments:

- Production of extracellular materials like lipides, lipopeptides, proteins and carbohydrates have the properties of surface active agents (biosurfactants).
- Contrary to other filaments and floc-formers, the cell walls of foam-forming microorganism are strongly hydrophobic.

The formation of stable foams in the aeration basins of activated sludge plants can create a wide range of operation problems:

(1) Aesthetic problems, slippery path along the aeration basins covered by escaping foam
(2) Floating biomass in secondary clarifier deteriorating the final effluent quality
(3) Accumulation of significant amount of biomass into the foam which is not available for treatment processes; loss of possibility to control activated sludge age

### EVALUATION OF SETTLING AND FOAMING PROPERTIES OF ACTIVATED SLUDGES

### Zone Settling Velocity

The conditions during settling in secondary settling tanks can be best simulated by the measurement of zone settling velocity. The measurement should be performed in transparent cylinders so that the speed of movement of the sludge-supernatant interface can be read. The higher and wider the cylinder is, the more realistic figures we obtain. The graph plotting sludge layer height vs. time is called settling curve which has usually three distinct settling phases: I—reflocculation; II—zone settling; and III—transition and compaction. The zone settling velocity is calculated from the slope of the curve in the phase of zone settling (II). ZSV depends on sludge concentration. This is treated by solids flux theory [17].

### Volume of Settled Sludge, Sludge Volume Index

In wastewater treatment practice, however, the measurement of so-called sludge volume index SVI is the most common way how to characterize activated sludge settleability. The standard sludge volume index is defined as follows:

$$SVI = \frac{V_{30}}{X} \tag{1.15}$$

where

SVI = sludge volume index [ml/g]
$V_{30}$ = volume of settled sludge after 30 minutes sedimentation
$X$ = concentration of activated sludge (mixed liquor suspended solids) [g/l]

The numerical value of SVI is affected by many factors, especially by sludge concentration or the volume of sludge after 30 minutes sedimentation and by so-called wall effect in the settlometer. Therefore, in some countries the conditions of the settling test are standardized in a certain way. The common modifications are:

- stirred sludge volume index SSVI; the measuring cylinder is equipped with slowly rotating stirring device
- SVI at standard concentration; the test is performed under defined sludge concentration, for instance 3.5 g/l ($SVI_{3.5}$)
- diluted sludge volume index DSVI; the volume of settled sludge after 30 minutes sedimentation should not exceed 200 ml/l, if yes, the test is repeated with diluted sludge

Activated sludges can be classified according to the zone settling velocity and the sludge volume index as shown in Table 1.1.

TABLE 1.1.  **Types of Activated Sludge According to Its Settleability.**

| Type of Sludge | SVI, ml/g | ZSV, m/h |
|---|---|---|
| Well settling | < 100 | > 3 |
| "Light" | 100–200 | 2–3 |
| Bulking | > 200 | < 1.2 |

## Length of Filaments

When the sludge separation problems are caused by filamentous micro-organisms, it is necessary to quantify the amount of filamentous microorganisms in activated sludge. For that purpose the measurement of total extended filament length (TEFL) was developed [18]. The total length of filaments protruding from flocs or freely floating in the bulk liquid is measured. The TEFL values of $10^7$ $\mu$m/ml or $10^4$ m/g are considered a boundary between nonbulking and bulking activated sludges.

The manual measurement of TEFL can be replaced today by using computer image analysis when the dimension of microscopic objects are measured in a processed computer image of the sample. However, both hardware and software equipment for this technique is rather expensive and thus the computer image analysis is still used more for research purposes than for routine measurements.

## Abundance

The manual measurement of TEFL is rather laborious and time consuming. For a routine examination of activated sludges in wastewater treatment plants, a rapid and simple method of subjective scoring of filament abundance was developed [19]. The abundance of filaments is classified according to scale given in Table 1.2 from none to excessive.

In case of increased abundance of foam-forming nocardioform actinomycetes, which are of rather irregular shape and frequently branched, Pitt and Jenkins [42] recommend a modified method based on counting the number of intersections of nocardia filaments with a hairline in the eyepiece.

## Scum Index

Probably the most exact method how to predict the amount of foam formed in aeration basins and the extent of resulting problemswas suggested in South Africa by Pretorius and Laubscher [20]. The scum index SI [%] is calculated from the following formula:

$$SI = \frac{\text{mass of biomass in foam}}{\text{total mass of biomass}} \times 100 \qquad (1.16)$$

The portion of biomass in the foam is estimated by means of fractionary floatation performed with a standard aeration rate of 10 m$^3$/m$^3$ · h in the batch floatation cell (80 mm in diameter; 500 mm high). The floatation is repeated with the sediments after separating the scum from settleable

TABLE 1.2. Subjective Scoring of Filament Abundance [19].

| Numerical Value | Abundance | Description of Microscopic Picture |
|---|---|---|
| 0 | None | |
| 1 | Few | Filaments present, but only observed in an occasional floc |
| 2 | Some | Filaments commonly observed, but not present in all flocs |
| 3 | Common | Filaments observed in all flocs, but at low density (e.g., 1–5 filaments per floc) |
| 4 | Very common | Filaments observed in all flocs at medium density (e.g., 5–10 filaments per floc) |
| 5 | Abundant | Filaments observed in all flocs at high density (e.g., >20 filaments per floc) |
| 6 | Excessive | Filaments present in all flocs—appears more filaments than flocs and or filaments growing in high abundance in bulk solution |

biomass several times until all foam-forming microorganisms are transferred into the scum. The scum index can be used for predicting the operational problems which can be expected as a result of the presence of foam-forming microorganisms in activated sludge. The extent of problems according to the scum index values is classified in Table 1.3.

The plant operators may also follow the tendency of activated sludge in a particular activated sludge plant by using the method suggested in England by Kocianova et al. [57]. The percentage of surface of aeration basins and secondary settling tanks covered by foam is regularly recorded. The readings should be taken at the same time of the day when loading conditions and operating procedures are similar. The authors found a good correlation between cover percentage and other parameters reflecting the presence of foam-formers like foam accumulation rate and hydrophobicity of activated sludge.

TABLE 1.3. Scum Index and Expected Operational Problems [20].

| SI % | The Extent of Problems |
|---|---|
| 0–0.5 | Negligible |
| 0.5–6 | Small |
| 6–10 | Medium |
| 10–15 | Serious |
| >15 | Catastrophic |

## FILAMENTOUS MICROORGANISMS

Filamentous microorganisms are a specific morphological group of organotrophic (mostly) microorganisms growing in activated sludge. At present we distinguish about 30 different kinds of filamentous microorganisms.

### Role of Filamentous Microorganisms

The role of filamentous microorganisms in the biocenosis of activated sludge can be evaluated from different viewpoints, but the main aspects are as follows:

(1) The filamentous microorganisms are believed to form a ''backbone'' of activated sludge flocs on which floc-forming bacteria are fixed by means of extracellular polymers [19].
(2) The deterioration of activated sludge properties is caused by a high occurrence of filamentous microorganisms in the biocenosis (usually abundance 4 and more according to Table 1.2).
(3) The increased occurrence of filamentous microorganisms in the biocenosis of activated sludge indicates that the activated sludge system is not designed or operated properly.

It was suggested [19,21] that the presence of certain filaments can indicate the conditions causing activated sludge bulking. Jenkins et al. [19] correlated the occurrence of individual filamentous types with technological parameters, operational conditions and wastewater characteristics. Five groups of filamentous microorganisms were established according to the causative conditions:

- low DO (dissolved oxygen)
- low *F/M* (low sludge loadings; high solids retention times)
- septic wastewaters (increased concentration of sulphide)
- nutrient deficiency (limiting concentrations of N,P, other nutrients?)
- low pH in mixed liquors.

Sometimes it is not easy to define exactly the range of the above causative conditions. For instance, the limiting concentration of $O_2$ depends on the actual floc loading by substrate (simultaneous diffusion and consumption in the floc interior). Is the solids retention time of 15 days high at winter temperatures ($<10°C$)? For this and many other questions the indicative role of filamentous microorganisms must be considered with caution.

### Needs for Identification and Classification

A correct finding of the taxonomic position is necessary for the documentation of the microorganism and for its preservation in microbiological collections. Once the microorganism is positioned taxonomically, the identification of isolates from activated sludge becomes much faster and easier. The knowledge of the taxonomic position will also help in recognizing growth and nutrition requirements of the given microorganism by comparison to other strains within a genus or family. Such correct taxonomic identification is still very difficult and expensive. Fortunately, the real needs of engineers are more prosaic. A microbiological examination of the bulking and/or foaming activated sludge must result in a kind of information that enables the engineers:

(1) To find out or predict the extent of filamentous bulking or foaming, i.e., not only to quantify the number or length of filaments but also to specify the type of filaments present in a particular sludge: It is a recognizable fact that all filamentous microrganisms do not deteriorate settling properties of activated sludge in the same manner. Much more severe bulking problems are to be expected when *Sphaerotilus spp.*, Type 021N, and *Thiothrix* are the dominating filaments in comparison with, for example, *Microthrix parvicella, Haliscomenobacter hydrossis,* and *Nostocoida limicola* bulking. The foaming problems threaten only when particular filaments are present.

(2) To estimate the causes of the presence of filamentous microorganisms in the biocenosis: By making such an estimate other factors should also be considered, especially wastewater character and composition, aeration basin configuration, and all operational parameters. It is evident that neither a microbiologist nor an engineer, however experienced, can trace the causes himself. A close cooperation between the specialists is of extreme importance.

(3) To rectify the bulking and foaming problems: This is in fact the ultimate objective of our effort. The information resulting from the identification of filamentous microorganisms must help in selecting proper measures against excessive growth.

### Possibilities of Identification

The conventional method of identification of bacteria or other microorganisms is based on their isolation from mixed cultures and on subsequent exhaustive tests of morphological, biochemical, and physiological features. The results of these tests are then compared with standard references given in the manual such as *Bergey's Manual of Determination Bacteriology*

[25]. The main advantage of the conventional method seems to be in the unambiguous taxonomic location obtained in this way. But at present the taxonomic position of a given microorganism without genetic studies can hardly be considered as "definite". For the purpose of isolation the filamentous microorganisms from the diverse biocenosis of activated sludge, the conventional methods are too cumbersome, time-consuming, and mostly unnecessary or even inappropriate because of the following reasons:

(1) There is a need for at least thirty stable features and as many as possible variable features (e.g., morphology of cells and cell agglomerates including trichomes) for a formal recognition of nomenclatoric taxon.

(2) There is still a severe lack of pure cultures of filamentous microorganisms originating from activated sludges that isolates can be compared with.

(3) There is a great probability that many morphological, physiological, and genetic changes may occur in the filamentous microorganisms within the pure cultures from bacteriological collections as a result of various undefinable influences on them in wastewater treatment plants.

(4) There are many difficulties in preparing a pure culture inoculum from the mixed culture of activated sludge, especially due to the removal of accompanying cells of other microorganisms.

(5) The low growth rates of most filamentous microorganisms cause the isolation and identification to take weeks or months. Thus, the results obtained do not correspond with the actual state of the biocenosis in the treatment plant the isolates originate from.

## Identification to Types

Because of high time and labour demands of conventional microbiological identification procedures, the activated sludge filamentous microorganisms are not identified to taxa but to the so-called types introduced to wastewater treatment practice by Eikelboom [58]. With few exceptions (e.g., *Nocardia* spp., *Thiothrix* spp., *Sphaerotilus natans*) the taxonomic position of most Eikelboom's types is still unknown and the types are characterized by numbers (e.g., Type 0041 or Type 021N) or by taxonomically invalid names like *Microthrix parvicella*.

The identification to types is based on some characteristic morphological features (cell and trichome size and shape, branching of trichomes, presence or absence of sheath and of sulphur, polyphosphate and PHB granules) and on reactions to Gram and Neisser staining [19,24]. Unfortunately, the morphology and staining reactions may vary significantly depending upon cultivation conditions and wastewater composition. For instance, an increased fraction of industrial wastes in treated wastewaters results in colours different from those described for the Gram stain in the above manuals. These problems cause that there is always some uncertainty in the results of microscopic identification. A typical example of this uncertainty was the most problematic filamentous microorganisms *Microthrix parvicella* when the scientists studying this filament could not agree until recently that all the laboratories all over the world work with the same microorganism. This situation led to the requirements for more precise but rapid and applicable in practice methods of identification and to the development of gene probes for the *in situ* identification and quantification of filamentous microorganisms.

### Identification by Using Gene Probes

This novel technique exploits recent findings of molecular biology. The molecular identification is based on specific sequences of ribosome 16S and 23S rRNA/DNA which can be detected by hybridization with fluorescently labelled oligonucleotides. With this technique Blackall showed that the strain of foam-forming microorganism *Nocardia pinensis* most probably belongs to the genus *Nocardia* although there were some doubts in the literature about this allocation [59]. On the other hand, the latest studies on "nocardioform" actinomycetes (*Nocardia amarae*—like organisms) showed that they do not belong to genus *Nocardia* but *Gordona*. Wagner et al. succeeded to identify *Sphaerotilus natans, Haliscomenobacter hydrossis* and *Thiothrix nivea* by using *in situ* hybridization technique with 16S rRNA targeted probes [60].

The number of (not only filamentous) activated sludge microorganisms which can be identified by this technique is increasing. Unfortunately, because of the high requirements on laboratory equipment and price of used chemicals, the Eikelboom's system of identification is still the basis for routine microbiological examination of activated sludges in practice. However, the gene probes can be used for a certain "calibration" or "verification" of a given microscopic laboratory or in cases of doubtful results of microscopic identification.

### Classification According to Morphology and Metabolic Properties

In conventional activated sludge systems the diversity of filamentous microorganism population was rather low. The most commonly found filamentous microorganisms up to the early 1980s were *Sphaerotilus natans* and Types 021N, 1701, 0041, i.e., the filamentous microorganisms mostly connected with low DO conditions or with low concentration gradients of readily biodegradable substrates in aeration basins. When nutrient removal activated sludge systems were introduced in wastewater treatment practice, the diversity of types of filamentous microorganisms in activated sludges increased. The filamentous microorganisms dominating in activated sludges from nutrient removal systems are different from those reported earlier for conventional (oxic) activated sludge systems. Recently, a new phenomenon of filamentous foaming has become typical exactly of nutrient removal activated sludge systems. It is thus logical to assume that the observed shift in filamentous population is connected with different properties of individual filamentous types.

Wanner and Grau summarized their own and literature data and proposed the classification of filamentous microorganisms into four groups [23]. The groups were set up on the basis of:

- morphological similarity (sheath, trichome, cells)
- similar staining reactions, intracellular deposits
- metabolic similarity, i.e., abilities to utilize substrates and gain energy under the same cultivation conditions
- occurrence in the same operational arrangements and conditions
- similarity in problems the filamentous microorganisms cause

The four groups are as follows:

(1) Group S: *Sphaerotilus*-like oxic zone growers—Sheathed filamentous microorganisms which are able to utilize organic substrates only in oxic conditions. The presence of polyphosphates and PHB granules in cells may be observed but the rate of their formation and exploitation is not technologically important. Their occurrence in activated sludges is connected with saccharidic and other readily biodegradable wastewaters, higher SRT, and low DO. Characteristic representatives are *Sphaerotilus natans* and Type 1701, then Types 0041 and 0675.

(2) Group C: *Cyanophycae*-like oxic zone growers—Group C includes filamentous microorganisms morphologically resembling colourless blue-green algae: Type 021N and *Thiothrix*. Although not included in Eikelboom's manual, the genus *Leucothrix* should be added to group C. The taxonomic position of Type 021N is not clear yet, it seems that Type 021N is not identical with either the genus *Leucothrix* or the genus *Thiothrix*. Type 021N is a very common filamentous microorganism in bulking activated sludges from conventional activated sludge plants treating domestic or other wastewaters containing readily biodegradable substrates. Contrary to oxic zone growers the microorganisms from group C are able to metabolize sulphur (with the exception of *Leucothrix*). Especially *Thiothrix* can take advantage of its mixotrophic way of life in systems with anaerobic zones [13].

(3) Group A: All zones growers—The term *all zones growers* means that these microorganisms are equipped with such a diverse enzymatic apparatus that they can utilize substrates under all cultivation conditions in activated sludge plants. In addition, the all zones growers may accumulate low molecular substrates under oxic and anoxic conditions and synthesize storage products with rates comparable to those of floc-formers. *Microthrix parvicella* and Type 0092 exhibit undoubtedly the characteristic features of the all zones growers. *Nostocoida limicola* was commonly found in our pilots and today in full-scale plants with anaerobic zones [23,61].

(4) Group F: Foam-forming microorganisms—The foam-formers are microorganisms which can produce biosurfactants enabling them to froth and create scum. The floating effect is supported by hydrophobic surfaces of cells. The hydrophobic cells stabilize the air bubbles or oil droplets entrapped in the foam. The formation of biological foam is mostly connected with actinomycetes, namely the so-called norcardioform actinomycetes (*Nocardia-Gordona*, *Rhodococcus*) or *Microthrix parvicella* which is probably also an actinomycete. These microorganisms can be considered as primary foam-formers together with Types 0092, 0041, and *Nostocoida limicola*. Other filamentous microorganisms commonly found in biological foams are probably only captured to already established scums.

The differences between individual groups S, C, A, and F can be illustrated by the response of filamentous microorganisms to systems combining the kinetic selection (i.e., systems with high concentrations gradients) with the metabolic selection under different cultivation conditions (for the selection mechanisms see Table 1.4). The characteristic representatives of individual groups are shown in Figures 1.5–1.8.

## FILAMENTOUS MICROORGANISMS GROWTH CONTROL

The methods of controlling an excessive growth of filamentous microorganisms can be divided into two groups:

(1) Biological methods selectively targeted on the suppression of filamentous microorganisms growth and on the support of floc-formers growth

(2) Non-specific control methods treating the consequences (symptoms) of the occurrence of filamentous microorganisms in activated sludge

The first group of methods are of a preventative nature and should always be considered in a design of newly constructed plants. On the other

TABLE 1.4.  Response of Filamentous Microorganisms to Systems Combining Kinetic Selection (High Concentration Gradients) with Metabolic Selection under Different Cultivation Conditions [34].

| Group of Filamentous Microorganisms | Cultivation Conditions | | |
|---|---|---|---|
| | Oxic | Anoxic | Anaerobic |
| S (e.g., *Sphaerotilus natans* and Type 1701) | – | – | – |
| C (e.g., Type 021N and *Thiothrix*) | – | – | –/+ (+ for sulphur filaments) |
| A (e.g., *Microthrix parvicella*, Type 0092, *Nostocoida limicola*) | 0 | 0 | 0/+ |
| F (e.g., nocardioform actinomycetes) | ? | ?/– | –/? |

– Suppression; + stimulation; 0 no effect; ? effect uncertain.

hand, the second group of methods are remedial (cure of symptoms) and cannot guarantee a long-term effect.

## BIOLOGICAL SELECTION MECHANISMS FOR FILAMENTOUS BULKING AND FOAMING CONTROL

The actual microbial composition of activated sludge is a result of strong competition for energy and carbon sources, nutrients and sometimes also for electron acceptors. The competition is influenced by numerous factors, both intrinsic to the activated sludge process like biomass retention time, actual substrate, oxygen and nutrients concentrations in reactor or cultivation conditions (oxic, anoxic, anaerobic) and factors from outside the activated sludge system. These external factors like wastewater composition, temperature, pH are difficult to control but their effect must be considered in wastewater treatment plant control strategies. The following basic mechanisms are decisive for the selection of microbial species in activated sludge [62].

### Kinetic Selection

This selective pressure results from differences in growth and substrate utilization rates. In case of balanced growth (readily biodegradable substrate uptake and cell growth occur simultaneously) two types of competition strategy are possible (in terms of saturation kinetics; see Figure 1.9):

- $\mu$-strategy: maximum specific growth and substrate utilization rates at high substrate concentrations

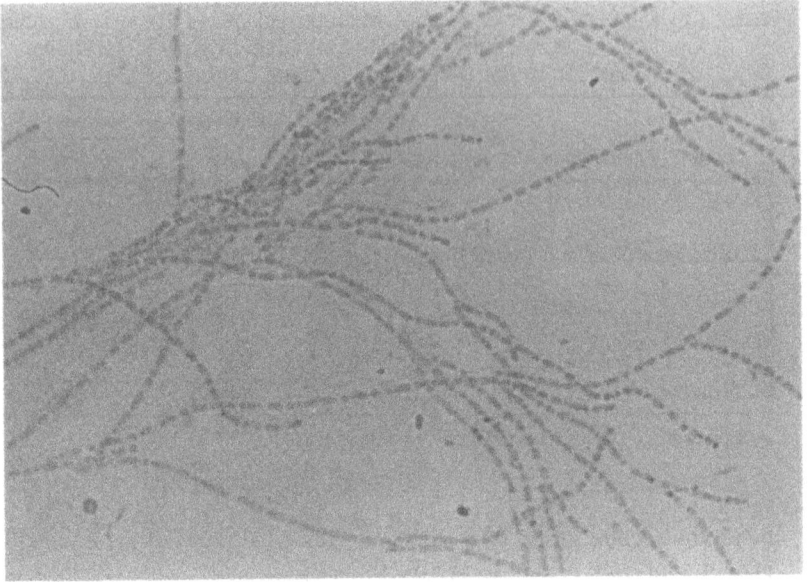

**Figure 1.5** Microphotograph of *Sphaerotilus natans* (1250×, Gram stain).

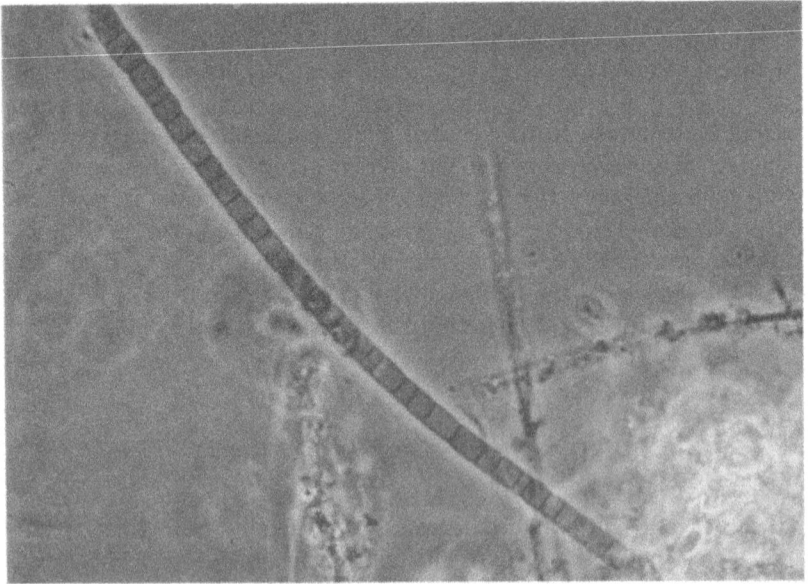

**Figure 1.6** Microphotograph of Type 021N (1250×, phase contrast).

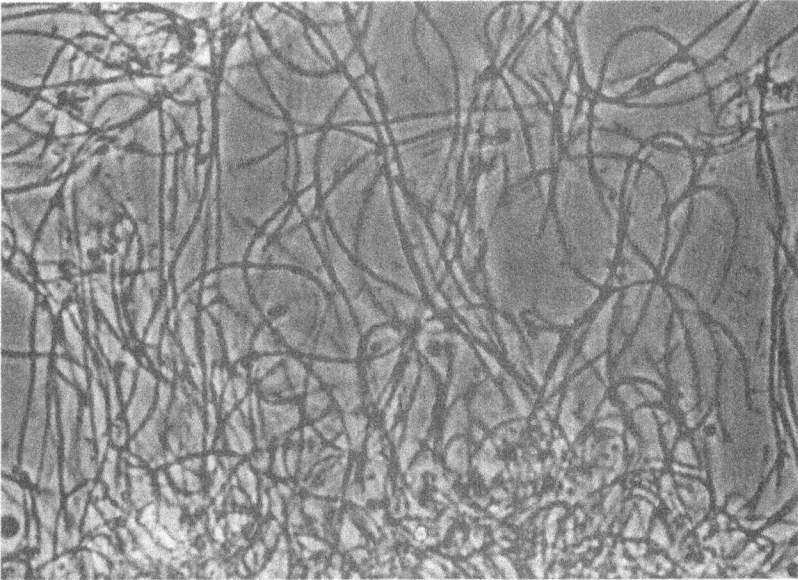

**Figure 1.7** Microphotograph of *Microthrix parvicella* (1250×, phase contrast).

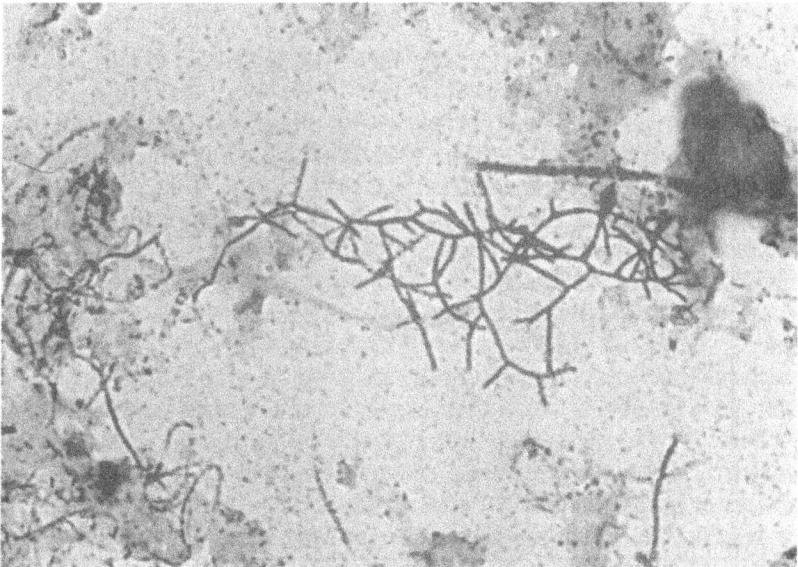

**Figure 1.8** Microphotograph of nocardioform actinomycetes (GALOs) (1250×, Gram stain).

**39**

- K-strategy: high substrate affinity at low substrate concentrations

Activated sludge floc-forming microorganisms are generally the $\mu$-strategists while most of activated sludge filamentous microorganisms belong to the K-strategists. However, the so-called unbalanced growth, when the phases of substrate (both soluble and particulate) or nutrients uptake and utilization are partially or fully separated, is a more common case in activated sludge process. In the competition under the conditions of unbalanced growth the microorganisms which can sequester substrate more rapidly from the bulk liquid will be selected provided that there is time enough for the regeneration of *accumulation/storage* capacity in the cells of these microbes [32]. In activated sludge the accumulation/storage and regeneration selection mechanism is typical of most floc-forming microorganisms. The process of substrate accumulation/storage takes place in the contact zone of activated sludge system (see below), which can be not only oxic but also anoxic or anaerobic, while the endogenous metabolism (regeneration) occurs in the main aeration basin or in a separate regeneration zone placed in the return sludge stream.

The results of competition of floc-forming and filamentous microorganisms based on kinetic selection are influenced by the following factors:

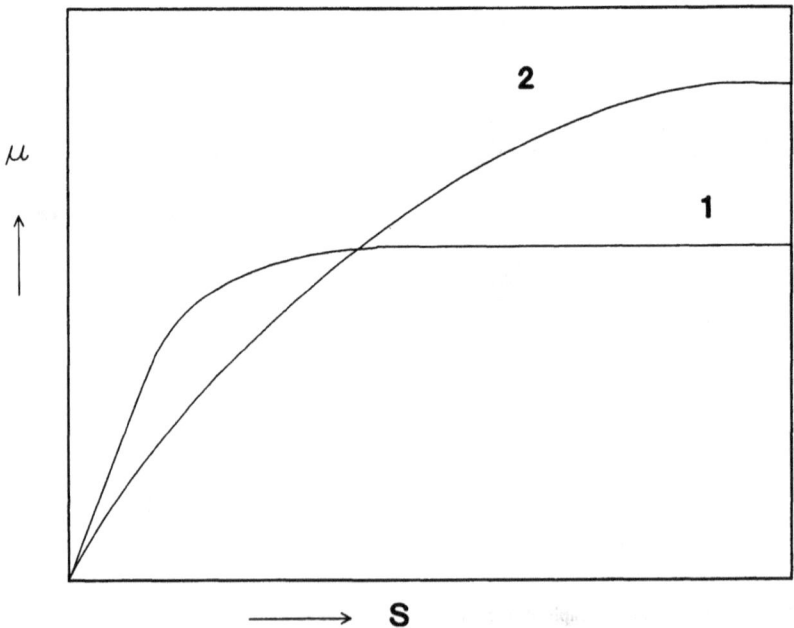

**Figure 1.9** Graphical presentation of $\mu$- and K-strategies: (1) K-strategists; (2) $\mu$-strategists.

## Wastewater Composition

Wastewater is a source of substrates, nutrients and micronutrients for the microbial consortium of activated sludge. In addition, wastewater continuously inoculates this consortium by bacteria growing in sewers.

### READILY BIODEGRADABLE SUBSTRATES

Low molecular weight organic compounds, i.e., compounds with simple molecules, can usually be directly utilized by bacterial cells. These compounds (e.g., monosaccharides, alcohols, volatile fatty acids and amino acids) represent approximately 10–20% of COD of common municipal wastewaters. Therefore their concentration in the mixed liquor of activated sludge is usually not very high, which is a favourable factor for the growth of some filamentous microorganisms (K-strategists). Especially Types 021N, *Sphaerotilus natans*/1701, 0041 and *H. hydrossis* can be expected in systems treating wastewaters with elevated concentrations of those compounds. A special attention has to be paid to the design of activated sludge systems for wastewaters with increased fraction of industrial wastewaters (food industry). Reduced sulphur compounds can also be easily metabolized and thus support the growth of sulphur-metabolizing filaments like Type 021N/*Thiothrix* or *Beggiatoa.*

### PARTICULATE, SLOWLY BIODEGRADABLE SUBSTRATES

Most of utilizable organic compounds in municipal wastewaters can be described as the so-called particulate substrate. The particulate substrate is formed by large-molecule organic compounds which are present in the wastewater either as colloids or as true suspended solids. In both cases those molecules have to be attacked by extracellular enzymes before they become available for cells. The mechanism of particulate substrate disintegration is generally termed *hydrolysis* and the products of this process are very similar from chemical point of view to readily biodegradable substrates originally present in wastewaters. However, the hydrolysis is a surface phenomenon and is connected with flocs. Therefore, if the main aeration basins, where the hydrolysis occurs, are not operated as *completely mixed tanks,* the readily biodegradable substrates released by hydrolysis will be more available to floc-formers than to filamentous microorganisms. However, some special kinds of particulate substrates can support the growth of filamentous microorganisms. This is especially the case of fats and grease which are selectively concentrated in foams and thus support the growth of nocardioform actinomycetes. The long-chain fatty acids are reported in the literature as specific substrate supporting the growth of

*Microthrix parvicella.* Oleic compounds in wastewater can also support the so-called viscous bulking caused by zoogloeal bacteria, especially in activated sludge systems with high substrate concentration gradient.

## INOCULATION OF ACTIVATED SLUDGE SYSTEMS FROM WASTEWATERS

The sewerage network and wastewater treatment plants form one system and the processes in sewers may significantly affect the composition of activated sludge consortia. Especially in extended sewer systems readily biodegradable substrates can be taken up by bacteria growing in inner slimes. These wall growths are continuously sloughed off and inoculate the treatment plant.

### Biomass Retention Time

Biomass retention time $\Theta_x$ ("sludge age") has a great impact on the distribution of individual microbial species in the consortium of activated sludge according to their growth and decay rates. Low biomass retention time $\Theta_x$ may result in a wash-out of slow growing species whose net growth rate is slower than the dilution rate $D = 1/\Theta_x$. On the other hand, high $\Theta_x$ values favour the slow-growers in the mixed culture of activated sludge. Therefore, the designer should realize that at different regions of $\Theta_x$ different filamentous microorganisms will predominate in activated sludge. Unfortunately, the "safe" values of $\Theta_x$, when all filamentous microorganisms are certainly washed-out, are too low to be applied as a common bulking control measure. The distribution of filamentous microorganisms with retention time is, moreover, strongly affected by temperature—at elevated temperatures the "wash-out" biomass retention time is lower. The group of filaments preferably growing at high $\Theta_x$ values is also termed "low *F/M*" filaments [19,26].

### Actual Substrate Concentration in Reactor

The activated sludge cultured in reactors with substrate concentration gradients gains certain features which were described by Gabb et al. as "selector" effect [26]:

(1) High rates of substrate consumption
(2) High oxygen (or generally: electron acceptor) uptake rate
(3) Enhanced growth of zoogloeal bacteria

When the activated sludge flocs are alternately exposed to higher and lower substrate concentrations, a significant stratification of the flocs may appear, which increases the diversity of metabolic processes occurring inside the flocs. Thus even in fully aerated activated sludge systems a significant fraction of available substrate can be utilized under anaerobic or anoxic conditions [27]. The substrate concentration gradient leads consequently to a combined effect of kinetic and *metabolic selection* (see below). Because of the floc stratification, the fourth feature can be added, viz.,

(4) Increased metabolic diversity

*Dissolved Oxygen, Nutrients, pH and Temperature in Aeration Basins*

DISSOLVED OXYGEN

Some filamentous microorganisms (e.g., *Sphaerotilus natans*/Type 1701, *H. hydrossis*) exhibit great affinity to dissolved oxygen at low concentrations because of low values of half-saturation constant $K_O$ [19]. The boundary between "bulking" and "non-bulking" DO concentrations is not fixed because this value depends on the actual value of activated sludge loading $B_X$ in the reactor. For the estimate of DO concentrations required for avoiding low DO bulking the data obtained by Palm et al. are summarized in Table 1.5 [28]. As it can be seen from this table, at higher sludge loadings the "safe" DO concentrations are quite high. This is important fact for the design of aeration systems. Especially in the head-end of aeration basins the oxygenation capacity of the aerators must be high enough to guarantee the DO concentrations given in Table 1.5. The aeration equipment in oxic zones of nutrient removal activated sludge systems should convert the mixed liquor from preceding anaerobic or anoxic conditions to oxic conditions as quickly as possible.

NUTRIENTS

Similarly to dissolved oxygen, some filaments exhibit higher affinity also to nutrients (i.e., nitrogen, phosphorus and micronutrients). However, the experiments with Type 021N carried out by Richard et al. [29] showed

TABLE 1.5. Relationship Between $B_X$ and "Safe" DO Concentrations in Oxic Contact Zones [28].

| $B_X$, kg/kg · d | DO, mg/l |
|---|---|
| 0.3 | 1.0 |
| 0.5 | 2.0 |
| 0.75 | 3.0 |
| 0.9 | 4.0 |

that the dominance of filamentous microorganisms under low nutrient concentrations is not only a matter of the K-strategy but that the filaments may exhibit a kind of "accumulation" capacity for nutrients, which gives them another advantage under the conditions of unbalanced growth. Therefore the addition of nutrients for controlling filamentous bulking should be sufficient to saturate this accumulation capacity. The exact doses of limiting nutrient(s) can only be assessed from pilot tests. When the tests cannot be performed, Richard [30] recommended to maintain these effluent concentrations: total inorganic nitrogen at least 1 mg/l, soluble orthophosphate phosphorus more than 0.2 mg/l.

The possibility of filamentous bulking by nutrient limitation stresses the requirement to know the composition of the treated wastewater before starting the design of activated sludge system, especially in case of important contribution of industrial wastewaters. A severe lack of nutrients may result also in serious problems with viscous bulking because an elevated synthesis of extracellular polymers is one of the reactions of microbial cells to such unfavourable conditions.

## pH

None of the most common filamentous microorganisms exhibits a special preference to extreme pH values. Some micromycetes prefer low pH values, however, such a decrease in pH in municipal wastewater treatment plants (e.g., by nitrification) leading to excessive fungal growth is improbable.

## TEMPERATURE

Temperature effects significantly the rates of all biochemical processes as well as the solubility of oxygen in the mixed liquor. Thus the elevated temperatures will support the growth of filamentous microorganisms connected with low DO concentrations. The oxygenation capacity of aeration systems should be calculated for the highest mixed liquor temperatures expected during the year to guarantee the DO concentrations from Table 1.5 also in warm weather.

There are significant seasonal shifts in the dominance of individual filamentous types. Typically, *Microthrix parvicella* which dominates the filamentous population in winters is replaced by nocardioform actinomycetes, Type 0041 or *Nostocoida limicola* in warm seasons. This shift is certainly connected with temperature changes but an exact explanation of this phenomenon is still not available. According to the recent practical experience, more severe *Microthrix parvicella* foaming problems are to be expected in winters than in summers (e.g., [31]).

## Metabolic Selection

In nutrient removal activated sludge systems the decisive fraction of readily biodegradable substrates from wastewater is utilized not under oxic but under anaerobic and/or anoxic conditions. Thus the microorganisms able to metabolize substrates under these conditions will be selected. This is the principle of metabolic selection.

The alternation of anaerobic and oxic cultivation conditions is a prerequisite for the enhanced biological phosphorus removal (EBPR) mechanism in nutrient removal. As the filamentous microorganisms from Groups S and C (see above) are not able to utilize substrate under alternating anaerobic/oxic cultivation conditions with a rate comparable to that of floc-forming microbes, their suppression in nutrient removal activated sludge systems is based on metabolic principles. While the Group S and C filamentous microorganisms were quite common in conventional activated sludge plants, recent surveys of filaments in nutrient removal systems done in Europe, South Africa, U.S.A. and in Australia indicate that these filaments are not important in systems with anaerobic and/or anoxic zones. The elimination of Group S and C filaments in systems with anoxic zones is based on different denitrification rate between filaments and floc-formers [35,36].

## Selective Predation

Protozoa and metazoa can significantly affect the results of competition between bacterial species. Experiments carried out in Japan indicated that there are some "spaghetti eaters" among predatory ciliates which specialize in consuming filamentous microorganisms [41]. The latest results from Sweden support this possibility [37].

## APPLICATION OF FILAMENTOUS MICROORGANISMS GROWTH CONTROL PRINCIPLES TO THE DESIGN AND OPERATION OF ACTIVATED SLUDGE PLANTS

## Plant Configuration

The design of activated sludge system has to combine all factors favouring the growth of floc-formers, viz.,

- substrate concentration gradient in the system (or at least in its head-end)
- rapid accumulation (*"biosorption"*) of substrate with subsequent regeneration of accumulation/storage capacity
- in case of nutrient removal systems, the utilization of most substrate under anaerobic and/or anoxic conditions.

There are principally four process configurations effectively supporting the growth of floc-formers. In these configurations the basic principles of kinetic selection are employed either alone or in combination with metabolic selection:

(1) Sequencing batch reactor with short filling period ("dump-fill") or with an inlet mixing zone

(2) Continuous plug-flow reactors; as even the long corridor-type reactors exhibit a high degree of longitudinal mixing, the reactor should be compartmentalized to approach the plug-flow hydraulic regime

(3) Completely mixed or compartmentalized continuous-flow reactor with a contact zone placed ahead the main reactor; the contact zone should preferably be divided into 2–4 compartments

(4) Activated sludge system with return sludge regeneration; the main reactor should be arranged in configurations 2 or 3.

In all the above configurations, the return activated sludge can be mixed with the treated wastewater either under oxic or under anaerobic/anoxic conditions. An example of activated sludge configuration combining most of the factors favouring the growth of floc-formers is given in Figure 1.10.

The arrangement utilizing the phenomenon of anoxic phosphate uptake is shown in Figure 1.11. This arrangement proved to be very efficient in controlling the growth of an *all zones* filament *Microthrix parvicella* [46].

### Design Parameters

#### Oxic Contact Zone Design

The design criteria for the oxic contact zone can also be applied to the

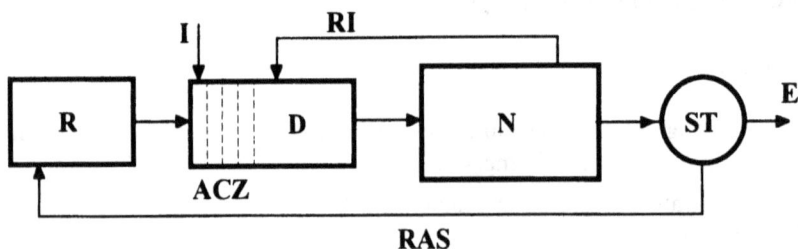

**Figure 1.10** Schematic of the so-called R-D-N process [33]: ACZ—compartmentalized anoxic contact zone; D—denitrification zone; N—nitrification zone; ST—settling tank; R—regeneration zone; I—influent; E—effluent; RAS—return activated sludge; WAS—waste activated sludge; RI—internal recirculation.

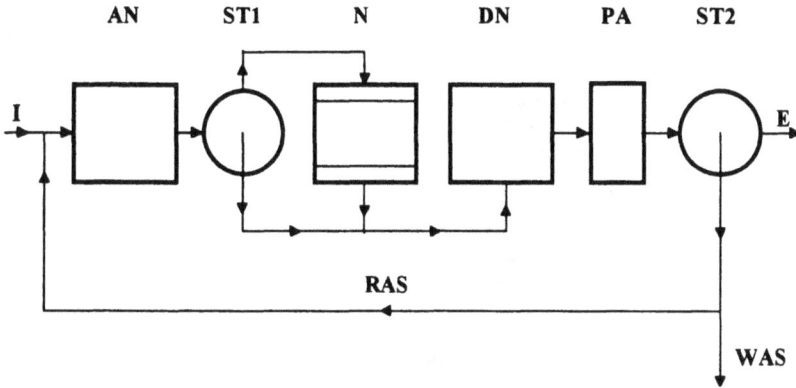

**Figure 1.11** Modified arrangement of nutrient removal activated sludge system [45]: I—influent, E—effluent, RAS—return activated sludge, WAS—waste activated sludge; AN—anaerobic zone, ST1—1st settling tank, N—nitrification of effluent from ST1 in fixed-film reactor, DN—denitrification, PA—post-aeration, ST2—final settling tank.

design of inlet part of compartmentalized plug-flow systems or to the design of SBRs. However, the oxic contact zone (originally described as the so-called selector in Czechoslovakia by Chudoba et al., [11]) can only be used in conventional activated sludge systems. This concept cannot be employed in biological nutrient removal (BNR) plants because the readily biodegradable substrate necessary for the EBPR and denitrification is "inefficiently" consumed under oxic conditions. In addition, the performance of "selectors" has been shown to be unstable in suppressing biological foaming [42]. The following parameters should be considered in the design of oxic contact zones:

CONTACT TIME

The contact time is based on the flow of mixed liquor and not only on the wastewater flow. The parameter is directly connected with the size of contact zone. When the contact time is insufficient, soluble substrate is not consumed in the contact zone and penetrates to the main reactor where it can support the growth of filaments. On the other hand, when the contact time is too long, the growth of filaments occurs right in the contact zone, especially when this is not compartmentalized. The recommended values of contact time are summarized in Table 1.6.

The recommended values of contact time lie in a range of 10–20 minutes. For conventional activated sludge systems this contact time results in the volume ratio $V_{CZ}/V_{TOTAL}$ of about 1/10. This ratio is also sufficient for a full restoration of accumulation/storage capacity in the main reactor. The

TABLE 1.6. **Recommended Contact Time in Oxic Contact Zones.**

| Source[a] | $\Theta_{CT}$, minutes |
|---|---|
| van Niekerk et al. (1987) | 12–18 |
| ATV Report 2.6.1 (1989) | 10–12 |
| Daigger et al. (1985) | 15 |
| Eikelboom (1991,b) | 10–15 |
| Pujol (1992) | 15 |
| Rensink and Donker (1993) | not less than 10 |

[a]The original references from this table can be found in Reference [62].

recirculation ratio of return activated sludge entering the contact zone should be less than one.

## ACTIVATED SLUDGE LOADING IN THE CONTACT ZONE

The activated sludge loading $B_X$ was selected as a design criterion instead of "floc loading." Although this parameter describes more precisely the relationship between actual substrate and biomass mass fluxes, for continuous-flow systems the activated sludge loading is more convenient for calculations. Table 1.7 brings the figures from the most successful case histories of bulking control found in the literature. The term "aerated" contact zone means that besides oxic conditions in the bulk liquid, anaerobic and anoxic conditions are possible in the flocs under such high loadings. All $B_X$ values are related to the whole volume of the contact zone, so that for compartmentalized contact zones the actual sludge loading in the first compartment is correspondingly higher. The asterisked values were calculated for MLVSS (organic fraction of X).

## CONDITIONS IN THE MAIN REACTOR

The selection of floc-formers does not occur only in the contact zone

TABLE 1.7. **Activated Sludge Loading in Aerated Contact Zones.**

| Source[a] | $B_X$, kg/kg · d; based on | |
|---|---|---|
| | BOD₅ | COD |
| Albertson (1991) | 3.0 | |
| ATV Report 2.6.1 (1989) | 3.0 | 20* |
| Chudoba and Wanner (1989) | > 3.0 | |
| Daigger and Nicholson (1990) | 3.2–4.9 | |
| Eikelboom (1991) | 2.0–5.0 | |
| Lee et al. (1982) | | > 20* |
| Linne et al. (1989) | 5.0–6.0 | |

[a]The original references from this table can be found in Reference [62].

and the conditions in the main reactor following the contact zone are part and parcel of the selection process (in this sense the term "selector" should cover the whole configuration and not only the contact zone). The activated sludge retention time in the main aeration basin is sufficient for a complete restoration of accumulation/storage capacity when the activated sludge loading in the whole system does not exceed 0.3–0.5 kg/kg · d (BOD$_5$, X). If the main reactor is a tank with high dispersion number, the soluble COD in the effluent from contact zone should not exceed the final effluent soluble COD by more than 20–30 mg/l.

## Design of Anaerobic and Anoxic Contact Zones of BNR Systems

The primary task of anaerobic and anoxic contact zones in BNR (biological nutrient removal) activated sludge systems is to create conditions for nitrogen and phosphorus elimination. Thus the design of these zones has to follow principally the guidelines for BNR systems design. However, the designer should always try to implement the bulking control strategies in the design although the conditions for bulking control can sometimes be contradictory to the requirements of BNR processes.

### ANAEROBIC ZONE

The anaerobic contact time sufficient for bulking control is about 0.5–1.0 h, which could be too short for the EBPR. The value of activated sludge loading and the existence of a substrate concentration gradient in anaerobic contact zones do not play such an important role in bulking control in comparison with oxic and anoxic contact zones. The floc-formers growing in systems with anaerobic zones differ from those in systems with oxic and anoxic zones and the competition between these floc-formers and filaments is based mostly on metabolic selection. For instance, the $B_X$ value of 1 kg/kg·d (BOD$_5$, X) in the anaerobic zone was enough for an efficient bulking (by Type 021N) control in a brewery wastewater treatment plant [43].

### ANOXIC ZONE

A certain analogy with the design of oxic contact zones can be used because the floc-formers supported by anoxic contact zones are mostly the same microbes which are favoured by oxic contact zones. From the successful case histories described in the available literature the activated sludge loading in the first compartment should be around $B_X = 6$ kg/kg · d (BOD$_5$, X).

To avoid a "secondary" growth of filamentous microorganisms (Oxic Zone Growers) in the oxic part of the biological nutrient removal system, the soluble COD in the effluent from anoxic zone should be as low as possible. The sizing of the anoxic contact zone can therefore be based on the $\Delta COD$ value necessary for achieving the effluent soluble COD of about 60 mg/l. The following typical figures are recommended for the calculation:

- specific COD consumption for nitrate nitrogen reduction 8 g/g (COD, $NO_3$–N)
- specific denitrification rate $r_{X,DEN} = 3$ mg/g·h ($NO_3$–N, X) or 4 mg/g · h ($NO_3$–N, $X_{Organize}$).

The volume of anoxic contact zone obtained under the above figures results in contact times between 10 and 30 minutes.

The plant operators should always be aware of the main principles of bulking control:

- In conventional plants the majority of organic substrate available in treated wastewater should be removed in the head part of the activated sludge system; the aeration basin should preferably be operated as a plug-flow reactor.
- In nutrient removal systems the majority of available organic substrate should be consumed in anaerobic and/or anoxic zone.

## NON-SPECIFIC CONTROL METHODS

### Bulking Control

The non-specific control methods treat the problem of an elevated presence of filamentous microorganisms in the activated sludge biocenosis. They have a short-term impact and should be applied repeatedly.

### Manipulation with the Sludge

A general operator's reaction to sludge bulking accompanied by losses of suspended solids over the clarifies weirs is to increase the sludge recycle ratio (in order to decrease the clarifier sludge blanket level) and to decrease the mixed liquor suspended solids concentration (in order to decrease the clarifier flux). Both measures are able to prevent loss of suspended solids. Reduction of biomass inventory in the system, however, can lead up to dispersed growth and a substantial drop in treatment efficiency. Both measures have a limited effect. If the clarifies overflow rate is high, not much can be achieved unless the SVI is dramatically lowered.

### Weighting the Sludge

The settling velocities of bulking sludge can be increased by increasing the specific weight of flocs. It is a well-known fact that activated sludge systems treating raw sewage suffer from bulking less frequently. One of the reasons (but not a complete explanation) is addition of well settling solids to the flocs. They are enmeshed in activated sludge and improve settling. In European practice inorganic coagulants (precipitant) such as lime, or ferrous or ferric salts are added into the aeration basins. The heavier precipitates sweep down the bulking sludge flocs and improve settling properties. The increased sludge production should be taken into account. It should be noted that iron is an important nutrient. Because of the formation of various precipitates, not all iron determined analytically maybe easily available to the microorganisms. In some cases, rather small doses of iron improve settling. Weighting of flocs is in such cases not the most probable mechanism of action.

### Polymer Addition

In an emergency case, when the activated sludge escapes from the clarifies, synthetic polymers can be dosed to the clarifier center well (flocculation zone). The dose and type of polymer (cationic alone or in combination with anionic) should be determinated in jar tests. Polymer addition generally improves clarifier efficiency in winter, when the density difference between the flocs and water is smaller and viscosity of water higher.

### Damaging the Filamentous Microorganisms

The filaments protruding from the flocs into the bulk liquid can be selectively damaged by strong oxidants. The filaments are exposed to lethal concentrations of chemical while the concentration inside flocs is sublethal thanks to the transport limitation and consumption of the toxicants during the transport through the floc.

In practice, chlorine and hydrogen peroxide have been used successfully [19]. The effective chlorine dosages are in the range of 1–10 g $Cl_2$/kg · d (MLSS). The use of chlorine for this purpose is widespread in the U.S.A., while in Europe the method is applied only occasionally because of the fear that addition of chlorine involves the risk of uncontrollable formation of chlorinated organic compounds which are discharged into the receiving waters.

In South Africa, ozonation of activated sludge was studied for bulking control, with excellent results [38].

The filaments can also damaged or their length decreased mechanically by a shear stress. Under such conditions the growth of filamentous microorganisms in hindered. Various pieces of mechanical equipment such as blender, high speed pumps or aerators can be used for this purpose.

Surprisingly, a kind of shear stress can also be evoked by a continuous contact of filaments with small inert particles added to the mixed liquor, which are kept in suspension together with the flocs. A sawdust was successfully used for the suppression of filamentous bulking even in completely mixed tanks [39,40].

## Foaming Control

### Manipulation of Biomass Retention Time $\Theta_x$

This method is based on the fact that most of foam-forming filamentous microorganisms exhibit lower net growth rates in comparison with floc-forming bacteria. Thus by lowering the sludge age the slow growing filaments can be washed out of the microbial community of activated sludge. However, the practical applicability of this method is rather limited because the values of sludge age at which the washout occurs are very low (1.5–3 days) and thus the washout of filaments may be accompanied by the washout of nitrification bacteria in nutrient removal plants.

### Use of Chemicals and Antifoam Agents

Contrary to bulking control, chlorination of return activated sludge is ineffective in foaming control because the foam-forming microorganisms are selectively retained in floating biomass, and their concentration in thickened return sludge is reduced. Much more promising results were achieved by spraying a chlorine solution, or sprinkling powdered hypochlorite directly onto the foam. The application of commercial antifoam agents is rather costly and uncertain method. Once the thick, viscous biological foam is established, it cannot be collapsed by common antifoam agents.

### Physical Removal of Biological Foams

#### SKIMMING THE FOAM

One of the most common methods of foaming control is a mechanical removal of the foam once it is established on the water surface. The selective removal of floating biomass with increased concentration of foam-forming filaments reduces the solids retention time of the filaments. The shortened solids retention time of foam contributes to the elimination of slow-growing foam-formers from the biomass of activated sludge. The mechanical skimming is usually performed in secondary settling tanks with effluent weirs protected by baffles against escaping floating biomass. The skimmers in both rectangular and circular secondary settling tanks should act over the whole water surface. If possible, the foam removed by skimmers should be handled separately from the waste activated sludge, especially when the waste sludge is stabilized in anaerobic digesters (to avoid foaming of anaerobic digesters). The experience from Czech wastewater treatment plants shows the floating biomass can be stabilized by lime and then mechanically dewatered before its final disposal.

## SELECTIVE FLOATATION

Selective floatation is based on similar principles as the previous method. However, the floatation does not occur spontaneously in the secondary settling tank but in a special aeration chamber placed either between the main aeration basin and the secondary settling tank or in the stream of return activated sludge.

## WATER SPRAYS

Biological foams can be destabilized by substantial dilution. The dilution opens the dams of filamentous microorganisms in liquid films surrounding the gas bubbles, so that the bubbles collapse more quickly. However, the amount of water sprayed over the foam surface in aeration basins or in secondary settling tanks needs to be much higher than in conventional water sprays used for detergent foam control. The intensity of spraying should be adjusted empirically for each particular plant. The use of water sprays does not eliminate the primary cause of foaming, i.e., the presence of foam-forming filamentous microorganisms in activated sludge. Filaments from foams collapsed with water sprays are returned to the mixed liquor or to the return activated sludge. In fact, this method helps keep the filaments inside the activated sludge system. Thus, the water sprays should only be applied in emergency cases.

## REFERENCES

1  Grau, P., M. Dohanyos and J. Chudoba. 1975. "Kinetics of Multicomponent Substrate Removal by Activated Sludge", *Wat. Res.,* 9:637–642.

2  Adams, C. E., W. W. Eckenfelder and J. C. Hovious. 1975. "A Kinetic Model for Design of Completely Mixed Activated Sludge Treating Variable-Strength Industrial Wastewater", *Wat. Res.,* 9:37–42.

3  Benefield, L. D. and C. W. Randall. 1987. "Evaluation of a Comprehensive Model for the Activated Sludge Process", *Journ. Wat. Poll. Control Fed.,* 49(7):1636–1641.

4  Dold, P. L. and G. v. R. Marais. 1986. "Evaluation of the General Model for the Activated Sludge Process", *Prog. Wat. Tech.,* 12(6):47–77.

5   Grady, C. P. L., W. Gujer, M. Henze, G. v. R. Marais and T. Matsuo. 1986. "A Model for Single-Sludge Wastewater Treatment Systems", *Wat. Sci. Tech.*, 18(6):47–61.

6   Grau, P., P. M. Sutton, M. Henze, S. Elmaleh, C. P. L. Grady, W. Gujer and J. Koller. 1982. "Recommended Notation for Use in the Description of Biological Wastewater Processes", *Wat. Res.*, 16(11):1501–1505.

7   Eckenfelder, W. W. 1980. *Principles of Water Quality Management*. Boston: CBI Pub. Co.

8   Cech, J. S., J. Chudoba and P. Grau. 1985. "Determination of Kinetic Constants of Activated Sludge Organisms", *Wat. Sci. Tech.*, 17(2/3):259–272.

9   Tchobanoglous, G. and E. D. Schroeder. 1985. *Water Quality*. Addison-Wesley Pub. Co.

10  Middlebrooks, E. J. and D. B. Porcella. 1971. "Rational Multivariate Algal Growth Kinetics", *Journ. ASCE*, 97(SA1):135–140.

11  Chudoba, J., P. Grau and V. Ottova. 1973. "Control of Activated Sludge Filamentous Bulking—II. Selection of Microorganisms by Means of a Selector", *Wat. Res.*, 7(10):1389–1406.

12  Henze, M. and C. Mladenovski. 1991. "Hydrolysis of Particulate Substrate by Activated Sludge under Aerobic, Anoxic and Anaerobic Conditions", *Wat. Res.*, 25(1):61–64.

13  Wanner, J., K. Kucman, V. Ottova and P. Grau. 1987. "Effect of Anaerobic Conditions on Activated Sludge Filamentous Bulking in Laboratory Systems", *Wat. Res.*, 21(12):1541–1546.

14  Yamammoto, R. I., T. Komori and S. Matsui. 1991. "Filamentous Bulking and Hindrance of Polyphosphate Removal Due to Sulfate Reduction in Activated Sludge", *Wat. Sci. Tech.*, 23(4/6):927–935.

15  Teichgräber, B. 1988. "Operational Problems of Two Stage Activated Sludge with Nitrification", *Wat. Supply*, 6:125–132.

16  Konicek, Z. and J. Burdych. 1988. "Effect of Activated Sludge Processes on Secondary Settling Tank Efficiencies", *Wat. Sci. Tech.*, 20(4/5):153–163.

17  Vesilind, P. A. 1968. "Theoretical Considerations: Design of Prototype Thickeners from Batch Settling Data Tests", *Wat. Sew. Wks.*, 115:302–306.

18  Sezgin, M., D. Jenkins and D. S. Parker. 1978. "A Unified Theory of Filamentous Activated Sludge Bulking", *J. Wat. Pollut. Control Fed.*, 50:362–381.

19  Jenkins, D., M. G. Richard and G. T. Daigger. 1986. *Manual on the Causes and Control of Activated Sludge Bulking and Foaming*. Pretoria: WRC; Cincinati: U.S. EPA.

20  Pretorius, W. A. and C. J. P. Läubscher. 1987. "Control of Biological Scum in Activated Sludge Plant by Means of Selective Flotation", *Wat. Sci. Tech.*, 19(5/6):1003–1011.

21  Wagner, F. 1982. *Ursachen, Verhinderung und Bekämpfung der Blähschlammbildung in Belebungsanlagen*, Munich, FRG: Kommissionverlag R. Oldenbourg, pp. 40–59.

22  Williams, T. M. and R. F. Unz. 1985. "Isolation and Characterization of Filamentous Bacteria Present in Bulking Activated Sludge", *Appl. Microbiol. Biotechnol.*, 22:273–282.

23  Wanner, J. and P. Grau. 1989. "Identification of Filamentous Microorganisms from Activated Sludge: A Compromise between Wishes, Needs and Possibilities", *Wat. Res.*, 23(7):883–891.

24  Eikelboom, D. H. and H. J. J. van Buijsen. 1981. Microscopic Sludge Investigation Manual. Report A94a, IMG-TNO, Delft.

**25** Buchanan, R. E. and N. E. Gibbons, eds. 1974. *Bergey's Manual of Determinative Bacteriology.* Baltimore: Williams and Wilkins.

**26** Gabb, D. M. O., D. A. Still, G. A. Ekama, D. Jenkins, M. C. Wentzel and G. v. R. Marais. 1988. "Development and Full-Scale Evaluation of Preventive and Remedial Methods for Control of Activated Sludge Bulking", Final Research Report W62, University of Cape Town.

**27** Albertson, O. E. 1987. "The Control of Bulking Sludges: From the Early Innovators to Current Practice", *J. Wat. Pollut. Control Fed.,* 59(4):172–182.

**28** Palm, J. C., D. Jenkins and D. S. Parker. 1980. "Relationship between Organic Loading, Dissolved Oxygen Concentration and Sludge Settleability in the Completely-Mixed Activated Sludge Process", *J. Water Pollut. Control Fed.,* 52(10):2484–2506.

**29** Richard, M. G., G. P. Shimizu and D. Jenkins. 1985. "The Growth Physiology of the Filamentous Organism Type 021N and its Significance to Activated Sludge Bulking", *J. Water Pollut. Control Fed.,* 57(12):1152–1162.

**30** Richard, M. 1989. *Activated Sludge Microbiology.* Alexandria, VA: The Water Pollution Control Federation.

**31** Foot, R. J. 1992. Effects of Process Control Parameters on the Composition and Stability of Activated Sludge. *J. Inst. Water Environ. Manage.,* 6(2):215–228.

**32** Grau, P., J. Chudoba and M. Dohanyos. 1982. "Theory and Practice of Accumulation-Regeneration Approach to the Control of Activated Sludge Filamentous Bulking", in *Bulking of Activated Sludge,* B. Chambers and E. J. Tomlinson, eds., Ellis Horwood Ltd.

**33** Wanner, J., M. Kos and P. Grau. 1989. "An Innovative Technology for Upgrading Nutrient-Removal Activated Sludge Plants", *Water. Sci. Tech.,* 22(7/8):9–20.

**34** Wanner, J. 1992. "Comparison of Biocenosis from Continuous and Sequencing Batch Reactor", *Water Sci. Tech.,* 25(6):239–249.

**35** Wanner, J., J. Chudoba, K. Kucman and L. Proske. 1987. "Control of Activated Sludge Filamentous Bulking—VII. Effect of Anoxic Conditions", *Wat. Res.,* 21(12):1447–1451.

**36** Shao, Y. J. and D. Jenkins. 1989. "The Use of Anoxic Selectors for the Control of Low F/M Activated Sludge Bulking", *Wat. Sci. Tech.* 21:609–619.

**37** Welander, T. 1996. "Microbial Interactions in the Activated Sludge and their Influence on Sludge Separability," Paper presented at the *6th Stockholm Water Symposium,* Stockholm, 5–9 August.

**38** van Leeuwen, J. and W. A. Pretorius. 1988. "Sludge Bulking Control with Ozone", *J. Inst. Water. and Environ. Manag.,* 2(2):223–227.

**39** Wanner, J., J. Zak and M. Hlavicova. 1990. CS Patent No. 269182.

**40** Sturzova, I. 1990. "Controlling Activated Sludge Filamentous Bulking", M.Sc. Thesis, Inst. of Chemical. Technol., Prague (in Czech).

**41** Inamori, Y., Y. Kuniyasu, R. Sudo and M. Koga. 1991. "Control of the Growth of Filamentous Microorganisms Using Predacious Ciliated Protozoa", *Wat. Sci. Tech.,* 23(4/6):963–971.

**42** Pitt, P. A. and D. Jenkins. 1990. "Causes and Control of Nocardia in Activated Sludge", *Research J. Wat. Poll. Control. Fed.,* 62:143–150.

**43** Wanner J. and J. Roskota. 1993. "Metabolic Selection in Full-Scale Plant", *Newsletter of the IAWQ Specialist Group on Activated Sludge Population Dynamics,* 5(1):4–9.

**44** Vlekke, G. J. F. M. 1988. "Biological Phosphate Removal from Wastewater with

Oxygen or Nitrate in Sequencing Batch Reactors'', *Environ. Techn. Letters,* 9:791–796.

**45** Wanner, J., J. S. Cech., M. Kos. 1992. ''New Process Design for Biological Nutrient Removal'', *Wat. Sci. Tech.,* 25(45):445–448.

**46** Sorm, R., Bortone, G., Saltarelli, R., Jenicek P., Wanner J., Tilche A. (1996) ''Phosphate Uptake under Anoxic Conditions and Fixed-Film Nitrification in Nutrient Removal Activated Sludge Systems'', *Water Res.,* 30(7):1573–1584.

**47** Sorm, R., J. Wanner, R. Saltarelli, G. Bortone, A. Tilche. 1996. ''Verification of Anoxic Phosphate Uptake as the main Biochemical Mechanism of the DEPHANOX, Process'', *Proc. IAWQ-NVA Conf. Advanced Wastewater Treatment,* Amsterdam, September, pp. 143–152.

**48** Toerien, D. F., A. Gerber, L. H. Lotter and T. E. Cloete. 1990. ''Enhanced Biological Phosphorus Removal in Activated Sludge Systems'', in *Advances in Microbial Ecology,* Vol. 11, K. C. Marshall, ed. New York, NY: Plenum Press, pp. 173–230.

**49** Painter, H. A. 1970. ''A Review of Literature on Inorganic Nitrogen Metabolism in Microorganisms'', *Water Res.,* 4(5):393–450.

**50** Rheinheimer, G., W. Hegemann, J. Raff and I. Sekoulov. 1988. *Stickstoffkreislauf im Wasser,* Muenchen, FRG: R. Oldenbourg Verlag.

**51** Stensel, H. D. and J. L. Barnard. 1992. ''Principles of Biological Nutrient Removal'', in *Design and Retrofit of Wastewater Treatment Plants for Biological Nutrient Removal,* C. W. Randall et al., eds. Lancaster, PA: Technomic Publishing Co., Inc., pp. 25–84.

**52** Grabinska-Loniewska, A. 1991. ''Denitrification Unit Biocenosis'', *Water Res.,* 25(12):1565–1573.

**53** Schlegel, H. G. 1985. *Allgemeine Mikrobiologie,* Stuttgart, FRG: Georg Thieme Verlag.

**54** Cloete, T. E. and P. L. Steyn. 1988. ''A Combined Membrane Filter-Immunofluorescent Technique for the In-Situ Identification and Enumeration of Acinetobacter in Activated Sludge'', *Water Res.,* 22(8):961–969.

**55** Eikelboom, D. H. 1982. ''Biosorption and Prevention of Bulking Sludge by means of a High Floc Loading'', in *Bulking of Activated Sludge: Preventative and Remedial Methods,* B. Chambers and E. J. Tomlinson, eds. Chichester, UK: Ellis Horwood Limited, pp. 224–242.

**56** Casey, T. G., M. C. Wentzel, R. E. Loewenthal, G. A. Ekama and G. v. R. Marais. 1992. ''A Hypothesis for the Cause of Low F/M Fialment Bulking in Nutrient Removal Activated Sludge Systems'', *Water Res.,* 26(6):867–869.

**57** Kocianova, E., R. J. Foot and C. F. Forster. 1992. ''Physicochemical Aspects of Activated Sludge in Relation to Stable Foam Formation'', *J. Inst. Water Environ Manage.,* 6(3):342–350.

**58** Eikelboom, D. H. 1975. ''Filamentous Organisms Observed in Activated Sludge'', *Water Res.,* 9:365–388.

**59** Blackall, L. L. 1994. ''Molecular Identification of Activated Sludge Foaming Bacteria'', *Water Sci. Tech.,* 29(7):35–42.

**60** Wagner, M., R. Amann, H. Lemmer, W. Manz, K. H. Schleifer. 1994. ''Probing Activated Sludge with Fluorescently Labeled rRNA Targeted Oligonucleotides'', *Water Sci. Tech.,* 29(7):15–23.

**61** Wanner, J., I. Ruzickova, P. Jetmarova, O. Krhutkova, J. Paraniakova. 1997. ''A National Survey of Activated Sludge Separation Problems in the Czech Republic:

Filaments, Floc Characteristics and Activated Sludge Metabolic Properties'', Paper presented at the *2nd IAWQ Intl. Conf. on Microorganisms in Activated Sludge and Biofilm Processes,* Berkeley, CA, 21–23 July.

**62** Wanner, J. 1994. *Activated Sludge Bulking and Foaming Control.* Technomic Publishing Co., Inc., Lancaster, PA.

# CHAPTER 2

# Activated Sludge Treatment of Municipal Wastewater— U.S.A. Practice

## GENERAL APPROACH

### ROLE OF DESIGNERS

IN the United States most municipal wastewater treatment plants are owned and operated by cities, counties, or Water Reclamation (Sanitary) districts. Some of the largest of these employ full-time design staffs to do some of their designs. Most, however, typically contract with an outside professional engineer, or engineering firm, to do the design.

Sizing of process units is done by the design engineers, but is typically restricted by state design standards or guidelines. The Ten State Standards [1] is the most commonly used of these and most state standards/guidelines either explicitly or implicitly refer to these. Typically these standards allow the designer some latitude, generally allowing exceptions if need is well-enough documented. In practice, however, few design firms deviate significantly from these standards. The Water Environment Federation (WEF, formerly WPCF) Manual of Practice 8 (MOP 8) is an often-used guide [2].

After the design engineers have prepared biddable plans and specifications, the project is bid upon by contractors. The contractors construct the project usually with observation, but not inspection (for legal reasons), by the design engineer. The owner typically supplies a construction inspector.

Mark Hsu (deceased) and Thomas E. Wilson, Rust Environment and Infrastructure, Oak Brook, IL.

## EFFLUENT CRITERIA

Designs are to meet effluent standards set up by the local and state regulatory agencies and reflect current U.S.EPA standards (Table 2.1). Secondary treatment is generally the minimum level of treatment required. EPA toxic regulations often require greater degrees of treatment in specific cases. Activated sludge plants, designed according to the principles described in this chapter, routinely produce an effluent containing less than 15 mg/l CBOD5 and SS; many actually produce less than 10 mg/l CBOD5 and SS.

## WASTEWATER CHARACTERIZATION

Wastewater characterization is generally the responsibility of the design engineer. For existing plants, this often can be done by analyzing 3–10 years' worth of plant data. For new plants, characteristics and flows often must be estimated, based upon expected population and the nature of industrial discharges. Table 2.2 presents typical U.S. characteristics. Phosphorus levels in plants where the community has limited the phosphate content in detergents are now typically:

Total P: 2–3 mg/l
Soluble P:0.5–1.5 mg/l

It has become common to consider the in-plant recycle steams. Recycle steams from sludge processing units may typically increase these raw wastewater BOD5, SS, TKN, and P concentrations and flow to aeration by 5–30 percent.

## PRETREATMENT

Pretreatment of raw wastewater precedes virtually all U.S. municipal activated sludge plants. This can range from simple screening and commi-

TABLE 2.1.  Current U.S.EPA Effluent Standards.[a,b]

| Type of Treatment Plant | CBOD$_5$ mg/l | BOD$_5$ mg/l | SS mg/l | N % Removal | pH Std Unit |
|---|---|---|---|---|---|
| Secondary treatment | 25 | 30 | 30 | — | 6–9 |
| Advanced secondary treatment (AST) | — | 10–29 | 10–29 | — | — |
| Advanced wastewater treatment (AWT) | — | 10 | 10 | 50 | — |

[a]30-day average.
[b]Reference 40 CFR 133, as of June, 1991; 44 FR 29534, May 21, 1979.

TABLE 2.2. U.S. Wastewater Characteristics.

| Item | Concentration | | |
| --- | --- | --- | --- |
| | Weak | Medium | Strong |
| Biochemical oxygen de-mand (BOD$_5$), mg/l | 110 | 220 | 400 |
| Chemical oxygen demand (COD), mg/l | 250 | 500 | 1000 |
| Suspended solids, mg/l | | | |
|   Total | 100 | 220 | 350 |
|   Volatile | 80 | 165 | 275 |
| Settleable solids, mg/l | 5 | 10 | 20 |
| Nitrogen (as N), mg/l | | | |
|   Organic | 8 | 15 | 35 |
|   Ammonia | 12 | 25 | 50 |
|   Total | 20 | 40 | 85 |
| Phosphorus (as P), mg/l | | | |
|   Organic | 1 | 3 | 5 |
|   Inorganic | 3 | 5 | 10 |
|   Total | 4 | 8 | 15 |

*Source:* Metcalf & Eddy. 1991 *Wastewater Engineering, Treatment, Disposal, Reuse,* 3rd Edition. New York: McGraw Hill, Inc.

nuters in the very small plants to series of screens, grit removal, and primary sedimentation in large plants. The trend is to use finer screens, with 0.25 inch screens becoming common. Primary treatment typically removes 15–45 percent of the BOD5 and 45–60 percent of the suspended solids. Higher removals are achievable with chemical additions such as alum and ferric chloride. The reader is referred elsewhere [2] for more detailed discussion.

## DESIGN APPROACH

This chapter is concerned only with designs for BOD5 and SS removal and nitrification. Many U.S. plants use this same design approach for full nutrient removal, by adding tertiary denitrification and/or chemical removal of phosphorus.

Activated sludge processes have two principal components:

- aeration tanks (reactors)
- clarifiers

Most of the following will discuss prevailing approaches to the design of these. Some briefer discussion of appurtenances is also included. Appurtenances include:

- aeration diffusers

- return sludge pumps
- waste sludge pumps
- scum removal equipment

The general approach presented in the following follows these basic principles:

- Carefully characterize wastewater with respect to flow, temperature, $BOD_5$, SS, N, P, alkalinity, and the variability (as load and concentration).
- Size the clarifiers first—based upon state standards or CRTC study results.
- Select MLSS settling characteristics, calculate MLSS based upon these, design flow, design flow variability and clarifier size.
- Select a design SRT (Solids Retention Time) based upon kinetic considerations. Factors should include temperature and need for nitrification.
- Estimate activated sludge solids production.
- Size aeration basins, based upon design MLSS, SRT, solids production, and reactor configuration.
- Determine need for and size selectors.
- Size appurtenances.
- Do layout on site; make dimensional decisions.

## CLARIFIER DESIGN

### SURFACE OVERFLOW RATE

In the U.S., surface overflow rate (SOR) is an important design parameter—but not by itself. It must be considered along with MLSS concentrations and settling characteristics. In the CRTC tests, circular clarifiers failed at a relatively low SOR because the clarifiers were overloaded on a solids loading basis. Other clarifiers did not fail at SORs more than 2000 gpd/sf because they were not overloaded on a solids loading basis [34].

The approach used in this chapter is to use the SOR dictated by local state design standards and to calculate the other design parameters. Most states typically require that the SOR of final clarifiers not exceed 800 gpd/sf at average flow or 1200 gpd/sf at maximum flow [3]. Since the CRTC tests have shown that well designed systems can safely taker higher loadings, some engineers have started routinely using maximum SORs as high as 1600 gpd/sf. Typically, where site restrictions allow, using the smallest clarifier allowed within these limits gives a cost-effective design. Virtually all standards require the same SOR for rectangular as for circular clarifiers. Some designers favor lower SORs for rectangular clarifiers, but recent studies suggest this is not necessary [4] and this practice is disappearing in the US.

## DEPTH

In the U.S. clarifier depth depends upon other clarifier design features and the designer's preferences Clarifier depth in modern U.S. designs tends to be getting deeper with each passing decade. Older designs often could be found to have 8–10 ft SWD (meaning "side water depth" for circular clarifiers, and "shallowest water depth" for rectangular clarifiers). Modern designs rarely are less than 12 ft SWD with some designers advocating as deep as 18–20 ft for circular clarifiers [5,6].

There is a good body of information that shows how well deep clarifiers perform (5.6). The real issue here turns out to be: how deep is deep enough? Everyone agrees that if a clarifier is too shallow it will fail, but how deep is deep enough?. Everyone seems to agree that one usually needs about a minimum of 10 ft in any case. Beyond that it depends on:

(1) How you isolate MLSS from the sludge hopper: If the influent MLSS is not separated from the sludge hopper, you need to add an extra 3–5 ft of depth to isolate it. This means that most center feed circular clarifiers with sludge scraped to hoppers need to be at least 13, probably 15 ft deep. It also means that rectangular "Gould Tanks" (which by definition isolate the sludge hopper—see [36]) can be as shallow as 10 ft.

(2) Whether you want store MLSS in clarifier or not: In stormwater influenced systems, for example, many designers like to add 2–4 ft to store MLSS washed into clarifiers. Others prefer to keep solids in aeration basins by using step feed approaches instead. Some designers like to provide both depth and step feed.

(3) Whether or not you want to thicken in clarifiers: Most wwtps have separate WAS thickening and designing final clarifiers to thicken is probably an expensive an unnecessary redundancy. Add 1–3 ft if you want to thicken.

For those who like a more formalized approach to calculating depth of clarifiers, one of the best are is Albertson's approach [33]; the reader is encouraged to review it.

## CLARIFIER SHAPE

In the US, circular and rectangular clarifiers are preferred. There are some significant numbers of other configurations, particularly octagonal with circular collectors, but they are becoming less common. Circular clarifiers are preferred [7,8], and asserted to be superior, by some consultants for any size and type plant. The assertion is often supported with reasonable sounding scientific reasons and supported by personal good experience with circular clarifiers and/or bad experience with rectangular clarifiers. Some designers routinely size rectangular clarifiers 25% larger than circular clarifiers. However CRTC studies have shown that well designed rectangular clarifiers can be expected to perform at least as well as well designed circular clarifiers of the same surface area. In fact, of the clarifiers studied by the CRTC, rectangular clarifiers performed the best—producing less than 10 mg/L ESS at surface overflow rates more than 2000 gpd/sf during stress tests.

Many new plants—particularly large ones—use rectangular clarifiers. Table 2.3 lists some of these. At similar loadings well-designed circular and rectangular clarifiers can be expected to perform similarly. Both type clarifiers have their place, with rectangular clarifiers probably being the best choice for large plants and on tight sites [4,28].

Several types of rectangular and circular clarifier designs are commonly used in the U.S.; two of the best documented clarifiers are "deep circular clarifiers with a center flocculation well" and "Gould" rectangular clarifiers. The key features of these are summarized in Table 2.4 and Figures 2.1 and 2.2. Table 2.4 also presents a comparison of these two types of clarifiers.

TABLE 2.3.  **Example U.S. Plants Using Rectangular Final Clarifiers.**

| Name | Date Built | Design Flow, mgd |
|---|---|---|
| Waukegan, NSSD, IL[a] | 1960, 1976 | 10, 20 |
| Whittier Narrows, LAC, CA | 1961 | 15 |
| Metropolitan WWTP, St. Paul, MN[b] | 1966, 1985 | 218, 250 |
| San Jose Creek, LAC, CA | 1969, 1982 | 37.5, 62.5 |
| Richmond, VA | 1974 | 70 |
| Yonker JTP, NY | 1976 | 92 |
| Tampa, FL[b] | 1978, 1990 | 60, 90 |
| Blue Plains, DC[b] | 1980, 1988 | 310, 370 |
| Carson, LAC, CA[a] | 1984 | 200 |
| Phoenix, AZ | 1984 | 120 |
| Philadelphia, PA (3 plants) | 1980–1986 | 120–250 |
| Def. #1, Camden, NJ[a] | 1987 | 40 |
| Little Rock, AR | 1989 | 12.5 |
| North River, NYC, NY | 1991 | 170 |
| Boston, MA[a,c] | In design | 540 |
| Newtown Creek, NYC, NY | In design | 370 |

[a]Oxygen activated sludge.
[b]Nitrifying system.
[c]Double deck.

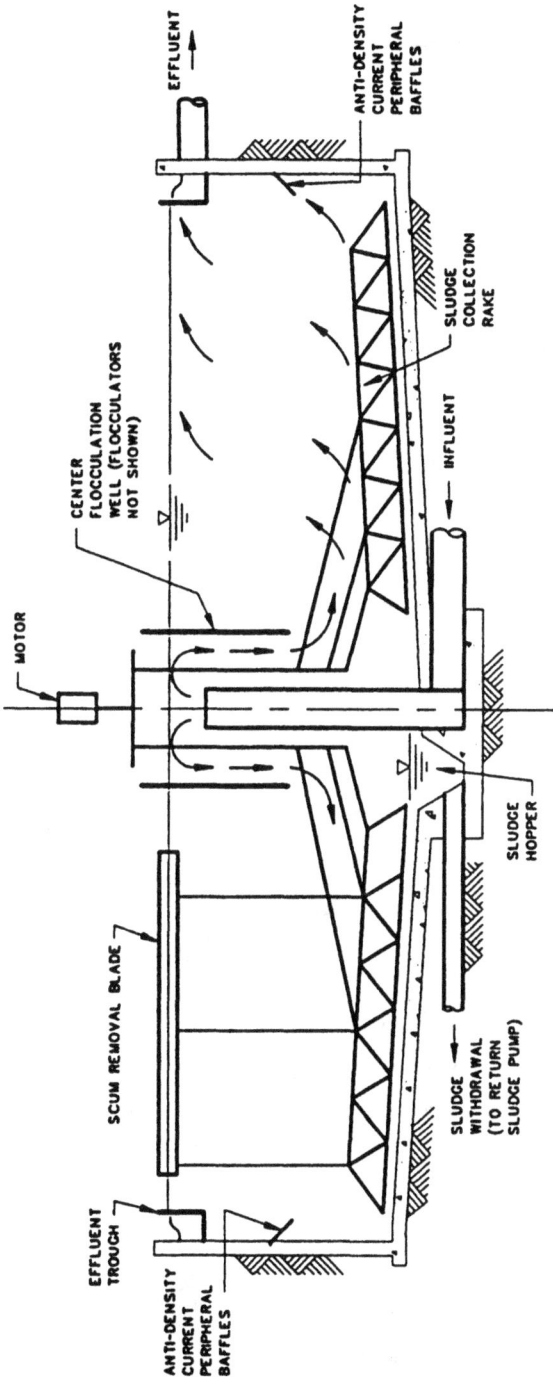

**Figure 2.1** Deep circular clarifier with center flocculation well.

65

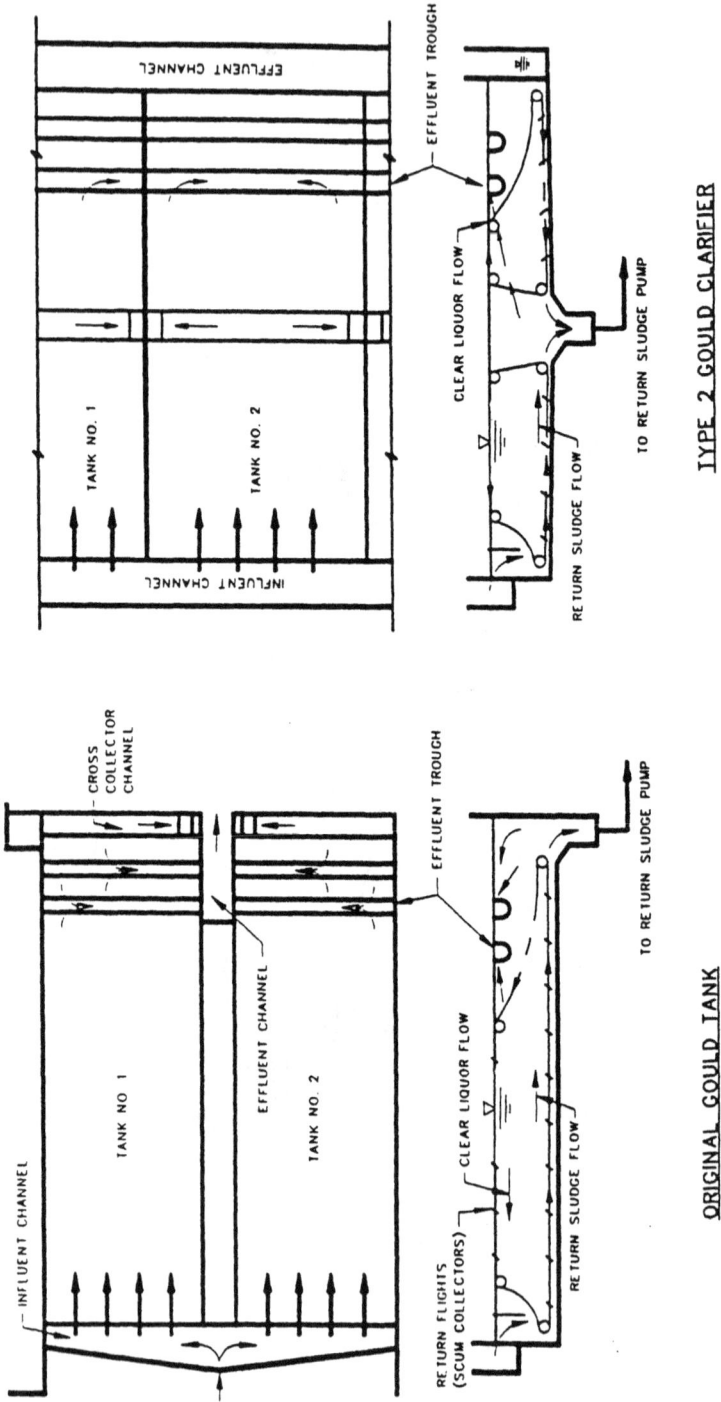

**TYPE 2 GOULD CLARIFIER**

**ORIGINAL GOULD TANK**

**Figure 2.2** "Gould" rectangular clarifiers. Reference: Wilson, T.E. and E.F. Ballotti, 1988. "Gould Tanks: Rectanuglar Clarifiers that Work," paper presented at *The 61st WPCF Annual Conference, October, 1988.*

66

TABLE 2.4. **Key Features and a Comparison of Clarifiers.**

| Deep, Center Floc Circular[a] | Item | Gould Rectangular[b] |
|---|---|---|
| 1890 (24 hr)[d] <br><br> 1500 (3 hr) <br> 2050 (30 min)[c] | Peak SORs demonstrated without failure | 72 000 gpd/st (24 hr) |
| < 10 mg/l | Typical ESS at < 1000 gpd/sf | < 10 mg/l |
| 15–20 ft | SW depth | 12 ft |
| Difficult, erratic | Scum/floating solids removal | Easy, complete |
| Good | Ease of covering for odor/VOC control | Excellent |
| High | Head loss for flow distribution among clarifiers | Low |
| High | Site requirement[5] | Low |
| No | Common wall construction | Yes |

*Source:* References [4,7,8,28,34].
[a]15'–18' deep, 90'–125' diameter.
[b]12' deep, 180 ft long.
[c]TF/SC sludge.
[d]Per square foot of clarifier area.

## SOLIDS LOADING—CSF/RSF

Solids loading is generally recognized as an important design parameter in the U.S., but is not a criterion used in most state standards. Many designers prefer to use a fixed number—typically 50 lb/d/sf at maximum loading. This has been shown to often undersize clarifiers [40]. A more rational approach takes mixed liquor settling characteristics into consideration. A simple way to determine if a clarifier is overloaded is to calculate the CSF and RSF:

$$CSF = V_i/SOR \qquad (2.1a)$$

$$RSF = R/R_{min} \qquad (2.1b)$$

where:

CSF = Clarifier Safety Factor
RSF = Return Safety Factor
$V_i$ = initial settling velocity (gpd/sf)
$R$ = return sludge flow rate / wastewater flow rate
$R_{min}$ = MLSS/(4/k—MLSS)

When both RSF and CSF exceed 1.0, the clarifier is not overloaded. To allow for differences between ideal and real world clarifiers, values higher than 1.0 are typically used for design. $V_i$ can be estimated as:

$$V_i = V_{max} \times \exp(-k \times MLSS \times 10^{-10}) \tag{2.1c}$$

where:

$V_{max}$, $k$ = settling constants of mixed liquor
MLSS = mixed liquor SS concentration, mg/l

Recently it has been found that $k$ is approximately equal to SVI × 4 [35, 41].

$$V_i = V_{max} \times \exp(-4SVI \times MLSS \times 10^{-6}) \tag{2.1d}$$

or, noting that, by definition:

$$SSV = (MLSS/1000) \times SVI \tag{2.1e}$$

where SSV is the 30 minute settled volume, cc / l.

$$V_i = V_{max} \times \exp(-4SSV/10^3) \tag{2.1f}$$

$R_{min}$ can be estimated as:

$$R_{min} = MLSS / (10^6 / SVI - MLSS) \tag{2.1g}$$

$$R_{min} = 1/(10^3 / SSV - 1) = SSV/(10^3 - SSV) \tag{2.1h}$$

Figure 2.3(a) shows the relationship between SSV and $V_i$. $V_{max}$ is typically near 23 fph (the dark, black, unlabeled curve) but specific for a particular MLSS, typically increasing with increasing water temperature and chemical addition (Wilson et al). The temperature dependency has been estimated to be:

$$V_{max}(fph) = 2.23T - 16.6 \tag{2.1i}$$

where $T$ is temperature in degrees C.

Typically an SVI of 150 is chosen for design. This leads to a $k$ of 600 l/mg. Most U.S. designers recognize 150 as the breakpoint between bulking and non-bulking sludges. Since, as discussed in Chapter 1, it is possible in design to prevent bulking, it is reasonable, but somewhat arbitrary, to

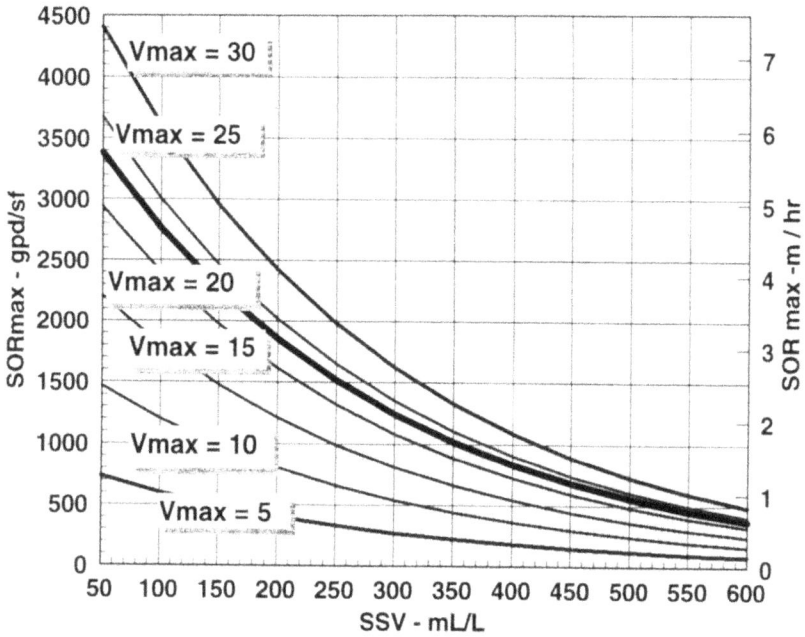

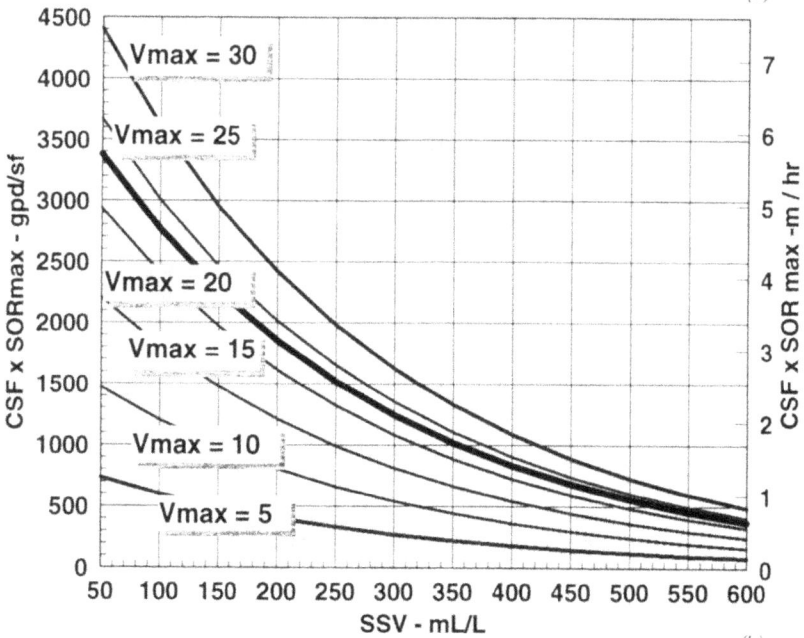

**Figure 2.3** (a,b) Effect of SSV on maximum allowable SOR

select 150. If a designer has confidence that the sludge will always settle better (as might be the case when expanding an existing plant, and if a good data base exists) a better design settling characteristic would be appropriate. Similarly, if experience suggests that a much bulkier sludge is to be expected, then a poorer characteristic would be appropriate.

The CSF is selected typically following guidance as shown in Table 2.5. Note that this approach allows the designer to rationally account for flow variability as well as other factors. CSF is defined as:

$$CSF = V_i \text{ (design)} \times 179.5/SOR \text{ (design)} \tag{2.2}$$

where 179.5 is a factor converting ft/hr to gpd/sf.

Selection of RSF is less well defined. Preliminary information suggests that type of clarifier may make a difference. Typically a value of at least 1.5 should be used.

## DESIGN MLSS

Since SOR is fixed as above, the usefulness of this approach lies in selection of the design MLSS (X). This is the concentration of the MLSS applied to the final clarifiers.

Combining Equations (2.1c) and (2.2) and rearranging

$$X = \frac{10^6}{K} \ln \frac{V_{max} \times 179.5}{CSF \times SOR} \tag{2.3}$$

TABLE 2.5. **Guidance for Choosing Design Clarifier Safety Factor (CSF).**

$$CSF = f_c \times f_m \times f_s \times f_u \times (Q_{msf}/Q_a)$$

where

$f_c$ = configuration factor; usually 1.0 for step aeration, 1.5 for all other configurations

$f_m$ = multiple clarifier factor; allows for uneven split among clarifiers, usually 1.0–1.25, increasing as number of clarifier increases; use 1.25 if in doubt

$f_s$ = series treatment factor; 1.0 if another activated sludge process, filter, clarifier, or lagoon follows; 1.5 if this is the last process

$f_u$ = uncertainty factor; 1.0 if settling characteristics, flow variability and clarifier performance known with a high degree of confidence; 1.5 or higher if uncertainties exist

$Q_{msf}/Q_a$ = maximum sustained flow divided by design average flow; definition varies with designer; typically maximum daily or monthly flow; 1.6 is a common value

Alternatively, Figure 2(a) can be modified to include CSF [see Figure 2(b)] and used by selecting $V_{max}$, SOR, CSF and SVI; then solving Equation (2.1e) for $X$:

$$X = 1000 \times SSV \, / \, SVI$$

### Example 2.1

For a design SOR of 800 gpd/sf and temperature of 17.8°C, what MLSS can be applied to the final clarifier, at a CSF of 2? Assume design SVI of 150. What should design R be if an RSF of 1.5 is used?
*Answer:*

$$V_{max}(fph) = 2.23T - 16.6 = 2.23 \times 17.8 - 16.6 = 23$$

$$k = 4SVI = 4 \times 150 = 600$$

$$X = \frac{10^6}{600} \ln \frac{23 \times 179.5}{2 \times 800} = 1580 \text{ mg} \, / \, 1$$

Alternatively:

$$SOR \times CSF = 800 \times 2 = 1600 \text{ gpd} \, / \, sf$$

Using Figure 2(b) read SSV = 240

$$X = 1000 \times SSV \, / \, SVI = 1000 \times 240/150 = 1600 \text{ mg/L}$$

$$R_{min} = 1/ \, (10^3 \, / \, SSV - 1) = SSV \, / \, (10^3 - SSV) = (MLSS/1000 \\ \times SVI) \, / \, (10^3 - MLSS/1000 \times SVI)$$

$$= 1580/1000 \times 150 \, /(10^3 - 1580/1000 \times 150) = 0.31$$

Design $R$ = RSF $\times R_{min}$ = 1.5 $\times$ .31 = 0.46 (say 50%)

The effect of choice of K and SOR on MLSS is shown in Figure 2.4.

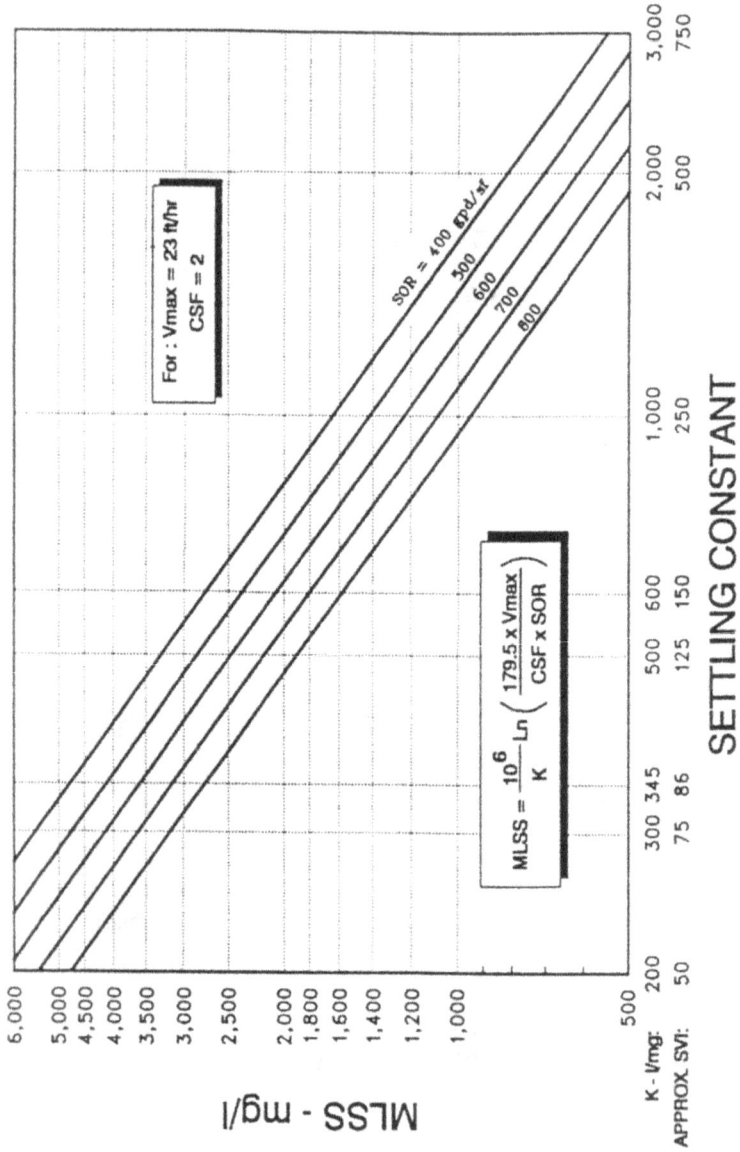

**Figure 2.4** Effect of settling constant on allowable MLSS.

## AERATION TANK (REACTOR) DESIGN

Traditionally, most designers and textbook authors have emphasized a kinetic approach to design aeration tanks. Many authors present "design" equations that suggest tank sizing is dependent upon the effluent BOD5 desired. The trend in the U.S., and this author's approach is to use kinetics only to assure the proper microbiology in order to:

- distinguish between $BOD_5$ removal (only) and nitrifying systems
- reflect temperature effects
- assure a good settling sludge

It is now becoming apparent that kinetics are of little importance in predicting effluent SS and $BOD_5$ concentrations for municipal wastewater treatment plants; the design of the clarifier and MLSS settling characteristics are more important for this. Some very complicated models which require long computation times on even the fastest computers are becoming available, but are not yet commonly used.

Kinetic models are, however, more useful for design of biological nutrient removal systems.

### SRT

The principal parameter used to size aeration basins is SRT (solids retention time). There is no 100 percent agreement in the U.S. as to how this is calculated, and how it differs from MCRT (mean cell retention time) and other, "sludge age" calculations. Many designers prefer the following definition and it is the one that shall be used in this chapter:

$$SRT = \theta = 8.34 \ X_a V_a/W \qquad (2.4)$$

where

$X_a$ = average mixed liquor solids concentration, mg/l
$V_a$ = aeration tank volume, mg
$W$ = solids wasted, lbs/day

The solids wasted, $W$, includes effluent suspended solids as well as intentional sludge wastage.

### SOLIDS YIELD

Solids yield is a function of the wastewater characteristics, the SRT, and other factors, including degree of oxygenation. It is probably the single most important design parameter used in the design of aeration tanks. Discussion of yield is simplified by rewriting Equation (2.4):

$$\theta = X_a V_d / (X_w Q_a) \qquad (2.5)$$

where

$Q_a$ = wastewater flow (exclusive of return sludge), mgd
$X_w$ = solids wasted per unit design flow, mg/l = $W/(8.34\ Qa)$
Now the apparent yield, $Y$, can be defined:

$$Y = X_w / C_i \qquad (2.6)$$

where
$C_i$ = total $BOD_5$ to aeration, mg/l
$Y$ = yield, lb SS/lb $BOD_5$ applied

There are numerous ways of calculating or estimating $Y$. Because of the typically high degree of variability in this parameter, there generally is no reason to use anything more sophisticated than the following:

| | Yield, lb SS / lb $BOD_5$ Applied | |
| --- | --- | --- |
| Type of System | With Primaries | Without Primaries |
| High rate, non-nitrifying | 0.9 (0.8–1.1) | 1.15 (1.0–1.3) |
| Nitrifying | 0.65 (0.5–0.75) | 0.85 (0.7–1.0) |
| Extended aeration | Not applicable | 0.60 (0.5–0.7) |

The higher the influent SS/$BOD_5$ ratio, the higher the yield that should be chosen. Degree of aeration is also important and the following correction factors are recommended:

| Degree of Aeration lb $O_2$/lb $BOD_5$ Applied[*] | Yield Correction Factor |
| --- | --- |
| < 0.8 (Low) | 1.2 |
| 1.0 (Average) | 1.0 |
| ≥ 1.2 (High) | 0.8 |

## REACTOR SIZING

With the values of SRT ($\theta$), MLSS ($X$), and solids yield determined, the volume of the reactor can be calculated by rewriting Equation (2.4) with the value of W expressed in terms of $Q_a$, $C_i$, and $Y$:

[*] A per lb. $O_2$ required for nitrifying systems; see oxygen requirements section of this chapter.

$$V_a = (\theta) \ (Q_a) \ (C_i) \ (Y)/X_a \qquad (2.7)$$

where $X_a$ is the average MLSS.

For plug-flow or CMAS (completely mixed activated sludge) systems, $X_a$ is equal to $X$, the MLSS concentration to the clarifier (see "Design MLSS Section" for calculation). For step aeration plants (and contact stabilization, which is a variant of step aeration), $X_a$ may be calculated:

$$X_a = Z_x \qquad (2.8)$$

where $Z$ is the "step feed factor."

$Z$ is a function of reactor configuration, point of wastewater and return sludge feed, and the return sludge fraction ($R$). Typically, a value of 1.5 is used for step feed design. Detailed discussion of the value of $Z$ can be found elsewhere [27].

## DEPTH

Most U.S. aeration tanks have a water depth of 15 ft. The most common exceptions are in very small plants (below 1 mgd) which typically are shallower, and on tight sites where deeper tanks are becoming more common (see the section on "Configuration," later in this chapter).

## SECONDARY TREATMENT—BOD$_5$ REMOVAL

If only BOD$_5$ removal is needed, very small aeration tanks can be used, with a theoretical minimum SRT of only about 0.2 days. Design texts suggest a kinetic approach, but warn that designing (or operating) at this minimum "washout" SRT is not possible because of the poor settling characteristics of the mixed liquor at such low SRTs. Typically their suggested design SRTs are 2–20 times washout values, and this will result in aeration tank design BOD$_5$ loadings of 35–60 lbs/d-1000 cf. Using the IAWPRC design model [9], the washout SRT at 20–25°C is about 0.3 days. Recently, however [18], it has been demonstrated at full plant scale in a 60 mgd wastewater treatment plant, that plants can be operated successfully at near-washout conditions. This plant operated in the 0.3–0.4 day SRT range and was still able to maintain an effluent with less than 20 mg/l BOD$_5$ and SS in it, and with SVIs near their normal 120–150 values. This suggests that kinetics of BOD$_5$ need not be limiting, and that aeration tank design should be set by some other parameters. As discussed later, this parameter is related to the type of oxygen transfer devices used.

Since most U.S. review agencies do not accept aeration designs above

60 lbs $BOD_5/d$-1000 cf (except for oxygen-activated sludge systems), a typical design practice is to assure that the design $BOD_5$ loading, at average conditions, is below 60 lbs/d-1000 cf, while selecting an SRT of at least 3 days in cold wastewaters (below 20°C) and of at least 2 days in warm (above 20°C) wastewater. Exceptions to this are in the design of the first stage of two-stage nitrification plants and oxygen-activated sludge plants, which can be designed at one-third to one-half of these numbers (i.e., 0.7–1.5 days).

### Example 2.2

For the clarifier and mixed liquor described in Example 2.1, size a reactor system for a 10 mgd treatment plant with a 120 mg/l $BOD_5$, 100 mg/l SS primary effluent fed into it. Assume effluent standards of 20/20 (mg/l $BOD_5$/mg/l ss) and no N standard. Assume design water temperature is 14°C and 1.2 lb $O_2$/lb $BOD_5$ applied will be supplied.

*a.* Size as a plug-flow reactor.
*b.* Size as a step feed plant, with $Z = 1.5$.

*Answer:*
a. (1) $V_a = (\theta)(Q_a)(C_i)(Y)/X_a$
   let $\theta = 3$ days ($T < 20°C$)
     $Y = 0.8 \times 0.9 = 0.72$ (high aeration)
     $V_a = 3 \times 10 \times 120 \times 0.72/1580$
       $= 1.64$ mg
  (2) Check $BOD_5$ loading:
      $= 120 \times 8.34 \times 10/[1.64 \times 10^6/(7.48 \times 10^3)]$
      $= 45.6 < 60$ lbs/d-1000 cf (O.K.)
   Use 1.64 mg
b. (1) $V_a = 3 \times 10 \times 120 \times 0.72/(1580 \times 1.5)$
     $= 1.09$ mg
  (2) Check $BOD_5$ loading:
      $= 120 \times 8.34 \times 10/[1.09 \times 10^6/(7.48 \times 10^3)]$
      $= 68.7 > 60$ lbs/d-1000 cf (too high)
   Use $68.7/60 \times 1.09 = 1.25$ mg

## NITRIFICATION

If nitrification is required, higher SRTs are needed. Both one- or two-stage nitrification plants are used in the U.S. Figure 2.4 illustrates these approaches. Generally, two-stage systems are only found in large (above 10 mgd) plants with moderate- to high-strength wastewater (above 120

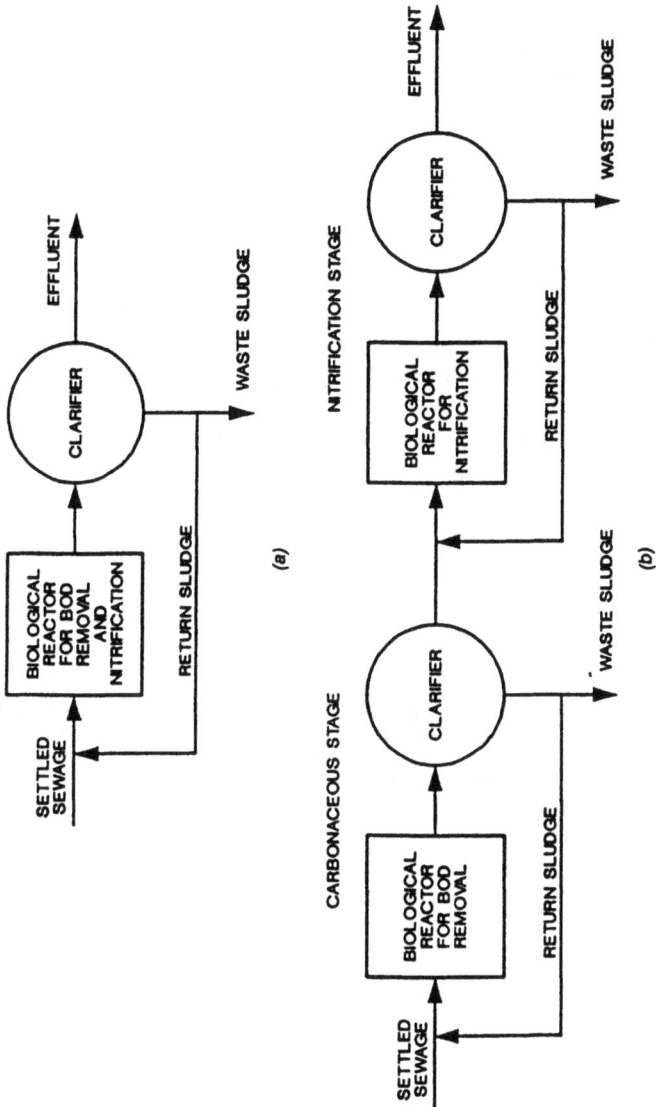

**Figure 2.5** Biological nitrification systems: (a) single-stage nitrification process; (b) two-stage nitrification process.

77

TABLE 2.6.  Conditions Favoring Selection of Two-Stage Nitrification.

- Wastewater strength is high.
- Wastewater temperature is low.
- High inert solids are in activated sludge influent.
- Uncontrollable toxic loads are expected.
- Flexibility is important.
- Highly competent full-time operators are available.
- Pure oxygen is used instead of air.
- Sludge disposal costs are low.
- Energy costs are high.
- Post denitrification is required.

mg/l $BOD_5$ to aeration system) which have primary treatment. Table 2.6 lists conditions favoring selection of each type of these processes. Table 2.7 lists the U.S. plants known to the author to be two-stage plants. The more common design, particularly among smaller plants, is often single-stage—with or without primary treatment—and, most recently, with some sort of selector.

Nitrification design requires using a design minimum SRT that is a function of the minimum monthly temperature. The U.S.EPA model seems to work well [10]:

$$\theta_{min} = 5.56 \exp [0.116 \times (15 - T)] \tag{2.9}$$

where $T$ is water temperature in °C.

TABLE 2.7.  Example Two-Stage Biological Nitrification Plants in the U.S.

| Location | Average Design Flow, mgd |
|---|---|
| Marlborough, Massachusetts | 5.5 |
| MWRDCGC, John Egan Plant, | |
|    Chicago, Illinois | 30 |
| NSSD | |
|    Waukegan, Illinois | 19.9 |
|    Gurnee, Illinois | 12.9 |
|    Clavey Road (Highland Park, Illinois) | 17.8 |
| Blue Plains, Washington, DC | 370 |
| Tampa, Florida | 96 |
| Jackson, Michigan | 13.5 |
| Flint, Michigan | 56 |
| Whittier Narrows, LACSD, California | 12 |
| Loxahatchee, Florida | 4 |
| Mahoning, Ohio | 4 |
| Amherst, New York | 24 |
| Pensacola, Florida | 32 |
| Empire, Minnesota | 9.0 |

This equation usually works well to produce an effluent with less than 0.5 mg/l ammonia N in it. Higher SRTs may be needed to consistently reach lower values. A typical, more rigorous, approach (USEPA) uses the equation:

$$\theta_{design} = (1 + N) / [N \times 0.47 \times \exp (0.098 (T - 15))]$$

where $N$ is desired effluent ammonia nitrogen, mg/L

Besides SRT, the other important factor to be considered in the design of a nitrification plant is the alkalinity requirement. This is the case at Blue Plains WWTP, Washington, D.C., for example. Alkalinity requirements in a nitrification process can be estimated as:

$$\text{Alk. required} = 7.2 \times (NO_3-N)_{formed} + 50 - (Alk)_{in} \qquad (2.10)$$

where alkalinity is expressed in terms of mg/l of $CaCO_3$ and nitrate formed in mg/l nitrogen. The maximum hour requirements of alkalinity are usually used in design. The calculation of $(NO_3-N)$formed is presented in the next section.

## WET WEATHER CONSIDERATIONS

In much of the U.S. short term flows can often reach several times dry weather flows during and just after storms. Historically many treatment plants just bypassed, ahead of the plant, all flows over some predetermined quantity—typically 1.5–2 times dry weather flow, but sometimes as low as just one times average design flow. Modern practice is to provide at least primary treatment and disinfection to all flow and to treat as much as possible through as much as possible of the full treatment plant. This requires using design and operating approaches which allow as much flow as possible to be treated through the activated sludge process without losing biomass.

Using Step Feed capability—moving feed point towards and ultimately to last pass is one very effective way to handle wet weather flows. Another method that is becoming increasingly popular in North America is to provide a partial bypass around the aeration tank, but not clarifiers. This is illustrated on Figure 2.6a. Typically all flow *over* a fixed flow (usually 1–1.5 times average dry weather flow, $Q_D$) is bypassed. This reduces the solids loading to the clarifiers. This is illustrated on Figure 2.6b where it is shown how bypassing all flow over design flow ($Q_D$) reduces MLSS to clarifiers. This in turn slows the decrease in clarifier safety factor (CSF) as flow increases. This is illustrated for two cases:

- a under loaded case when CSF is 2 at start of storm [Figure 2.6c] and
- a stressed case where CSF is just above 1 at start of storm [Figure 2.6d

As demonstrated on these figures, this approach allows flows of up to 3.5–4.9 times design flow before failure (as defined by CSF dropping below 1.0) occurs.

## OXYGEN REQUIREMENTS

There are many methods used in the U.S. to estimate the oxygen requirements in the activated sludge process. Generally it is adequate to use the following:

| System | Oxygen Requirements |
|---|---|
| Low SRT, Non-nitrifying | 1 lb $O_2$/lb $BOD_5$ applied |
| High SRT, Nitrifying | 1.2 lb $O_2$/lb $BOD_5$ applied + 4.6 lb $O_2$/lb $NO_3$–N formed |

In most cases it is adequate to assume that $NO_3$–N formed is equal to $NH_3$–N applied to the aeration tank. For unusual wastewaters, an estimate of $NO_3$–N formed is recommended. The general expression for $NO_3$–N formed in mg/l is:

$$(NO_3–N) = (TKN)_{in} - (f_N)(X_w) - (NH_3–N)_{eff} - N_{soe} \qquad (2.11)$$

where

$\quad (TKN)_{in}$ = total kjeldahl nitrogen (including ammonia nitrogen and organic nitrogen) applied to the aeration tank, mg/l

$\quad f_N$ = fraction of nitrogen in waste sludge, typically 0.06 to 0.1

$\quad X_w$ = solids wasted per unit design flow, mg/l

$\quad (NH_3–N)_{eff}$ = ammonia nitrogen in effluent, typically 0.0–1.0 mg/l

$\quad N_{soe}$ = soluble organic nitrogen in effluent, typically 1.0–1.5 mg/l

Typically a firm capacity for maximum day loading is provided or, more simply, two times the average requirements.

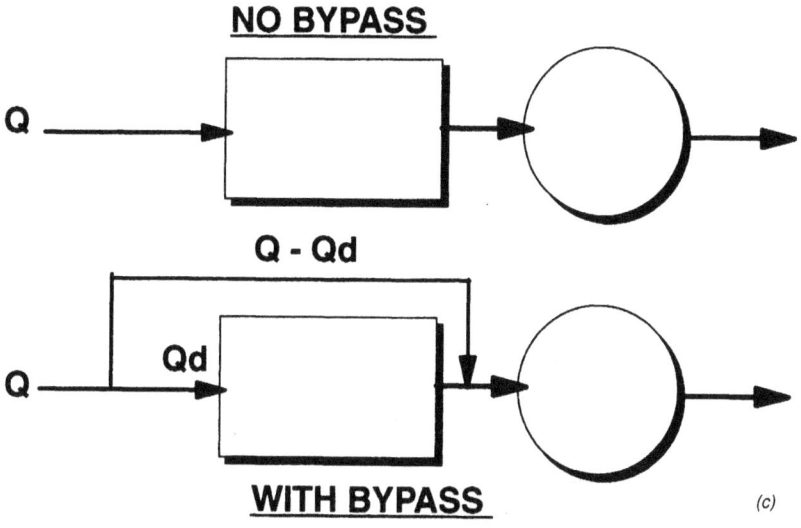

**NO BYPASS**

**WITH BYPASS**

**EFFECT OFSTORMWATER BYPASS**

MLSS TO CLARIFIER

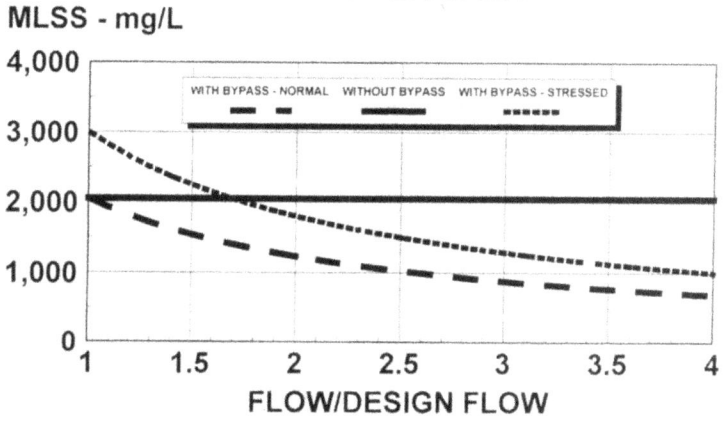

**Figure 2.6(a)** Wet weather partial bypass concept; **(b)** effect of MLSS on relative capacity;

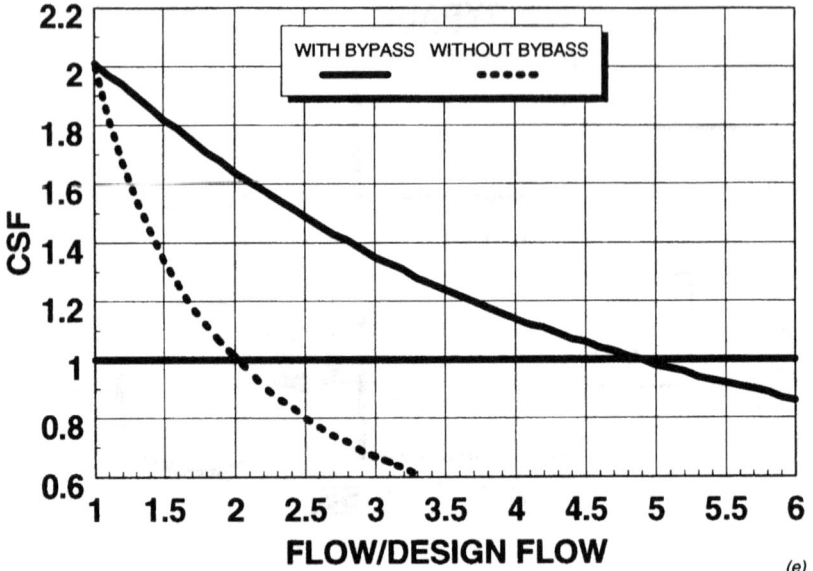

(e)

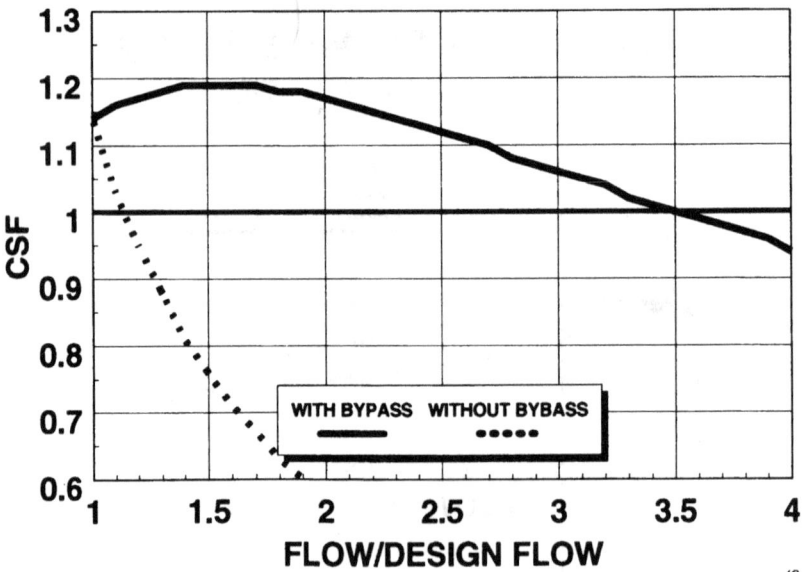

(f)

**Figure 2.6(c) (Continued)** effect of bypass on CSF; (d) effect of bypass on CSF: stressed case.

## Example 2.3

For the same wastewater clarifier and mixed liquor as in Example 2.2, size a reactor system for full nitrification of the wastewater at 14°C. Use a single-stage system.

*Answer*
a. As a plug-flow reactor
$V_a = (\theta)(Q_a)/(C_i)(Y)/X_a$
$\theta_{min} = 5.56 \exp [0.116 \times (15 - 14)]$
$\quad = 6.24$ d
   Assuming a safety factor of 1.5:
$\theta = 6.24 \times 1.5 = 9.4$ d
$Y = 0.65$ (nitrifying)
$V_a = 9.4 \times 10 \times 120 \times 0.65/1580$
$\quad = 4.64$ mg
b. As a step feed plant
$V_a = 9.4 \times 10 \times 120 \times 0.65/(1580 \times 1.5)$
$\quad = 3.09$ mg

## Example 2.4

How much alkalinity must be added to the aeration basin in Example 2.3 if the following are in the feed to the aeration system?

Alkalinity: 100 mg/l (as $CaCO_3$)
TKN: 20 mg/l
$(NO_2 + NO_3)$—N: 0 mg/l

Assume 0.5 mg/l $NH_3$—N, and 1.5 mg/l soluble organic N, in plant effluent, and $f_N = 0.08$.

*Answer*
$(NO_3-N)_{formed} = (TKN)_{in} - (f_N)(X_w) - (NH_3 - N)_{eff} - N_{soe}$
$\quad = 20 - 0.08 \times (0.65 \times 120) - 0.5 - 1.5$
$\quad = 11.8$ mg/l
Alk. required $= 7.2 \times (NO_3 - N)_{formed} + 50 - (Alk)_{in}$
$\quad = 7.2 \times 11.8 + 50 - 100$
$\quad = 35$ mg/l as $CaCO_3$

## PERCENT OXYGEN TRANSFER

To satisfy the requirements as determined above, enough oxygen must be supplied but only a fraction of the amount supplied can be absorbed and become available to the microorganisms. The oxygen transfer efficiency is in general lower in the aeration tank mixed liquor than it is in the clean water; it can also decreases with the time of service due to the continuous fouling of the diffusers in the diffused air systems. The general expression [16] is:

$$\text{OTE}_f = \alpha F \, (\text{SOTE}) \theta^{(T-20)} \, (\Omega \, \tau \beta \, C^*\infty_{20} - C)/C^*\infty_{20} \qquad (2.12)$$

where:

$\text{OTE}_f$ = oxygen transfer efficiency under process conditions, %

$\alpha$ = (process water $K_L a$ of a new diffuser)/(clean water $K_L a$ of a new diffuser)

$F$ = (process water $K_L$ of diffuser after a given time in service)/($K_L a$ of a new diffuser in the same process water)

$\text{SOTE}$ = oxygen transfer efficiency under standard conditions (20°C, 1 atm, C = 0 mg/L), %

$\theta$ = 1.024

$T$ = process water temperature, C

$\theta^{(T-20)} = K_L a/K_L a_{20}$

$K_L a$ = apparent volumetric mass transfer coefficient in clean water at temperature $T$, 1/hr

$\Omega$ = pressure correction for $C^*\infty = P_b/P_s$ (approximation for water depths of 20 ft or less)

$P_b$ = field atmospheric pressure, psia

$P_s$ = standard atmospheric pressure (14.7 psia or 1.0 atm at 100 percent relative humidity), psia

$\tau$ = temperature correction for $C^*\infty = C^*\infty/C^*\infty_{20} = C^*s/C^*s_{20}$

$\beta$ = (process water $C^*s$)/(clean water $C^*s$)

$C^*\infty_{20}$ = steady state DO saturation concentration attained at infinite time for a given diffuser in clean water at 20°C and 1 atm, mg/L

$C$ = process water DO concentration, mg/L

$C^*\infty$ = steady state DO saturation concentration under process conditions at T°C and field atmospheric pressure $P_b$, mg/l

$C^*\infty_s$ = tablular value of dissolved oxygen surface saturation concentration at process water temperature, standard atmospheric pressure and 100% relative humidity, mg/l.

$C^*s_{20}$ = tablular value of dissolved oxygen surface saturation concentration at 20°C, standard atmospheric pressure and 100% relative humidity, mg/l.

Before converting $\text{OTE}_f$ values to SOTE values, a preliminary selection of the fine pore diffuser must be made. Information specific to that diffuser can then be obtained, including values for certain of the coefficients in Equation (2.12).

The fouling factor is dependent upon the cleaning frequency of the diffusers. Diffuser fouling can be due to the deposition of airborne particles on the inside surfaces of the diffusers and inorganic precipitates, as well as the accumulation of biological slime on the wetted surface of the diffusers. Frequent cleaning is often necessary during plant operation. Air side fouling should not be a problem as long as proper filtration is provided with the blowers. In design, a fouling factor of 0.9 is typically used with fine pore aerators. Detailed discussions on diffuser fouling, the change of fouling factor with time, and methods for diffuser cleaning are given elsewhere [16].

The value of SOTE is dependent upon the aeration equipment selected, water depth, diffuser placement density, airflow rate/diffuser, diffuser configuration (i.e., full floor, spiral roll, etc.) and other factors, and is usually available from the manufacturers. Typical values for 15 foot deep aeration basins are 20 to 40 percent for fine pore diffusers and 5 to 15 percent for coarse pore diffusers. The value of $\alpha$ varies and is a function of diffuser type, wastewater characteristics, and whether or not the system is nitrifying. Typically $\alpha$ for fine pore systems is in the range of 0.3 to 0.6 with higher values corresponding to nitrifying systems. Coarse pore $\alpha$ values are typically about 50 percent higher. It is highly desirable that $\alpha$ be determined from field tests. The off-gas oxygen transfer test is often used to determine the value of $\alpha$ [24].

Typical values of $OTE_f$ for diffused air aeration systems in 15 ft deep tanks are as follows:

| System | Diffusers | $OTE_f$, % |
|---|---|---|
| Low SRT, non-nitrifying | Coarse | 4–7 |
| | Fine | 6–10 |
| High- SRT, nitrifying | Coarse | 4–7 |
| | Fine | 9–15 |

For mechanical aeration systems, Equation (2.12) is normally written using the amount of oxygen transfer in terms of lb $O_2$/hp-hr instead of the percent transfer, and the value of a is typically in the range of 0.8 to 0.9. For activated sludge mixed liquors, the actual transfer is typically in the range of 1.0 to 2.0 lbs. $O_2$/hp-hr (wire horsepower) at 20°C.

## AIR REQUIREMENTS

The air requirements, which are used to blower capacity for the select diffused air aeration system, is determined by using the expression:

$$Q_{air} = 100 \times W_{oxygen}/(OTE_f \times 0.2315 \times 1440 \times 0.075) \quad (2.13)$$
$$= 4 \times W_{oxygen}/OTE_f$$

where

$Q_{air}$ = air requirements, scfm
$W_{oxygen}$ = oxygen requirements, lbs/day
0.2315= percent of oxygen in air by weight
 1440 = conversion factor, min/day
0.075 = specific weight of air at 20°C, 1 atmospheric pressure, lb/cf

This amount of air supply is then checked with that required to bring about an adequate mixing in the aeration tank. The power required for mixing typically ranges from 0.4 to 1.5 hp/1,000 cf (delivered horsepower) [25]. The following expression is used to calculate the delivered horsepower (HP) at an air supply of Qair (scfm) [26]:

$$HP = 0.227 \, (Q_{air}) \, [(P_2/P_1)^{0.283} - 1] \quad (2.14)$$

where

$P_1$ = atmospheric pressure—blower suction piping loss, typically: $14.7 - 1.0 = 13.7$ psia
$P_2$ = atmospheric pressure + diffuser static head + measured diffuser head loss + blower discharge piping loss, typically: $14.7 + 0.433 \times 14 + 1.0 + 0.1 = 21.86$ psia for fine pore diffusers and 21.36 psia for coarse pore diffusers, assuming an aeration tank water depth of 15 ft.

## PURE OXYGEN-ACTIVATED SLUDGE PLANTS

Oxygen is supplied to the activated sludge process through aeration with either air or pure oxygen. The design of a pure oxygen plant is, in principle, the same as for an air aeration system. Municipal pure oxygen-activated sludge plants in the U.S. typically consist of four covered, completely mixed reactor stages in series. Surface aerators, with draft tubes or bottom impellers in deeper tanks, are commonly used to mix and provide oxygen transfer. High-purity oxygen is introduced into the first stage, along with return sludge and wastewater, and a gas containing 20–60 percent oxygen is vented from the last stage. DO concentrations vary, but typically are in the 5–10 mg/l range. In the U.S. the process is marketed by Lotepro (UNOX®) and Kruger (OASES®).

## CONTROL OF SETTLING PROPERTIES

To achieve a satisfactory degree of treatment, adequate settling of the biological sludge is essential. Sludges having good settling properties should contain a well-flocculated biological floc with a limited number of filamentous organisms. Sludges with poor settling properties are most commonly associated with large numbers of filamentous organisms.

Three methods are commonly used to control the excessive growth of the filamentous organisms:

- selectors
- return sludge chlorination
- step feed operation

### Selectors

Selectors are small tanks where the aeration tank influent and the return sludge are mixed. The function of a selector is to prevent the massive growth of filamentous organisms by providing a short period of detention in a high *F/M* environment. Depending upon the design of each particular wastewater treatment plant and the type of filamentous organisms present, selectors can be aerobic, anoxic, or anaerobic. Selectors may be covered by one or more patents [11]. Anoxic selectors generally can be designed in nitrifying systems, in the range of 10–30 minutes detention time, with satisfactory results. A good design usually provides plug-flow characteristics, often by compartmentalizing into at least three chambers in series. Selectors are discussed in more detail elsewhere [38].

### Return Sludge Chlorination

This is another technique commonly used to control the excessive growth of filamentous organisms in the activated sludge process. Most larger U.S. plants include the capability to chlorinate return sludge. While other oxidants, such as hydrogen peroxide, could also be used, chlorine generally is the most popular chemical selected. Filamentous organisms are usually more susceptible to chlorine. When properly dosed, chlorine is not as toxic to the floc-forming organisms normally existing in the biomass but discourages the development of filaments. Chlorine is typically added to the return sludge line at a dose in the range of 5 to 15 lbs/1000 lb of solids in the aeration tank [29].

### Step Feed Operation

Strictly speaking, step feed operation does not actually control or suppress the growth of filaments—rather, it accommodates them. According to Equation (2.1), the settling velocity of a bulkier sludge can be improved by lowering the mixed liquor suspended solids concentration (the value of $X$ in the equation). The function of step feed is to decrease the aeration tank effluent MLSS by shifting the solids inventory to the front portion of the aeration tank. In a step feed plant (typically called a "step aeration" plant in the U.S.), the aeration tank is separated into 3 to 5 (typically 4) passes in series. Generally each pass is a plug-flow type reactor. Return sludge is always fed to the first pass while the aeration tank influent can be optionally directed to one or more of the passes. By doing so, the solids content in the aeration tank effluent being applied to the clarifier is significantly decreased, but the total inventory of the biological solids, and thus the SRT of the treatment system, remains the same. Figure 2.7 shows an example of step feed operation. By feeding the plant influent to pass C, the MLSS of the aeration tank effluent can theoretically be decreased by 50 percent, from 2000 to 1000 mg/l. For a sludge with a $V_{max}$ of 23 ft/hr at a settling constant of 600 l/mg, this reduced MLSS would increase the settling velocity of the activated sludge by 82 percent (from 6.93 to 12.6 ft/hr).

It may be noted, however, that a step feed option can only be considered a temporary measure, allowing the treatment system to accommodate a sudden increase in filamentous organisms. To provide room for unexpected plant upsets, operation should be brought back to the plug-flow mode as soon as there is a sufficient decrease of the filaments. Permanent solutions such as provision of an increased air supply, plant influent control to ensure an adequate nutrient balance, or the construction of selectors, should be instituted for actual control of the filamentous organisms.

## NOCARDIA CONTROL

According to the *Bergey's Manual of Determinative Bacteriology* [12], *Nocardia* belongs to Order Actinomycetales, the mold-like bacteria. They are filamentous and branched, characterized by short, net-like branches sometimes described as "aerial hyphae." They are strictly aerobic and differ from other species in containing a high percent, as much as 11 percent, of lipids in cell by weight. Lipids are found in both the cytoplasm and the cell wall. In an activated sludge process, Nocardia is unique in that it is almost always associated with the heavy foam production in the aeration tank. This is probably due to the high lipid content in its cells. Heavy foam production is a nuisance since it [13,30]:

- is unaesthetic

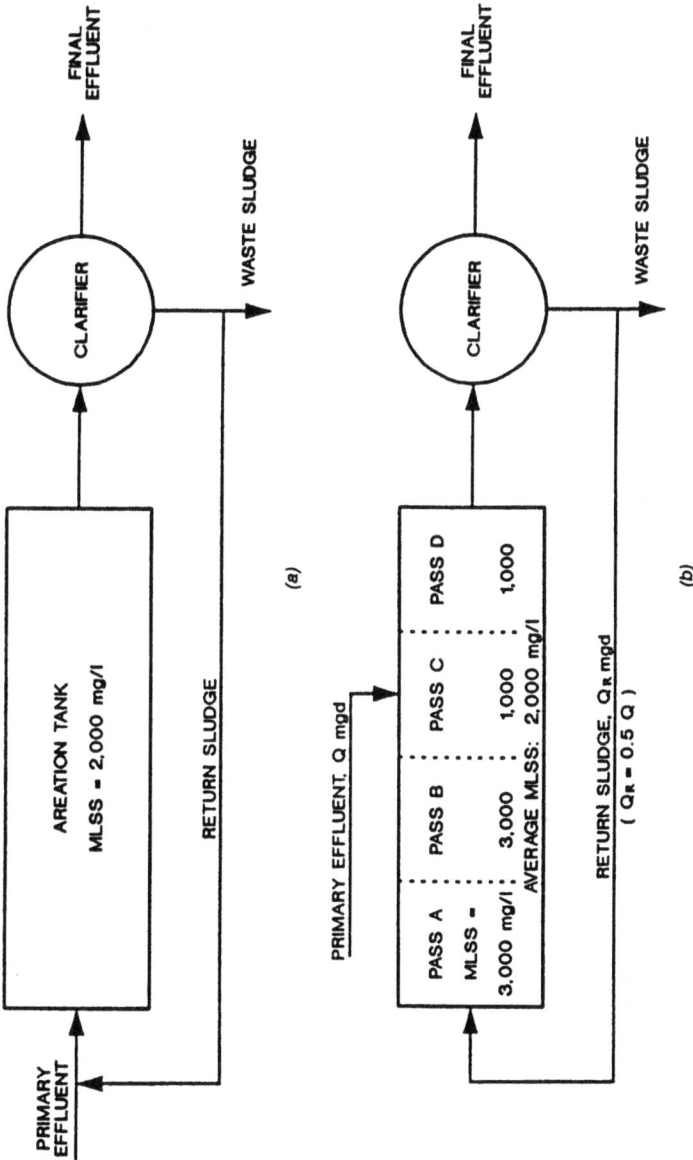

**Figure 2.7** Effect of step feed options: (a) plug-flow operation mode (conventional); (b) mode C step feed operation mode.

**89**

- creates safety problems by spilling the tank contents onto the walkways
- increases effluent BOD and SS
- causes difficulties in controlling solids inventory in the aeration tank

Most recently, it has been found that what has been historically referred to as *Nocardia* are actually a group of organisms now referred to as Nocardioforms [39]. It has also been reported that Nocardia forms in the waste activated sludge has caused operational problems in the anaerobic sludge digesters [13].

Various studies have been conducted to determine the causes of excessive growth of *Nocardia* forms and the general conclusions are that *Nocardia* forms predominate at:

- high SRT
- low *F/M*
- high temperature

High SRT and low *F/M* are actually interrelated since the higher the MLSS, the longer the SRT and the lower the *F/M* will be. It has also been reported that the presence of surfactants may significantly enhance foaming in *Nocardia* form-containing mixed liquors [14].

Control measures that have been developed include decreasing SRT through the increased rate of sludge wastage. In warm waters, the SRT may have to be reduced to as low as 0.7–1 day [31,32]. In warm water using pure oxygen sludge systems, values as low as 0.4 day have been required [18].

The use of anoxic selectors in the nitrification plants has also been found to be effective in controlling *Nocardia* forms [15].

Another way of controlling Nocardia form is to avoid trapping it in the aeration tank. In terms of tank design, this means using weirs instead of submerged ports to transfer mixed liquor between aeration tanks and clarifiers. This is becoming the preferred method of tank design in the U.S.

Other methods have been developed depending upon the specific design of each plant. These methods may include [13,14,15,39]:

- increasing the return sludge flow rate
- return sludge chlorination
- addition of surfactant degrading organisms
- addition of lipid and hydrocarbon degrading organisms
- addition of anaerobic digester supernatant
- trapping and spraying foam with hypochlorite
- use of polyelectrolytes

## APPURTENANCE DESIGN

### AERATION DIFFUSERS

The most important function of an aeration tank is to provide enough oxygen (e.g., air) to support the growth of microorganisms and through which the organic content, i.e., $BOD_5$ of the influent wastewater, is removed. Sufficient air supply is also one of the practical ways to avoid the excessive growth of filamentous organisms.

In the activated sludge process, air is introduced to the mixed liquor via two major types of devices—mechanical aerators and diffused air aerators. Mechanical aerators can be:

- surface aerators (with air or oxygen)
- turbine aerators (with air or oxygen)
- rotor-brush aerators (with air)

Diffused air aerators may include:

- static tube aerators
- jet aerators
- diffusers

Among these, diffusers are currently the most popular, except perhaps in large plants where mechanical surface aerators with pure oxygen are common (see the section on "Configuration for Large Plants"). Diffusers can be either fine pore or coarse pore depending upon the size of air bubbles generated—about 2 to 5 mm diameter for fine pore diffusers, and 6 to 10 mm diameter for coarse pore diffusers [16].

In the U.S. there is a steady trend toward the use of fine pore diffusers because of their higher transfer efficiency and thus lower energy consumption. In diffused air aeration activated sludge plants, aeration normally represents 50 to 90 percent of the total plant energy requirements; about 50 percent of aeration energy can be saved by using fine pore instead of coarse pore diffusers [16].

In applications where the mixing requirements dictate the oxygen transfer, such as those in the extended aeration plants or in aerobic digesters, coarse pore diffusers or mechanical aerators are generally the choice.

Another trend in the U.S. is to use the membrane diffusers, either tubes or discs, with fine pore slots, to replace the currently prevailing rigid ceramic-type fine pore diffusers. The major advantage of flexible membrane diffusers is their so-called self-cleaning mechanism. Vibration of the membrane during service can discourage the growth and accumulation of biological slime and chemical precipitates (scale) around the pore and thus can reduce maintenance cost for the diffusers.

Depending upon the type of diffusers selected, the number of diffusers needed can be determined through the allowable air flow rate for each unit. This information is usually available from the diffuser manufacturers. The arrangement of diffusers in plug-flow tanks is typically tapered corresponding to the air (oxygen) demand along the tank. For step feed plants, care should be taken to have enough diffusers provided in the passes where aeration influent can be added. Diffusers in each pass typically should be tapered.

## RETURN SLUDGE PUMPS

Return sludge is provided to supply microorganisms to the incoming wastewater. Enough return flow is needed to maintain the desired MLSS concentration. Since return sludge is withdrawn from the final clarifier, the required flow rate is also determined by the settling characteristics of the activated sludge. The settled sludge should be drawn fast enough to prevent the sludge blanket from rising—a condition that may eventually lead to a solids washout from the final clarifiers. The following expression has been developed to incorporate the settling characteristics of the sludge in the determination of the minimum required return sludge flow rate for an activated sludge process [17]:

$$(Q_r)_{min} = (Q_a X)/(4,000,000/K - X) \qquad (2.15)$$

where

$Q_a$ = wastewater flow (exclusive of return sludge), mgd
$X$ = MLSS to clarifiers, mg/l
$K$ = settling constant, l/mg (600 typically used)

It is seen that the higher the value of K, and thus the slower the settling of the sludge, the faster the settled sludge should be withdrawn and recycled. Historically, typical return sludge flow rates were designed at about 25 percent of the plant flow. In more recent designs, return sludge flow rates have typically been increased to 100% or more of the plant flow.

## WASTE SLUDGE PUMPS

This is probably the most important process control element in an activated sludge plant. SRT is controlled through sludge wastage. Sludge is typically wasted from the return sludge line. Sludge can also be wasted directly as mixed liquor from the aeration tank. A separate thickener is usually required to concentrate such wasted mixed liquor before sludge treatment and disposal. This is a common option where gravity thickeners are already in place for thickening of primary sludges.

Due to the possible variation in the settling properties of the sludge (and therefore, the return sludge concentration), the capacity of the waste sludge pumps should be chosen so that the desired SRT can be obtained at the lowest return sludge concentration possible. To achieve this, pumps with a firm capacity of four times the design value at average operating conditions are typically desirable.

## SCUM REMOVAL EQUIPMENT

While scum removal has always been common in primary treatment, it had not been practiced much in final clarifiers until the 1970s. The current practice in the U.S. is to provide scum removal facilities in both primary and final clarifiers. Skimmers in the primary clarifiers are used mainly for the removal of oily and greasy substances present in the raw sewage. Scum in the final clarifier is quite different, however, consisting mostly of foams containing biological solids such as Nocardia. In general, rectangular final clarifiers are more amenable to scum removal. The flight used for plowing sludge may also serve as the skimmer on return trips (see Figure 2.2). Special attachments are needed for the removal of scum in circular clarifiers (see Figure 2.1) and the removal may not be as complete as that in rectangular tanks (Table 2.4). Due to the possible accumulation of Nocardia forms as addressed above, scum from secondary clarifiers should not be recycled to the aeration tank, nor should it be discharged to the digesters. It is desirable that secondary scum skimmings be thickened and followed by disposal through means such as incineration or landfill along with the grit.

## CONFIGURATIONS

The choice of configuration depends upon many factors including designer's and regional preference. Plant capacity and function probably are among the most common other factors used in selecting configuration. The choice of single versus two-stage configurations for nitrifying plants was addressed in the ''Nitrification'' subsection. In the following are what appear to be the current trends in the U.S. for other types of municipal plants.

## SMALL PLANTS

For small (approximately 3 mgd or less) plants, the most common choices are oxidation ditches and sequencing batch reactors (SBRs). Some extended aeration and contact stabilization plants are being built, particularly in pre-engineered plants, but oxidation ditches and SBRs seem to be the most popular current choices. Commonly no primary treatment is used with these types of plants. Selectors are commonly used in the front of the ditches, and SBRs are commonly operated in a mode that includes a selector type treatment.

## MEDIUM-SIZED PLANTS

For medium-sized plants (3–50 mgd), the "conventional" activated sludge process seems to be the choice. In this size range primary treatment is usually employed and aeration generally uses fine pore diffusers. Simple plug flow or completely mixed activated sludge (CMAS) plants are common conventional plants, with CMAS possibly falling out of favor—unless preceded by a selector. Many conventional plant designers prefer the flexibility of step aeration (see Figure 2.5) and this type of plant appears to be gaining in popularity in this size range.

## LARGE PLANTS AND TIGHT SITE DESIGNS

Plants over 50 mgd tend to use the same configurations as medium-sized plants with step aeration being a common choice. Site limitations can lead to the use of deep aeration tanks; the 330 mgd Blue Plains Plant in Washington DC has 30-foot deep aeration tanks, for example. Care must be taken when designing tanks deeper than about 18 feet to provide for release of super-saturated gases before clarification [19]. This may be done as simply as by providing a 12–24 inch mixed liquor drop over a weir [20], or, in tanks over about 100 feet deep, providing gas strippers [21,22].

Another way of saving site space is to use oxygen-activated sludge. Oxygen-activated sludge often becomes economically attractive, even when site space is not tight, in this size range. For secondary treatment, loadings in excess of 100 lb $BOD_5$/(day—1000 cf) are not uncommon for oxygen activated sludge plants. These can be run at very low SRTs—approaching theoretical washout values, and still maintain good effluent quality [18]. Settling characteristics vary, but typically are similar to those recommended for air activated sludge plants (see section on "Clarifier Design"). Sludge production is less only if more oxygen is used (see subsection on "Solids Production"). There are large oxygen plants in most U.S. metropolitan areas including: Philadelphia, Boston, Baltimore, Newark, Miami, Tampa, Detroit, Houston, Denver, San Francisco, Los Angeles, and Seattle.

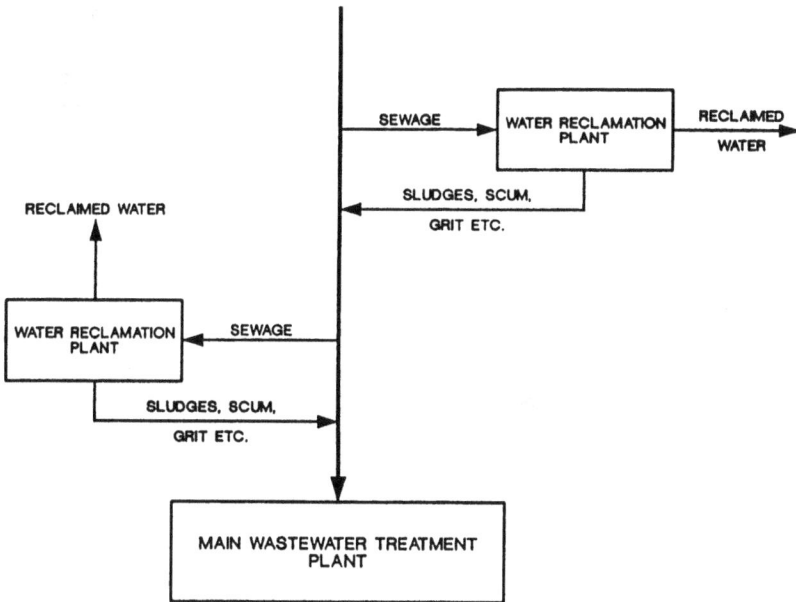

**Figure 2.8** Water reclamation plant concept.

Yet another way of fitting an activated sludge plant on a tight site is to use multi-deck clarifiers. Double-deck clarifier systems are used at Westchester County, NY and Boston, MA.

## WATER RECLAMATION PLANTS

In parts of the U.S., particularly in the arid Southwest, the term "Water Reclamation Plant" is used to describe a plant that treats a portion of a municipal wastewater for local reclamation use. Typically this use is primarily for irrigation of public lands and golf courses. These plants typically do not treat or dewater residual solids, but return all sludges, scum, grit, etc., back to the main flow of the wastewater for processing at a larger, or a main plant, downstream (see Figure 2.6). These plants can be of almost any size, although most fall in the "small plant" category. Their configuration is generally chosen the same as any other plant in their size range.

## REFERENCES

1  Great Lakes-Upper Mississippi River Board of State Sanitary Engineers. 1997. "Recommended Standards for Sewage Works," Health Education Services, Albany, NY.

**2** WEF, 1992. *Design of Municipal Wastewater Treatment Plants,* WEF MOP 8, Book Press Inc.

**3** Wilson, T. E. and T. M. Keineth. 1990. "Secondary Clarifier Design Standards: An Issue Paper of the ASCE Clarifier Research Technical Committee." ASCE Environmental Division Conference.

**4** Wilson, T. E. 1991. "Rectangular Clarifiers Should Be Considered," *Water/Engineering and Management,* 138(4):20.

**5** Parker, D. S. and R. Stanquist. 1986. "Flocculator Clarifier Performance," *J. WPCF,* 58:214.

**6** Parker, D. S. 1983. "Assessment of Secondary Clarification Design Concepts," *J. WPCF,* 55:349.

**7** Parker, D. S. 1991. "The Case for Circular Clarifiers," *Water/Engineering and Management,* 138(4):23.

**8** Parker, D. S. 1991. Rebuttal to "Rectangular Clarifiers Should Be Considered," *Water/Engineering and Management,* 138(4):22.

**9** IAWPRC. 1986. "Activated Sludge Model No. 1," Scientific and Technical Reports No. 1.

**10** U.S.EPA. 1975. "Nitrogen Control," EPA Process Design Manual.

**11** Daigger, G. T. and G. A. Nicholson. 1990. "Performance of Four Full-Scale Nitrifying Wastewater Treatment Plants Incorporating Selectors," *WPCF Research Journal,* 62:676.

**12** Buchanan, R. E. and N. E. Gibbons, ed. 1974. *Bergey's Manual of Determinative Bacteriology,* 8th Ed. Baltimore: Williams and Wilkins.

**13** Pitt, P. A. and D. Jenkins. 1988. "Causes and Control of Nocardia in an Activated Sludge," paper presented at the *61st WPCF Annual Conference,* October, 1988.

**14** Ho, C. F. and D. Jenkins. 1991. "The Effect of Surfactants on *Nocardia* Foaming in Activated Sludge," *Water Sci. Tech.,* 23:879.

**15** Blackall, L. L. et al. 1989. "The Physiology of *Nocardia amarae* and Its Implications for the Control of Nocardia Foaming in Activated Sludge," paper presented at the *62nd WPCF Annual Conference,* October, 1989.

**16** U.S.EPA. 1989. "Fine Pore Aeration Systems," *U.S.EPA Technology Transfer Design Manual,* EPA/625/1–89/023.

**17** Riddell, M. D. R. et al. 1983. "Method for Estimating the Capacity of an Activated Sludge Plant," *J. WPCF,* 55:360.

**18** Wilson, T. E. et al. 1989. "Operating Experience at Very Low SRTs," paper presented at the *62nd WPCF Annual Conference,* October, 1989.

**19** Lu, L., et al. 1990. "Deep Tank Aeration Experience at NYC's North River Plant," paper presented at the *63rd WPCF Annual Conference,* October, 1990.

**20** Associated Engineers, Greeley and Hansen. 1990. "Floating Sludge Study for North River Water Pollution Control Plant," NYC, WP-164.

**21** Daly, P. G. and C. C. Shen. 1989. "The Deep Shaft Biological Treatment Process," *Proceedings 43rd Purdue Industrial Waste Conference,* Lewis Publishers, p. 479.

**22** Sasser, L. W. et al. 1989. "Filamentous Microorganisms Control in the Deep Shaft Activated Sludge Process," paper presented at *62nd WPCF Annual Conference,* October, 1989.

**23** Eckenfelder, W. W., Jr. 1966. *Industrial Water Pollution Control.* New York: McGraw-Hill, Inc.

**24** Redmon, D. et al. 1983. "Oxygen Transfer Efficiency Measurements in Mixed Liquor Using Off-Gas Techniques," *J. WPCF,* 55:1338.

**25** Metcalf & Eddy, Inc. 1991. *Wastewater Engineering Treatment, Disposal, Reuse,* 3rd Edition. New York: McGraw-Hill, Inc.

**26** Yunt, F., et al. 1980. "An Evaluation of Submerged Aeration Equipment—Clean Water Test Results," paper presented at the *WWEMA Industrial Pollution Conference,* June, 1980.

**27** Buhr, H. O. et al. 1984. "Making Full Use of Step Feed Capacity," *J. WPCF,* 55:325.

**28** Wilson, T. E. 1991. Rebuttal to "The Case for Circular Clarifiers," *Water/Engineering and Management,* 138(4):26.

**29** Jenkins, D. et al. 1993. "Manual on the Causes and Control of Activated Sludge Bulking and Foaming," 2nd ed. Lewis Publishers MI

**30** Jenkins, D. and G. T. Daigger. 1990. "The Use of Selectors for Control of Activated Sludge Bulking," paper presented at *Michigan WPCF Conference February,* 1990.

**31** Steytler, R. B. et al. 1982. "Watching the 91st Avenue Plant in Phoenix," City of Phoenix and Greeley and Hansen.

**32** Personal communication with David Wendell, City of Phoenix, Arizona, 1985.

**33** Albertson, O. "Clarifier Design" *Design and Retrofit of Wastewater Treatment Plants for Biological Nutrient Removal,* Water Quality Management Library, Volume 5, Lancaster, PA: Technomic Publishing Co. Inc. (1992).

**34** Wahlburg, Augustas, Chapman, Chen, Esler, Keinath, Parker, Tekippe, and Wilson "Evaluating Activated Sludge Secondary Clarifier Performance Using CRTC Protocol: Four Case Studies" presented at the *Water Environment 67th Annual Conference & Exposition* (October 1994).

**35** Wilson, T. E. (1996) "A New Approach to Interpreting Settling Data," Proceedings of WEFTEC'96, the 69th Annual WEF Conference.

**36** Wilson, T. E. and Ballotti, E. F. "Gould Tanks: Rectangular Clarifiers That Work" presented at *61st Annual Conference,* WPCF (October 1988).

**37** USEPA *Manual—Nitrogen Control,* EPA/625/R-93/010 (September 1993).

**38** Albertson, O. "Control of Bulking and Foaming Organisms" Design and Retrofit of Wastewater Treatment Plants for Biological Nutrient Removal, Water Quality Management Library, Volume 5, Lancaster PA, Technomic Publishing Co., Inc. (1992)

**39** Jenkins, David "The Causes and Control of Activated Sludge Foaming," Proceedings. 3rd Annual CSWEA Education Seminar, Madison, WI, April 7, 1998.

**40** Wilson, T. E. "Status of Final Clarifier Design and Operation." Proceedings of 1st Annual CSWEA Education Seminar, Madison, WI, March 1996.

**41** Wilson, T. E. "Getting More Information with Less Work: Application of ISU Test to the Operation of Activated Sludge Plants." Proceedings, WEFTEC 97, Chicago, IL, October 1997.

# CHAPTER 3

# European Practices

## INTRODUCTION

EUROPE extends from Portugal in the West to Russia in the East and from Scandinavia in the North to the countries bordering the northern banks of the Mediterranean in the South. Since the break-up of the Soviet Union and former republic of Yugoslavia this area embraces about 50 countries. Fifteen of these countries (Austria, Belgium, Denmark, Finland, France, Germany, Greece, Holland, Ireland, Italy, Luxembourg, Portugal, Spain, Sweden, and the U.K.) now form the European Union (EU).

A European Water Pollution Control Association (EWPCA) was formed in 1981 and so far includes thirteen of the EU member states plus Hungary, Norway, and Yugoslavia. However it has not yet turned its attention to the issue of manuals of practice. The EU countries are all subject to Directives that set limits to the concentrations of certain quality determinands in municipal and industrial wastewaters and in receiving waters, but local pollution control authorities can and do set more rigorous standards in some cases. These authorities have local discretion to set standards for many determinands for which limits are not laid down by the EU and which differ from country to country. The EU Urban Waste Water Treatment Directive (1991) [1] is having a major impact upon treatment facilities in the 15 member states. This Directive sets framework specifications for plants serving different population equivalent (pe) sizes together with a

A. L. Downing, D.Sc., FIChem.E., Herts, England.
P. F. Cooper M.Sc. C.Eng. M.I. Chem. E., WRc Swindon, England.

timetable for compliance with the regulations. Considerable structural funding is also being given from the EU central funding to bring about improvement in sewers and wastewater treatment facilities in the member states such as Spain, Portugal, Greece and Ireland where these facilities were not as developed as in the others. The directive requires that by 2001 wastewater treatment plants serving pe's of >15,000 pe have to have secondary treatment to achieve effluent qualities of:

- 25 mgBOD$_5$/l
- 35 mgTSS/l
- 125 mgCOD/l

Population equivalent is defined as the organic biodegradable load having a five-day BOD of 60 g oxygen per day.

Treatment plants discharging to the sea may have a dispensation allowing them not to use secondary treatment but have to achieve at least primary treatment standards.

In addition where they discharge to areas which are sensitive to eutrophication plants with pe's >10,000 have to achieve, by 1999, the following nitrogen and phosphorus concentrations in the effluent:

$$\left.\begin{array}{l} 2 \text{ mg Total P/l} \\ 15 \text{ mg Total N/l} \end{array}\right\} \begin{array}{l} \text{Plants treating} \\ \text{>10,000 pe <100,000 pe} \end{array}$$

$$\left.\begin{array}{l} 1 \text{ mg Total P/l} \\ 10 \text{ mg Total N/l} \end{array}\right\} \begin{array}{l} \text{Plants treating} \\ \text{>100,000 pe} \end{array}$$

As an alternative it is possible under the Directive to designate an area (e.g. a river basin) in which rather than comply with the above N and P standards the whole area has to achieve a removal of 75% total P and 75% total N. Some of the EU member states have decided to designate the whole of the territory and go for the percentage removal option, others are approaching it on a plant by plant basis. These standards for N and P discharges have to be implemented by the start of 1999 for all plants of > 10,000 pe discharging to sensitive waters.

It should be remembered that the national regulators still have the freedom to set even higher standards to protect rivers, lakes and estuaries within their territory. Moreover, although EU directives influence the degree of treatment provided, they do not prescribe the means by which limits must be met, except in very broad terms; thus designs and operating practices are quite varied. The situation in Eastern Europe is even more diverse. Thus, since there is as yet no common code of practice to which all countries subscribe, one cannot strictly write of European practice in regard to sewage treatment other than by reference to the individual practices of the different countries.

In the face of this lack of uniformity, the object of this chapter is to describe the "state of the art" in Europe in terms of a representative selection of the best modern practice wherever it is to be found, though indicating local preferences for alternatives where appropriate. It is assumed that the reader has a basic familiarity with the activated sludge process and will already have read previous chapters in this series.

## HISTORICAL PERSPECTIVE

Since its development in the U.K. just before the First World War use of the activated sludge process (ASP) for treatment of domestic sewage has grown steadily in Europe. For about the last four decades in the wealthier, more densely populated countries such as Austria, Belgium, Denmark, France, Germany, Holland, Norway, Sweden, Switzerland, and the U.K., it has usually been the preferred method of treatment for towns of substantial size, gradually replacing the older, more space-demanding process of trickling filtration. Indeed in Austria, for example, the process has gained in popularity to such an extent that it now serves 90 percent of the population. Introduction on a substantial scale to many of the other countries has been more recent, and in several of those in Southern and Eastern Europe, application has not yet progressed very far. Use of the process for small communities began in the late 1950s with the introduction of "package" plants, and rather more effectively with that of the oxidation ditch developed in Holland. In the early days of application of the process, the requirement was mainly to meet standards for suspended solids (SS) and five-day BOD (determined by the classical unsuppressed test), the so-called "Royal Commission 30:20 Standard" becoming a norm in the U.K. for inland towns; similar standards were often applied on the European continent. By the 1950s, limits were also beginning to be placed on concentration of ammonia for works discharging into rivers drawn on for public supply. Applications of a limit on ammoniacal nitrogen have become steadily more widespread; moreover the limits, expressed usually as maximum acceptable concentrations or percentile values, have become more stringent. More recent advances have been in the application of the process to meet limits on the concentrations of nitrogen (N) and phosphorus (P) in effluents, especially in Denmark, Switzerland, and Scandinavia but also, in the latest developments, in Germany and some of the other wealthier European states.

To match these changing requirements, the design of plants has been steadily evolved and many new variants have been introduced. Indeed examples of nearly every new version of the process that has been invented is to be found in Europe. Nevertheless in probably more than about 98 percent of plants the main treatment units for removing carbonaceous matter and ammonia are of a traditional style in the sense that they comprise open aeration basins (usually rectangular in large works) coupled to circular secondary settlement tanks, and aeration is commonly by diffused air or mechanical agitation. However, the resemblance to the original process designs ends there, since there are a host of major refinements in detailed features such as in:

- the choice of loadings
- hydraulic conditions
- geometry of the units
- design of aerators
- control equipment

Additionally and perhaps most notably, the aeration units in many modern plants are interlinked with other process stages of new types in which, for example, anoxic and anaerobic conditions are maintained or chemicals are added either to improve efficiency or to meet new standards for N and P. It is these features of modern European design that form the main subject of the remainder of this chapter.

## PROCESS DESIGN

Design in Europe is based on the common currency of knowledge that has accumulated over the years in the light of research and operating experience conducted in many countries. In the past there was considerable bias towards exclusively European experience and equipment; but in recent times design has been rather more influenced by experience from non-European countries in which plants had to be designed to meet new requirements, such as for nutrient removal or water reclamation.

The main concepts of this common currency of established knowledge are briefly as follows:

(1)  The most important traditional function of the process is to bring about biochemical oxidation of the majority of the polluting constituents of sewage (principally organics and ammonia). This is accomplished by aerating the sewage (after preliminary treatment and most commonly primary sedimentation) with a flocculent heterogeneous microbial sludge (the activated sludge) followed by settlement of the sludge, discharge of a purified supernatant (possibly to polishing or other tertiary treatment stages) and recycle of the majority of the

settled sludge to the aeration units. To achieve this traditional function most effectively the content of dissolved oxygen (DO) in the aeration units must be maintained above about 0.5 mg/l. Because the process of bio-oxidation is accompanied by growth of micro-organisms and because of other phenomena, such as adsorption, the sludge mass tends to increase, so to maintain concentrations within a suitable range surplus sludge must be bled off for separate disposal. The two features having the greatest influence on the performance achievable are the rates of biochemical reaction within the aeration units and the settleability of the sludge. Both of these features are influenced by many factors and are interdependent.

(2) It is well-established that, while the main structures required to bring about these aerobic processes are an aeration tank coupled to a secondary clarifier together with facilities for the return of the settled sludge to the aeration tank, various configurations of tank can be used and the flows of both sewage and returned sludge can be distributed into the aeration units in various ways. Suffice it to record that among the main variants of the traditional aerobic process recognised to be effective are those involving operation with:

- uniformly mixed aeration units
- baffled or compartmentalised plug-flow aeration unit
- aeration units divided into "contact" and "stabilization" zones
- "canalized" aeration units (oxidation ditches)

with basic flow sheets as indicated in Figure 3.1.

(3) Since the evolution of the traditional process, it has been discovered that other useful functions can be accomplished by including supplementary units or zones characterised by the status of oxygen within them. Thus anoxic zones (i.e., those in which the mixed liquor contains less than 0.5 mg/l DO but with a supply of utilizable oxygen available in the form of nitrate) are included if it is desired to remove nitrogen (N) by denitrification and to obtain certain other benefits. Anaerobic zones containing no DO or oxidised nitrogen are included if it is desired to mobilise phosphorus (P) so that it can be incorporated in microbial cells in excess of normal requirements (so-called "luxury uptake") in subsequent aerobic stages.

(4) In the aerobic stages, arguably the most important of the many features determining performance is the rate of application of degradable substrate to these sludge organisms. This importance derives essentially from the fact that sludge loading controls the solids retention time (SR.) (or sludge age—the mass of activated sludge in the aeration tank divided by the rate of production of this sludge) which

COMPLETELY-MIXED

"PLUG" FLOW

CONTACT STABILIZATION

EXTENDED AERATION DITCH
(PLAN VIEW)

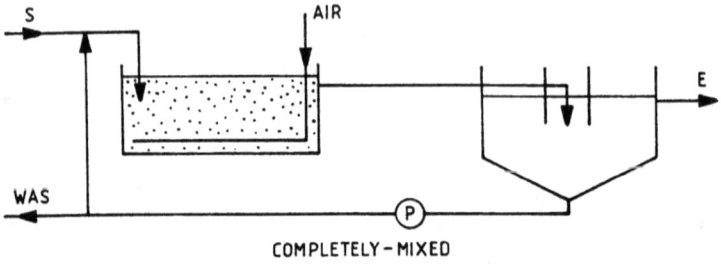

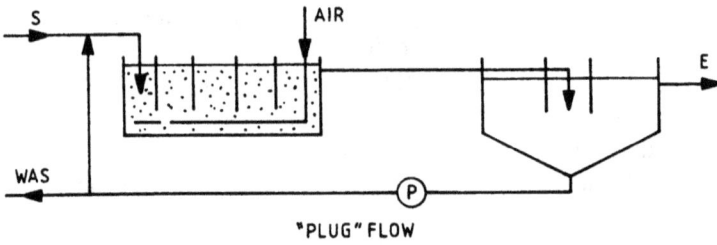

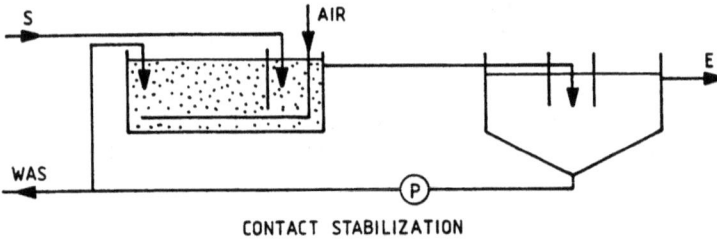

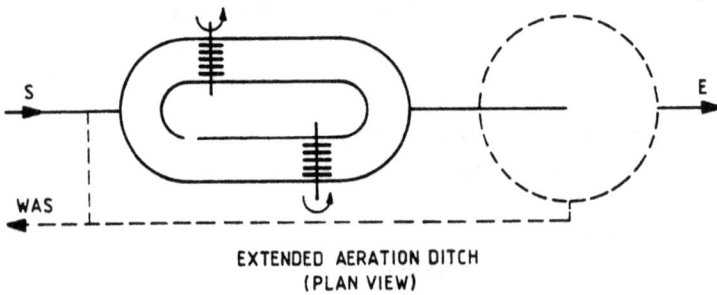

S: SETTLED SEWAGE    E: EFFLUENT
WAS: WASTE (SURPLUS) ACTIVATED SLUDGE

**Figure 3.1** Some modifications of the activated sludge process.

in turn controls the spectrum of micro-organisms that can be retained in the plant. Broadly, the lower the sludge loading and the longer the SRT, the better the quality of effluent, at least up to the point at which the loading is insufficient to sustain a coherent settleable sludge, which in turn is usually well below that at which it becomes more economical to remove remaining impurities by means other than reducing loading.

(5) However, the relationship between sludge loading or SRT and performance is not a simple continuous smooth trend but includes "step changes" that reflect the fact that as sludge loading is reduced and SRT increased the more slowly growing organisms with specifically useful properties that would be lost at low SRTs can be retained. This retention becomes possible once the percentage rate of increase in the sludge mass just exceeds the percentage rate of growth of these organisms. Two particularly important groups of such organisms are the nitrifying bacteria, which oxidize ammonia, and ciliated protozoa, which ingest freely suspended bacteria and other small non-settleable particles that give rise to turbidity in settled effluents. Thus low sludge loadings have to be adopted if well-nitrified non-turbid effluents are to be produced. Experience shows that under these circumstances, characterised by loadings in the range from about 0.1 to 0.2 g/g MLSS day and SRTs from about 30 to 15 days it is economically feasible to remove up to about 95 percent of BOD and all but a few mg/l of ammonia. Conversely if it is desired merely to provide partial purification (for instance, for removal of around two-thirds of BOD) then loadings can exceed 1 g/g day.

(6) Sludge loading and SRT also determine the rate of production of surplus sludge and sludge properties. Sludge production declines continuously with reducing loading, for example, from around 1 g/g BOD removed at loadings around 1 g/g day to a minimum of about 0.2 g/g at loadings around 0.05 g/g day. Also, the rate of uptake of DO by the sludge declines with decrease in loading, but in this case the trend includes a step change that marks the point beyond which nitrifying bacteria are retained in the plant. Usually the specific resistance of sludge (a measure of its dewaterability) also decreases with decreasing loading.

(7) The rate of settlement of sludge is also affected by sludge loading, but the trends appear to be to some extent site-specific. Many other factors affect rates of settlement. Thus under certain operating conditions, not as yet thoroughly understood, sludges can develop excessive populations of filamentous organisms such as *Sphaerotilus* spp., tending to cause bulking, or organisms such as *Nocardia*

spp. that tend to form scums on the surface of sedimentation tanks. While various remedies are available for such outbreaks the most useful positive step that can be taken by designers to minimize their occurrence appears to be to baffle or otherwise divide aeration units into "pockets" so as to reduce longitudinal mixing and render flow as "piston-like" as possible; additionally, if the plant is designed to nitrify, settleability can also benefit from inclusion of an anoxic zone. The reason for this beneficial effect is not wholly clear but may reflect the action of an "organism-selector mechanism."

(8) The previously mentioned features (5–7) are influenced by the composition of the sewage treated and by temperature. The composition of purely domestic sewage varies from place to place mainly in regard to "strength," determined by water usage, and its content of constituents introduced in the local water and those dependent on domestic practice particularly in the use of soaps, detergent and other cleansing agents. Of these features the two having the greatest significance are probably the strength of sewage, and its alkalinity, both of which can influence rates of nitrification. Rates of nitrification of ammonia appear to decrease with increasing strength of sewage, probably at least partly because ammonia begins to become mildly inhibitory at concentrations above about 10 mg/l. Alkalinity can have an influence because it reacts with the hydrogen ions liberated during nitrification, but if it is insufficient, pH value may fall into a range below pH 7 within which the rate of nitrification declines as pH value falls, becoming zero at around pH 6.

Often of greater importance is the content of constituents, originating from industrial sources, having inhibitory properties. A great deal is now known of substances having such properties, and although sludges can be adapted to tolerate or destroy many inhibitors it is a general practice to endeavour to limit them to acceptable levels at the source. Nevertheless such measures are not always effective and thus it may be necessary to conduct treatability tests to obtain reliable design parameters.

Lengthy detention of sewage in sewers or primary sedimentation tanks may lead to much greater than usual conversion of carbohydrates to lower fatty acids. While this may be useful in plants designed for nutrient removal it can also encourage the growth of filamentous organisms, leading to poor settlement.

Increasing temperature of sewage within the normal ambient range in Europe is in most respects beneficial in accelerating destruction of pollutants and increasing rates of settlement of sludge, with the reservation that in nitrifying plants the tendency of sludge is to rise in final clarifiers owing to denitrification increases with increasing temperature.

(9) It is accepted that aeration may be achieved satisfactorily in various ways, of which the two most common are by mechanical agitation and by diffusing air through the mixed liquor. Well-designed systems should normally be capable of achieving rates of oxygen input per unit power into fully deoxygenated mixed liquor between about 2 and 2.5 kg/kWh. Many factors can affect performance including particularly the composition of the sewage, the loading on the plant, and the geometry of the aeration units.

(10) In anoxic zones, the rate of denitrification at a given temperature is primarily dependent on the concentration and nature of the metabolizable substrate present and that of the active fraction of the MLSS. It is recognized that the highest rates are usually obtained when an external readily-degradable substrate such as methanol or acetic acid is added, though rates nearly as high can be obtained in systems in which organics released from hydrolysed primary sludge are fed into the zones. It is also accepted that average rates decline with increasing detention in the zone due to removal of the more readily biodegradable substrates before the more recalcitrant ones. Corresponding rates of oxidation of carbonaceous matter are no more than about 40 percent of those obtained when DO is utilised. For this reason it is not normally considered economic to provide anoxic zones having a retention of more than about 0.5 h when the main object is to recover energy and improve settleability of the sludge.

It is recognized that anoxic and aerobic zones can coexist in aeration units that are not formally divided into compartments when, in circulating back and forth between regions of relatively high and relatively low oxygen transfer, the oxygen demand of the sludge is high enough to reduce DO below 0.5 mg/l during transit. This is consistent with experience that it is feasible to create anoxic zones by aerating with diffused air at a low rate sufficient to keep sludge in suspension but not high enough to match the oxygen demand, although it is more usual to use slowly rotating paddles for this purpose.

(11) The kinetics of mobilization of phosphorus in anaerobic zones do not appear to have been conclusively unravelled, but on the basis of empirical evidence a residence time of about 1.5–2 h appears adequate.

(12) Satisfactory settlement of sludge demands maintaining settling velocities of the sludge particles in a range within which all but a small faction will separate within the detention time and at the overflow rates that

can economically be accommodated. It is widely accepted that settling velocity cannot be predicted precisely though given appropriate choice of loadings and other operating conditions, such as have been mentioned in items (6) and (7) previously, it can usually be relied on to fall within an acceptable range. In practice, this range is such that residence times of not more than 2 h at peak flow and corresponding overflow rates (flows per unit plan area of tank) of not more than 1 m/h are regarded as appropriate. It is also recognised that the flux of solids applied to the tank must not exceed the maximum mass rate of removal by settlement and return of sludge. Pumping capacity for return of sludge capable of delivering a flow not much more than that of the incoming sewage would also be regarded as appropriate. Other factors known to be important are the maximum diameter of tanks (to avoid wind effects and maintain overflow weirs' level), the amount of sludge storage capacity provided, the hydraulic loading per unit length of overflow weir, the height of sidewalls, the slopes of floors and the design of scrapers.

While the above features are part of the common knowledge forming a basis for design, detailed relationships employed by designers in Europe vary from country to country and within countries. However, often the differences amount to no more than the description of trends by alternative algebraic expressions whose numerical implications are nevertheless quite similar.

Generalizing broadly, designs are usually evolved by one or more of the following procedures, that is by:

- simple analogy with what has been successful in comparable situations in the past
- use of well-tried empirical or theoretical formulae or models relating to steady conditions and often applied to the expected ''worst case''
- use of dynamic models that have gained local popularity
- obtaining design parameters by conducting laboratory or pilot-scale trials particularly in the event of doubt about the influence of industrial constituents

The differences from practice outside Europe, which qualitatively are not great in the case of designs for a given duty, stem mainly from:

- a bias towards selection of locally invented processes and locally manufactured plant and equipment
- local requirements, which are uniquely site-specific in many cases or are country-specific
- a bias towards the use of design guides, manuals of practice, and relationships between design, operating conditions, and performance devised locally

It is becoming increasingly common for designers to use models to assist in the design process though, recognising that it is rarely possible to model all the features that may be important in the local context, final decisions normally demand the exercise of judgement and experience. Even when fully descriptive models are not formally used, many of the relationships that are built into them, for example as subroutines, are quite commonly employed on a more piecemeal basis. It is thus convenient to present examples of such relationships within the context of modelling.

## MODELS

No comprehensive statistics are available on the extent to which European designers make use of models. One major consulting engineering firm in the U.K. has made use of SWAT, a model developed by the U.S.EPA, for studies outside the U.K. [2]. Until 5 years ago greater use has certainly been made in the U.K. of a Sewage Treatment Optimisation Model (STOM) that was developed by collaboration of a large number of U.K. organizations acting under the auspices of the Construction Industry Research and Information Association [3]. This situation has now changed quite dramatically. In the last 5 years more groups have started to use the more sophisticated dynamic models (described later), and their use is now quite widespread. All significant large new plants being modelled and many older plants also being modelled to achieve more efficient operation.

The STOM model relates the performances and costs of well-established unit processes of sewage and sludge treatment to their capacities and operating conditions in such a way that the costs of alternative treatment sequences can be compared and the least-cost sequence for a given duty be identified. As structured in 1981, when the User Manual was published, it included some eighteen unit processes of which the activated sludge process and trickling biological filtration were the two main groups of alternatives for secondary treatment designed primarily to meet standards for SS, BOD, and ammonia.

However, within the last decade, processes other than those included have come to the fore as a result of innovations or new requirements, particularly aerated filters, biological nutrient removal systems, and (for sludge treatment) incineration. The last of these processes, though quite widely used in Europe, was not considered for STOM because in the U.K. it had become unpopular. However, it has since enjoyed a revival, mainly because of increasing restrictions on disposal of sludge at sea and on land. A second limitation of the model is that the relationships between performance and operating conditions are those for constant conditions; diurnal variations are not represented. Nevertheless, it remains useful for relatively rapid screening of possibilities to obtain a preliminary broad perspective against which to judge alternatives not included in the model, and to consider the implications of refinements such as can be made using dynamic models. Perhaps the chief value of STOM is that it necessitated the critical evaluation of the available scientific and engineering knowledge, and formation of a consensus U.K. view about the then most appropriate performance relationships and cost functions to use as a basis for process design.

In the particular case of ASP, the model predicts dissolved and suspended BOD in the effluent, the production of surplus sludge, and the rate of uptake of DO in the aeration tank. Influent to the aeration tank is assumed to contain suspended matter plus three types of biodegradable substrate, suspended and dissolved carbonaceous matter (SBC and DBC), and ammonia, which undergo the reactions indicated in Figure 3.2. Several usually less important pathways of interchange that have been included in more recent models are omitted in the interests of simplicity. In particular release of ammonia from nitrogenous organics is disregarded on the grounds that the quantity of nitrogen so released is usually about equal to that required for growth of new sludge.

Activated sludge is assumed to be composed of heterotrophs, nitrifiers, inert organics, and biodegradable but unreacted suspended organics. *Heterotrophic* bacteria are assumed to convert SBC to DBC and inert suspended matter, without uptake of DO, and to oxidize DBC utilising DO and synthesising new cell material in the process. Nitrifiers oxidize ammonia autotrophically with a cell yield per unit mass of substrate that is negligibly small by comparison with that of the heterotrophs. Both heterotrophs and nitrifiers respire endogenously in the absence of substrate, in the process consuming cell material. Nitrification is assumed to occur only when DO is above 20 percent of the air saturation value (though this is a rather crude simplification). Ammoniacal nitrogen required for synthesis of heterotrophic bacteria is taken to be assimilated without oxidation.

Average mass balances for the various components are written as

$$\text{Input} - \text{Output} = \text{Removal rate} - \text{Production rate}$$

For simplicity it is assumed that the concentrations of SBC are directly proportional to their carbonaceous $BOD_5$. The removal and production rates are defined by the expressions given in Table 3.1. The meaning of the symbols in Table 3.1 is as follows:

$\mu_{hm}$, $\mu_{nm}$ = maximum specific growth rates of, respectively, heterotrophic and nitrifying bacteria

$R_h$, $R_n$ = growth rates of heterotrophs and nitrifiers

$S_a$, $S_d$ = concentrations of SBC and DBC (expressed as $BOD_5$ in the model)

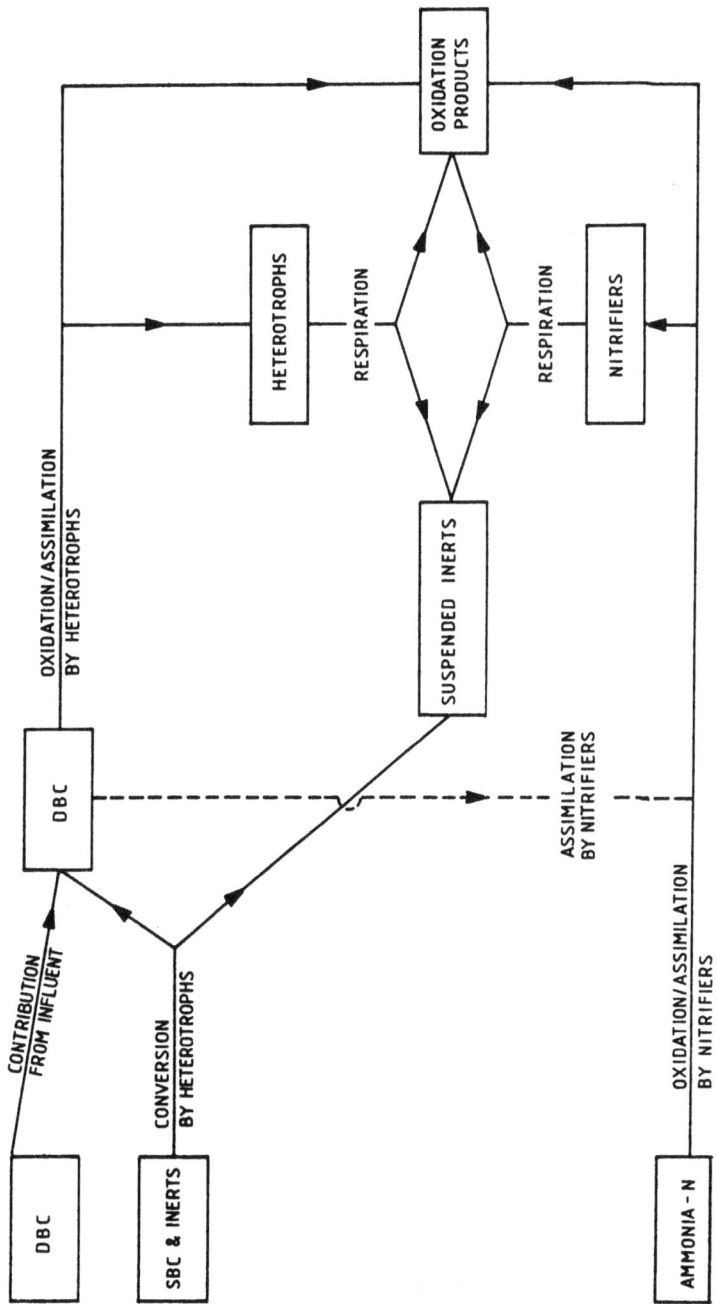

**Figure 3.2** Biochemical reactions assumed in STOM model of activated sludge process. (Reprinted with permission from STOM *User Manual* and Description TR 144, 1981.)

111

TABLE 3.1. **Rates Assumed for Interconversion of Constituents of Mixed Liquor in STOM.**

| Component | Rate of Formation | Rate of Removal |
|---|---|---|
| SBC | 0 | $\dfrac{f_h\mu_{hm}S_sX_h}{K_s + S_s}$ |
| DBC | Rate of conversion from SBC | $\dfrac{R_n}{Y_h}$ |
| Ammonia-N | 0 | $\dfrac{R_n}{Y_n}$ |
| Heterotrophs | $R_n = \dfrac{f_n\mu_{hm}S_d}{K_d + S_d} X_n$ | $f_hD_n\mu_{hm}X_n$ |
| Nitrifiers | $R_n = \dfrac{f_n\mu_{hm}S_n}{K_n + S_n} X_n$ | $f_nD_n\mu_{hm}X_n$ |
| Inert SS | Rate of release from SBC + 0.15 × sum of oxidation of heterotrophs and nitrifiers | 0 |

$K_s$, $K_d$ = saturation constants of SBC and DBC

$X_h$, $X_n$ = concentrations of heterotrophs and nitrifiers

$Y_h$, $Y_n$ = yield coefficients of heterotrophs and nitrifiers (i.e., mass of organisms produced per unit mass of substrate utilised)

$f_h$, $f_n$ = treatability coefficients representing the effect of the sewage on the growth rate of the organisms

$D_h$, $D_n$ = coefficients defining the rate of destruction of heterotrophs and nitrifiers by endogenous respiration

These are based on concepts of a type that have been widely recognised to be simplifications of real conditions, and some details of which have been specified differently, though not necessarily more accurately, in other work. These concepts are that rates of removal of dissolved substrates are proportional to the product of the concentration of the bacteria degrading them and a function of the substrate concentration according to the classical Monod equation; and a similar expression describes the conversion of SBC to DBC. The rates of endogenous degradation of the bacteria are constant and proportional to their maximum growth rates. Specific growth rates of the organisms are assumed to be dependent on the treatability of the sewage, a site-specific feature.

The overall yield, i.e., the mass of solids produced per unit of BOD removed, is derived from the loss of solids in the effluent, which is given by the difference between the rate of transfer of solids to the settlement tank and recycled in returned sludge.

The carbonaceous BOD per unit concentration of the suspended solids, a′, in the effluent is assumed to be given by the equation:

$$d = \frac{S_g + 0.50\ (X_h + X_n)}{(S_s/a) + X_i + X_h + X_n} \tag{3.1}$$

where
$d$ = the mass of oxygen absorbed per unit mass of SBC and DBC assimilated
$X_i$ = the concentration of inert suspended matter

The average rate of consumption of DO by heterotrophs, $F_h$, is taken as:

$$F_h = F_s - F_a + F_T \tag{3.2}$$

where
$F_e$ = the mass flow of oxygen that would be necessary for complete biochemical oxidation of the BCM removed
$F_a$ = the mass flow of oxygen that would be required for complete biochemical oxidation of BCM assimilated by heterotrophs
$F_r$ = rate of consumption of oxygen by respiration of heterotrophs

The model assumes that $F_s$ is approximately equal to:

$$\frac{1.5R_h}{Y_h}$$

where the coefficient 1.5 represents the approximate ratio between the five-day BOD and the ultimate oxygen demand (UOD).
$F_a$ is also proportional to the growth rate of heterotrophs.
$F_r$ is assumed to be proportional to the rate of destruction of organic matter by endogenous respiration.
The rate of consumption, $F_n$, of oxygen by the nitrifying micro-organisms is given by:

$$N_n = \frac{4.3R_n}{Y_n} \text{ (growth rate of nitrifier's biomass)} \tag{3.3}$$

The sum ($F$) of $F_h$ and $F_n$ equals the total average rate of consumption of oxygen by both types of micro-organism.
In agreement with general experimental observations, the model assumes that the rates of biochemical reactions that normally occur under ambient conditions increase by 7% for a 1°C rise in temperature. The rate constants in Table 3.1 are therefore multiplied by

$$1.07^{\text{(Temperature °C - 15)}}$$

to correct for seasonal changes in the temperature of the mixed liquor.

Examples of the output from the model for the treatment of roughly typical wholly domestic settled sewage containing 200 mg/l BOD, of which half is dissolved and half suspended, at a temperature of 15°C, are shown in Figure 3.3(a) and (b). The trends shown relating effluent BOD, ammonia, oxygen uptake, and sludge production to organic loading (g/ g day) are broadly consistent with experience. Particularly noteworthy features are:

- the low concentration of dissolved BOD escaping even up to quite high sludge loadings
- the necessity to maintain fairly low loadings in order to achieve substantial removal of ammonia
- the substantially higher oxygen demand in nitrifying compared with non-nitrifying plants

While these types of output are indeed useful for preliminary screening of process options, it was recognised from the outset that for the detailed design of modern plants, models would have to be extended to take account of such features as the effects of fluctuating loads on performance, to estimate the spatial variations in oxygen demand (and thus the oxygenation capacities to be provided) in plants in which the aeration units were not uniformly mixed. This would be required not only for traditional nitrifying and non-nitrifying plants but also for new generations of plants designed for N and P removal.

## DYNAMIC MODELS

### THE WRc-JONES MODEL

To take account of the fact that sewage arriving at municipal works normally varies substantially in both flow and strength, and that it is usually uneconomic to provide balancing capacity to reduce fluctuations to more than a moderate extent, dynamic models have been evolved to simulate the effects of these fluctuations and thus provide a better basis for design than steady-state models such as STOM. It was shown in early work in the U.K. that it was quite feasible to calculate the effects of longer-term variations in operating conditions on nitrifying performance in piston-flow plants by determining the changes in the populations of nitrifiers in successive passes of the sewage through the aeration units [4,5]. Then

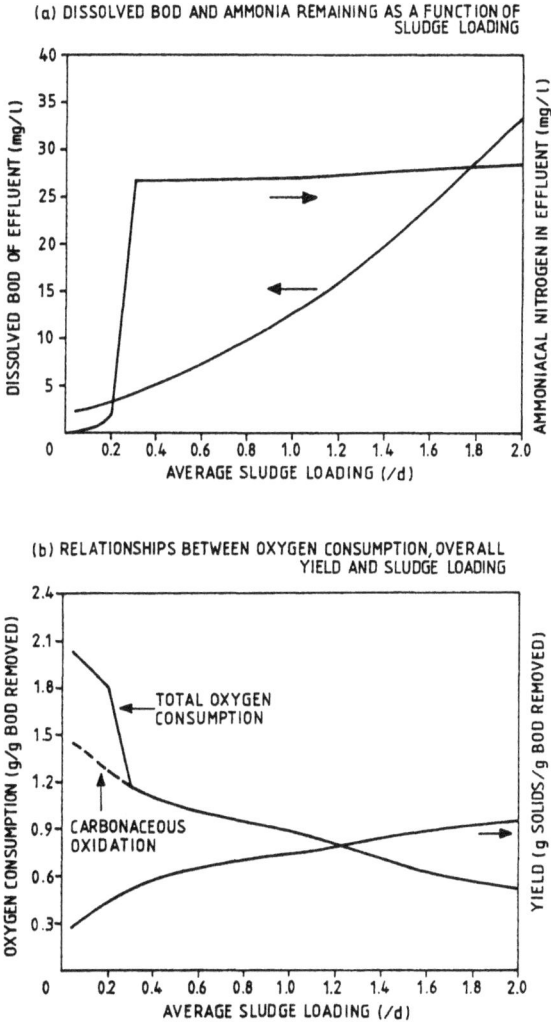

Figure 3.3 Solutions provided by activated sludge sub-model of STOM. (Reprinted with permission from STOM *User Manual* and Description TR 144, 1981.)

Gujer in Switzerland formulated a model in 1969 for simulating the effects of diurnal fluctuations on nitrification [6]. A little later, Jones in the U.K. developed a dynamic model embracing both nitrification and bio-oxidation of carbonaceous matter [7]. It was subsequently extended at the Water Research Centre (WRc), U.K., to form the basis of the current WRc model, which has recently been used as the basis of design for several new or upgraded works [8,9].

The essence of the Jones model is that it includes three mechanisms to account for the relatively low numbers of viable bacteria in activated sludge when determined by conventional microbiological techniques of enumeration and the fact that these numbers were always very much lower than the numbers that appeared to be necessary (for example from work in pure cultures) to account for the observed activity of the sludge. These mechanisms are that:

- Cell mass was consumed to satisfy the maintenance requirements of bacteria in the absence of sufficient substrate to support growth.
- Substrates could be co-metabolized by mixtures of bacterial species that could not individually degrade them.
- The enzyme systems of moribund cells could exert "posthumous" biochemical activity.

While it has not yet been feasible to describe the kinetics of removal of the complex mixtures of carbonaceous substrates present in sewage in terms of the individual behaviour of the substrates and the organisms metabolizing them in the terms of the above mechanisms, their broad implications have been incorporated into empirical generalizations describing the overall summation of their effects on dissolved and suspended BOD. Such generalizations for steady-state conditions indeed form the basis of the AS sub-model for STOM.

In the WRc model, the aeration units are treated as consisting of a number (up to twelve) of uniformly mixed compartments in series. If required, one or more compartments can be considered as anoxic zones in which denitrification takes place if the loading on the plant permits nitrification. When applied to analyze performance of existing works the number of uniformly mixed compartments is that which would give the same residence-time distribution curve of concentration in the effluent as that of a tracer passing through the actual units. The dynamic mass balances for biodegradable substrates, activated sludge, and DO are then written for each compartment using the concepts of Jones to allow for removal of substrate by both viable and non-viable bacteria, with viability being related to the concentration of substrate in the mixed liquor. Thus the dynamic mass balance for a substrate in a single compartment is

$$\frac{dS}{dt} = \frac{Q(S_o + rS_r)}{V} - \frac{\mu_m S K_v}{Y(K_s + S)} - \frac{kX_{nv}S}{K_m + S} - \frac{Q(1 + r)S}{V} \quad (3.4)$$

where

$X_v, X_{nv}$ = the concentrations of viable and non-viable cells, respectively

$S_o, S_r$ = the concentrations of substrate in the feed and sludge recycle

$r$ = recycle ratio

$Q$ = the feed flow

$V$ = the volume of the compartment

$\mu$ = the specific growth rate of viable cells

$k$ = maximum specific rate of conversion of substrate by non-viable cells

$K_m$ = the concentration of substrate at which the specific rate of conversion is $k/2$

The same form of equation is used for BOD and ammonia. It is assumed that organisms remain fully viable until concentration of substrate falls below that which would support a specific growth rate of not less than 10 percent of the maximum. Below this concentration, viability declines.

The mass balance equations for cells corresponding to Equation (3.4) allow for endogenous decay of cells at a rate proportional to their concentration.

Concentration of DO is assumed to affect rate of removal of substrates by both viable and non-viable cells by an expression comparable to the Michaelis-Menton equation describing the influence of substrate concentration.

Oxygen is utilized in the degradation of substrate and also by the endogenous respiration of cells.

If an anoxic zone is specified, a switching mechanism is brought into play when the rate of bio-oxidation supportable by nitrate oxygen is greater than that supportable by DO. Bio-oxidation using nitrate is assumed to take place according to the same forms of equation as those for aerobic degradation, but with much lower rate coefficients.

Settlement in final tanks is empirically assumed to be good, average, or poor, as characterized by SSVIs of 80, 100, and 150 ml/g, tank performance being subject to the overriding requirements that the applied mass-flux must not exceed the rate of sludge removal (by settlement and sludge return) as discussed in a subsequent section on secondary settlement.

An iterative procedure is used to obtain an optimum design. The diurnal patterns of variation in sewage characteristics (flow, temperature, BOD, SS, and ammonia) the rate of sludge recycle, the MLSS concentration thought likely to be appropriate and the mass-transfer coefficient for absorption of atmospheric oxygen ($k_La$) are all specified by the user. The sludge wastage rate is determined by trial and error to yield the desired MLSS concentration. The chosen values of $k_La$ are then adjusted until the DO profile is as desired, which usually would be less than 0.5 mg/l in the anoxic zone and around 1–2 mg/l in the aerobic units. The necessary $k_La$ values are translated into the corresponding equipment required, taking account of the known performances of aeration systems under standard conditions and the expected variations along the length of the aeration units in a-factors (the ratio of the oxygen transfer rate in mixed liquor to that in the standard conditions in which performance of aerators is normally measured, usually in clean water).

In this way, design is evolved iteratively until the simulated performance is optimally consistent with that required particularly in regard to the standards that have to be met and any other constraints such as there may be, for example, unavailability of land, or tank geometry. A typical example of the outcome of this procedure is given in the subsequent section on process sequences.

Since 1990 the WRc Jones Model has been included in WRc STOAT, which is a software package for process modelling for design and operation. STOAT (Sewage Treatment Optimisation over Time) is a dynamic simulation which includes many other models than the WRc activated sludge model but this is still the most important and developed one in the package. STOAT contains:

- storm tanks
- primary tanks
- trickling filters
- disinfection
- mesophilic anaerobic sludge digestion
- thermophilic aerobic sludge digestion
- sludge dewatering

The activated sludge variants included are:

- several variants of the WRc model
- oxidation ditches
- sequencing batch reactors
- N and P removal
- IAWQ activated sludge models No. 1 and No. 2
- activated sludge settlement tanks

The WRc STOAT model has been developed to the stage where for the past 4 years it has operated on an IBM-compatible PC and is now used throughout the world. It has been used in the design of more than

100 works in Europe, USA and the Middle East [10]. The STOAT package is capable of using both $BOD_5$ and COD models for activated sludge.

This software is capable of linking to sewerage and river quality models such as MOUSETRAP, MIKE II or HYDROWORKS.

## IAWQ AS MODELS NO. 1 AND NO. 2

In the last few years many Europeans have been members of the IAWQ (International Association on Water Quality (formerly known as IAWPRC)) Specialist Group developing a generalized dynamic model of the activated sludge process. This model is a synthesis of experience in many countries, notably including the Republic of South Africa, several European states, Japan, and the U.S.A. The model is now ready for use as a design tool, though because it has become available only very recently the extent of its application so far is very small. Nevertheless it has the potential to become a principal design tool for new works in the future. The model has been described in a number of publications, and a detailed account is beyond the scope of this chapter [11,12].

Suffice it to record that though the model represents many of the processes occurring in more detail than in the WRc-Jones model, several of its main features are similar. The more important differences are that it:

- represents carbonaceous matter, both biomass and substrate, in terms of COD rather than BOD
- takes separate account of the behaviour of carbonaceous matter not containing nitrogen and of organic nitrogen
- accounts for hydrolysis of organic nitrogen to ammoniacal nitrogen
- takes formal account of hydrolysis of slowly degradable organics to produce readily degradable species but assumes that no uptake of DO is involved in these reactions
- assumes decay of biomass results in formation of slowly degradable COD, although the assumed rate of decay is so low that the component formed is virtually inert
- does not adopt the concepts of co-metabolism and posthumous activity
- assumes that rapidly biodegradable matter is all dissolved and that slowly biodegradable matter is all suspended

The IAWQ AS Model No. 1 covered carbonaceous oxidation, nitrification and denitrification. The No. 2 model is essentially the same but has had biological phosphorus removal added to it.

There are a number of other modelling packages around now such as GPSX and BIOWIN. In most cases they use the IAWQ COD-based models.

## AERATION SYSTEMS

Modern designs of plants depend on many factors, but particularly the effluent quality required and the size of population served. However, all types of conventional activated sludge plants involve use of aeration systems and final clarifiers, and it is thus appropriate to review the characteristics of these units before proceeding to consider the influence of the effluent quality required and size of population served on plant design.

While examples of many types of aeration systems are to be found, in the majority of plants aeration is achieved either by diffusing air through the mixed liquor or by agitating it mechanically in open tanks. A valuable summary is presented in a *Manual of Practice* issued by the Institution of Water and Environmental Management in the U.K. [13].

### DIFFUSED AIR PLANTS

It is well established that the mass of oxygen that can be transferred into mixed liquor from a given volume of air under given conditions increases with decreasing bubble size. On the other hand, the difficulties and associated costs of producing bubbles tend to increase with decrease in the size sought. The practical compromise most frequently adopted in Europe is to provide fine-bubble aeration systems delivering bubbles usually of around 2 to 2.5 mm in equivalent diameter.

This is most commonly achieved by delivering filtered air through porous sintered materials, such as the Alundum domes popular in the U.K. (Figure 3.4), or porous plastic tiles or discs. In certain proprietary designs the plastic discs are sufficiently flexible to expand as back-pressure rises, thus increasing orifice size and providing opportunity for clogging particles to escape.

It is well established that in vertically sided tanks with diffusers located at the bottom on a flat floor, the percentage of oxygen absorbed from the air under given conditions increases linearly with the depth of mixed liquor. However, construction and other costs tend to increase often more rapidly with depth, especially if the water table is not far below the surface, and thus there is again a practical compromise that in Europe normally results in the preferred depth being about 3.5 to 4 m. It is recognized that aeration efficiency is greatest when the air diffusers required to achieve the desired rate of aeration are distributed as uniformly as possible over the floor.

However, because the oxygen demand to be satisfied declines from inlet to outlet in a ''plug-flow'' plant and the a-factor increases as the liquid phase becomes purified, the number of diffusers provided in each pocket is reduced (tapered) from inlet to outlet. A typical arrangement is shown in Figure 3.5.

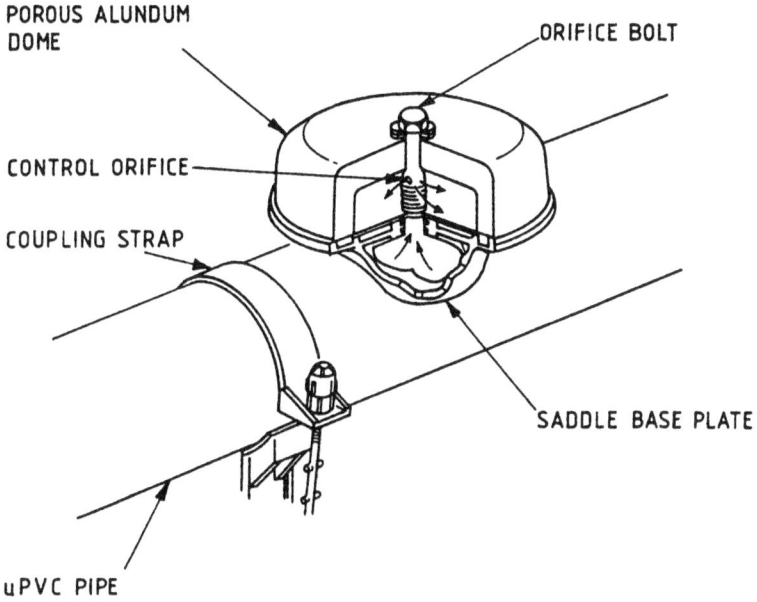

**Figure 3.4** Dome diffuser. (Figure reproduced with the kind permission of Water Engineering Ltd., England.)

**Figure 3.5** Diffused air activated sludge plant equipped with dome diffusers. (Figure reproduced with the kind permission of Water Engineering Ltd., England.)

121

One of the more notable variants that has been used in Europe, though mainly in Scandinavia, is the Inka System in which air is injected from a perforated grill, mounted about 0.6 m below the mixed liquor surface on one side only of a longitudinal baffle running the full length of the aeration tanks. A spiral flow is induced, which carries the bubbles for some distance around the tank and shears them to smaller sizes than would be produced from the relatively large orifices in still water.

Because of the small hydrostatic head to be overcome, fans are used rather than compressors and this offers some advantage in reducing unproductive dissipation of energy. However the oxygen input per unit energy under the most favourable conditions, probably around 2.2 kg/kWh is somewhat lower than attainable with fine-bubble porous diffusers, and the difference may well be even more unfavourable toward the outlet ends of plug-flow plants where the required oxygenation capacities are relatively low.


## MECHANICAL AERATORS

In some European countries a substantial proportion of works have been equipped with mechanical aerators. In the U.K., for example, this proportion is about two-thirds of the roughly 150 plants servicing populations greater than 10,000. However the proportion of the total flow treated in these mechanically aerated plants is somewhat lower than that treated in diffused air plants.

Mechanical aeration systems include several types of both vertical shaft and horizontal shaft aerators. Three types of vertical shaft aerators are shown in Figure 3.6. The single most popular type in the U.K. has been the Simplex aeration cone, which is mounted at the top of a draft tube as indicated in Figure 3.7. In compartmentalized plants, each pocket (usually square in the plan) is equipped with a cone at its center and often fillets are added to the base of the tank to assist vertical circulation; also, vertical baffles are sometimes used to prevent horizontal rotation of the tank contents. Depths are usually about 5.5 m and a freeboard of 1.5 m is provided to retain spray. With other types of vertical spindle aerators not employing a draft tube, the tanks are usually somewhat shallower.

On the European continent the proportion of mechanically aerated plants employing horizontal-spindle aerators is greater than in the U.K. A popular type is the "Mammoth" rotor developed in Germany (Figure 3.8). This type of rotor is manufactured in lengths up to 9 m and diameters up to 1 m, and typically operates at immersion depths up to 300 mm and speeds of rotation around 70–80 rev/min. Oxygen inputs per unit energy between 2 and 2.5 kg/kWh are achievable. The aerators are usually individually powered by weatherproof, fan-cooled motors (up to about 120 kW in power) acting through reduction gear boxes.

Adjustable weirs are often incorporated at the outlet ends of mechanically aerated units to enable the depth of immersion of the rotating impellers and thus the rate of aeration to be varied in response to demand. In addition, the aerators may be covered by acoustic shrouds to reduce noise. Normally one can expect mechanical aeration systems to deliver oxygen inputs per unit energy of around 2 kg/kWh when driving force is near the maximum.

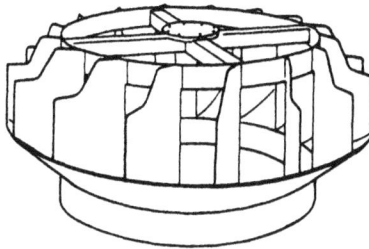

(a) SIMPLEX AERATOR

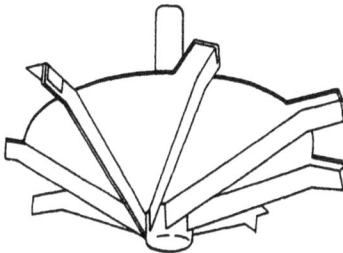

(b) SIMCAR SAL AERATOR

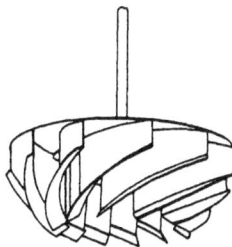

(c) LIGHTNIN AERATOR

**Figure 3.6** Three types of vertical shaft aerators (photograph courtesy of Biwater Treatment Limited).

**Figure 3.7** Simplex cone aeration unit (photograph courtesy of Biwater Treatment Limited).

**Figure 3.8** Mammoth aeration rotor (photograph courtesy of Biwater Treatment Limited).

124

## "HYBRID" SYSTEMS

In passing it is noteworthy that in aeration zones in which the organic load is relatively high (for example at the inlet ends of "plug-flow" plant or in uniformly mixed plants required to achieve only partial purification) fine-bubble diffusers are reported to undergo relatively rapid clogging requiring more intensive maintenance or replacement than those in more lightly loaded zones [14]. Also $\alpha$-factors for the heavily organically loaded zones tend to be quite low.

In contrast, mechanical aerators do not suffer from these drawbacks. It has thus been suggested that the optimum design for a plug-flow compartmentalized plant could involve the use of mechanical aerators in the first few compartments and a fine-bubble diffused air system in the remainder. A plant in which this configuration has been adopted is expected to be completed by about the end of 1991 at Blackburn Meadows in Sheffield, U.K.

## ALTERNATIVES TO THE MAIN TYPES OF AERATION UNITS

### THE ICI DEEP SHAFT

The ICI Deep Shaft process was developed in the early 1970s [15]. In the process, screened and degritted crude sewage is fed with returned activated sludge into a vertical shaft up to 150 m deep though usually between 50 and 100 m deep and divided into downcomer and riser sections. Air is injected into the riser initially to set up a recirculating flow, after which air is released into the downcomer at a point low enough to ensure that the density of the air-water mixture in the riser is lower than that in the downcomer, thus both maintaining circulation and to achieving aeration. A feature of this method of aeration is that high rates of oxygen transfer can be achieved, thus making it feasible to match the oxygen demand of particularly strong sewages. Experience has shown that because gas bubbles coming out of solution in the riser attach themselves to sludge particles, tending to make them difficult to settle, it is necessary either to render the sludge settleable before transferring mixed liquor to secondary clarifiers or to use an alternative means of separation such as flotation. Various means of rendering the sludge settleable have been used, including vacuum degasification, but that which now appears to be most commonly preferred is to aerate the mixed liquor using conventional porous diffusers usually for between 1 and 2 hours. The flowsheet for a major plant in the U.K., performance of which is described later, is given in Figure 3.9.

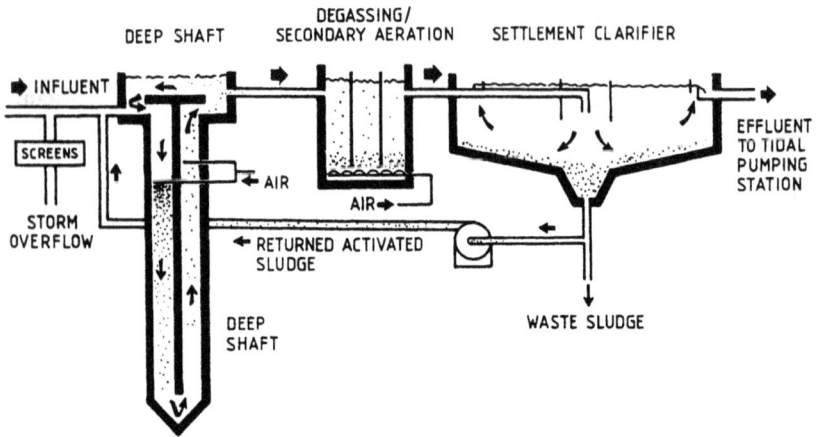

**Figure 3.9** Process diagram of Tilbury deep shaft sewage treatment works (Irwin, R. A., W. J. Brignal and M. A. Biss 1989. "Deep Shaft Process at Tilbury," *J. Inst. Wat. Envir. Mangt.*, 3:281).

Despite the commitment of capacity and thus of land area to degassing of sludge, the absence of primary tanks and the quite small area occupied by the shaft itself make it substantially less demanding on space than conventional AS plants.

## OXYGEN AS PLANTS (OASP)

Although a few notable successes have been reported in the use of oxygen for aeration rather than air in industrial effluent treatment, OASP has not often been preferred in European sewage treatment plants. There are a few plants of the "closed" UNOX type, developed in the U.S.A. by the Union Carbide Corporation, and some others using the "open" type such as the FMC Marox System and the British Oxygen Company's (BOC) Vitox System [16]. The main features of the Vitox System are shown in Figure 3.10.

In closed systems the oxygen is dispersed into covered tanks, the contents of which are mixed by vertical spindle vane impellers. In open systems the oxygen is injected into the throat of a Venturi tube in a sidestream through which mixed liquor is continuously recycled into otherwise conventional open aeration tanks.

Advantages claimed for the closed systems include lower sludge production, higher sludge densities, and greater reactivity per unit biomass than attained in conventional plants. However, probably the main reason for the limited use has been that any such advantages appeared to be insufficiently clear cut to overcome the drawbacks of greater complexity, a lower rate of stripping of $CO_2$, and thus the necessity usually to incorporate a second conventional stage to achieve nitrification when limits on ammonia or N in the effluent have to be met. Moreover, they would not appear to integrate very conveniently into modern nitrification and denitrification schemes for nutrient removal, control of sludge settleability, and energy recovery.

The open systems have been used mainly to uprate the oxygenation capacities of older conventional plants. Just such an application has recently been reported for one of the main plants at Dusseldorf in Germany where the oxygen injection system is of a proprietary design known as Lindesol-vax-B [17]. Applications of this type do not appear to have been many, but they may increase if the need for uprating outpaces construction of new works.

## SECONDARY SETTLEMENT

As with aeration units, there are only a few features that distinguish European practice significantly from that in the U.S.A. and other developed countries outside Europe. Thus, as noted earlier, detention times are usually chosen to be about 2 h at peak and 6 h at average flow, with the corresponding surface overflow rates being about 1.5 and 0.5 m/h.

Circular radical flow tanks with mechanical scrapers are the most popular design for larger works. For these it is usually considered advisable for:

- side walls to be not less than 2 m deep, to avoid carryover of

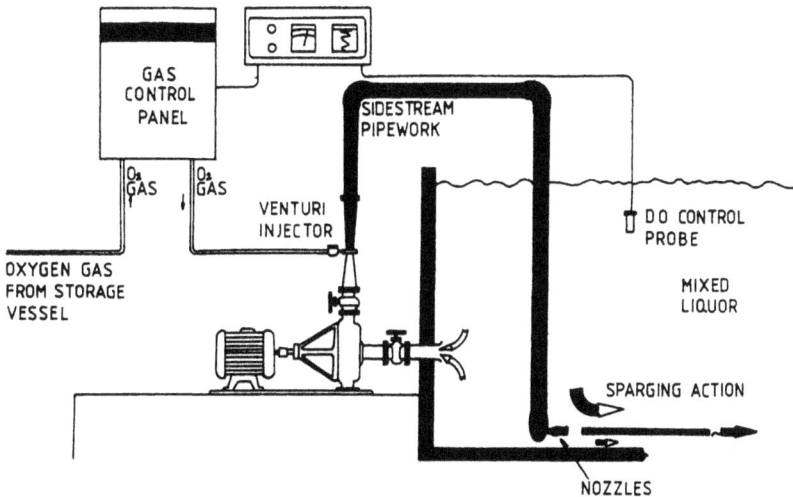

**Figure 3.10** Vitox high pressure sidestream dissolver.

solids picked up from the settled sludge in currents travelling to the overflow weirs
- the slopes of the conical floors to be not less than 5° to the horizontal, to facilitate transfer to the withdrawal sump, and not more than 30°, to limit depth and construction costs
- the maximum diameter of tanks to be not more than about 50 m
- weir loadings to be between 100 and 250 $m^{3/m\ d}$

Pyramidal tanks with walls in the lower section sloping at about 60° to the horizontal are often used in small works.

Perhaps the most notable development in Europe within the last decade has been the much wider recognition of the necessity to ensure that this mass-flux must not exceed the capacity of the units to remove and recycle them. This requirement was first brought into prominence by work in the U.S.A. by Dick and his co-workers [18,19]. Both Dick and Ford and Eckenfelder have suggested that a better indication of the settling velocity of sludge under the dynamic conditions of full-scale tanks than is afforded by the traditional sludge volume index (SVI), or its reciprocal the sludge density index (SDI), would be obtained by measuring velocity while gently stirring the sludge [20]. The local element of modern European practice has evolved from the development by White at the former WRc Stevenage Laboratory in the U.K. of such a new form of test to measure the stirred settled volume index (SSVI) [21].

In this test, mixed liquor is settled in a standard apparatus in which it is very gently stirred by a vertical ring impeller rotating at 1 rev/min as indicated in Figure 3.11. The impeller eliminates wall effects and induces a consistent degree of flocculation similar to that occurring under the dynamic conditions of full-scale tanks.

The SSVI is calculated in the same way as SVI:

$$SSVI = \frac{\%\ of\ total\ volume\ occupied\ by\ settled\ sludge\ after\ 30\ min}{concentration\ of\ MLSS\ (as\ \%)}$$

By analogy with SDI, the stirred sludge density index (SSD) is the reciprocal of SSVI. Because MLSS concentration influences settling velocity it is usual to measure SSVI at a concentration of 3.5% if possible so as to facilitate comparison with performances in other plants or in the same plant under different operating conditions.

White then deduced a relationship between the maximum mass-flow of suspended solids $R$ (kg/h) recycled from the base of a settlement tank to the downward velocity of the liquor, $U$ (m/h), and the average SSVI as:

$$R = 310A(SSVI)^{-0.77}\ U^{0.68} \tag{3.5}$$

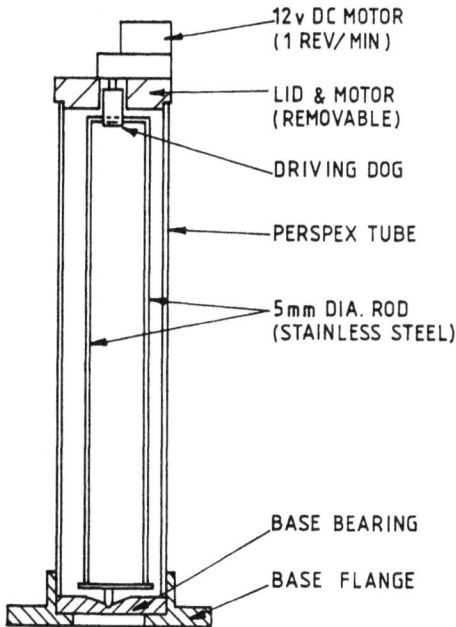

**Figure 3.11** WRc standard settling apparatus (White, M. J. D. 1976. "Design and Control of Secondary Settlement Tanks," *J. Inst. Wat. Pollut Control,* p. 461).

in which $A$ (m$^2$) is the cross-sectional area of the tank. This formed the basis of the convenient nomogram shown in Figure 3.12, which enables the applied solids loading to be compared with the predicted maximum acceptable loading for a given SSVI.

## PROCESS SEQUENCES

As indicated at the beginning of the preceding section, process selections are very dependent on effluent quality required and size of population to be served. The influence of effluent quality can adequately be illustrated by reference to five types of requirement, namely to reduce:

(1) Carbonaceous BOD and SS partially (e.g., by 60–85 percent) to meet limits in respect of SS and BOD of 60:40 mg/l as average values or better
(2) SS and BOD by 90–95 percent to meet limits of 30:20 as average values
(3) SS, BOD, and ammoniacal-nitrogen to meet limits of 10:10:5 (or lower in the last case sometimes down to 3 mg/l)

Figure content (nomograph scales and labels):

$\dfrac{Q_u}{A}$ (m/h)

UNDERFLOW RATE
(return) PER UNIT AREA

PREDICTED

APPLIED

$\dfrac{Q\,total}{A}$ (m/h)

TOTAL FLOW RATE (feed+return)
PER UNIT AREA

| STIRRED SLUDGE DENSITY (%) | STIRRED SPECIFIC VOLUME (ml/g) | Measured at S S concentration of 3.5g/l in WRC standard settling apparatus | SOLIDS LOADING (kg/m²h) ±20% | CONCENTRATION OF MIXED LIQUOR (g/l) |

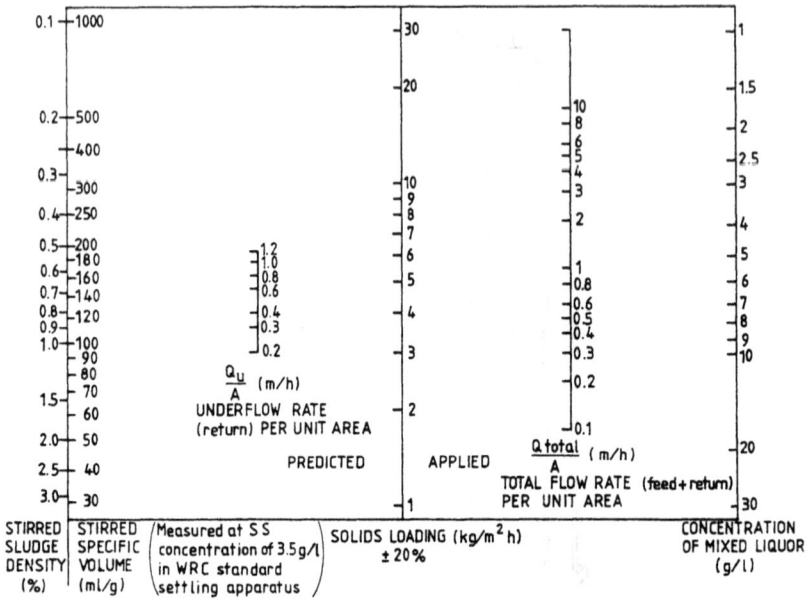

**Figure 3.12** Nomograph for the calculation of predicted and applied solids loading (White, M. J. D 1976. "Design and Control of Secondary Settlement Tanks," *J. Inst. Wat. Pollut. Control*, p. 464).

(4) SS, BOD, and ammoniacal-N as in (3) but in addition total nitrogen (Kjeldahl nitrogen plus oxidized nitrogen) to 10 mg/l (or lower down to around 2 mg/l)

(5) SS, BOD, ammoniacal nitrogen, total phosphorus (P) to 1 or 2 mg/l and often in addition total or total nitrogen as in (3)

These last two requirements [(4) and (5)] can conveniently be considered together under the heading of nutrient removal.

The sequences chosen to fulfil these requirements are considered below in relation to the size of population served. There is no sharp demarcation between this last feature and the types of plants installed, though broadly as populations served fall below 20,000 there would be a rapidly increasing proportion of plants either of relatively shallow depth and simple construction occupying relatively large areas of land per unit of load treated (such as the oxidation ditch) or of special simplified design (such as the Putox process used in Austria—see later section). For simplicity it is convenient to describe such plants in the context of facilities for "smaller populations," though recognizing that a few of them serve populations up to about 200,000 and there is some overlapping with more intensive elaborately engineered plants, with deeper reinforced concrete treatment units provided for "larger populations."

## LARGER POPULATIONS

### Partial Purification

The use of the activated sludge process solely for partial purification followed by release to a watercourse nowadays tends to be limited to discharges near the mouth of large estuaries or to coastal waters. The practice is not very widespread, but where it is employed the plants are usually of the uniformly mixed or contact stabilisation types with sludge loadings in the range from about 0.5 to 1 g/g day. Basic flowsheets are usually as in Figure 3.1(a) and (c). In principle, such plants could be used as the first stage of a two-stage process for producing final effluent of much higher quality, but it is more common to use a combination of high-rate biofilter followed by low-rate activated sludge plant for this purpose.

High-rate AS plants were particularly popular in Germany prior to 1970, sludge loadings usually being in the range of 1 to 3 g/g day. Thus, for example, the Kohlbrandhoft plant at Hamburg was designed to remove about two-thirds of the incoming BOD at a sludge loading of 3 g/g day [20].

An example of a large modern inland high-rate plant discharging to a large river (the Danube) is afforded by the main works for Vienna-Simmering, which treats the mainly domestic flow from more than 2 million people. Process design was by Von der Emde and his colleagues [22,23]. Settled sewage is treated in plug-flow mechanically aerated (vertical spindle) aeration units having a total residence time of 1.5 h and operating with a sludge loading of 1.1 g/g day and sludge age of 1.2 days (Figure 3.13). The aerators were powered with DC motors so that aeration intensity could be varied and matched to demand. Because of the high water table the treatment units were all rather shallow, the aeration basins for example being not more than 2.6 m deep. An unusual feature of the plant is that the final clarifiers are rectangular. The plant was required to achieve a 70 percent reduction in BOD but in practice reductions of 88 percent were obtained, except during periods when high loads of centrate from a separate sludge dewatering facility were returned to the works. Under more normal conditions, effluent BOD and SS averaged 31 and 21 mg/l, respectively, oxygen uptake per unit mass of BOD was 0.5 g/g, and sludge production 0.78 g/g BOD removed. These figures are reported to agree well with the theory of Marais and Ekama [24]. The sludge productions were also in reasonable agreement with an empirical equation derived by Hopwood and Downing [25].

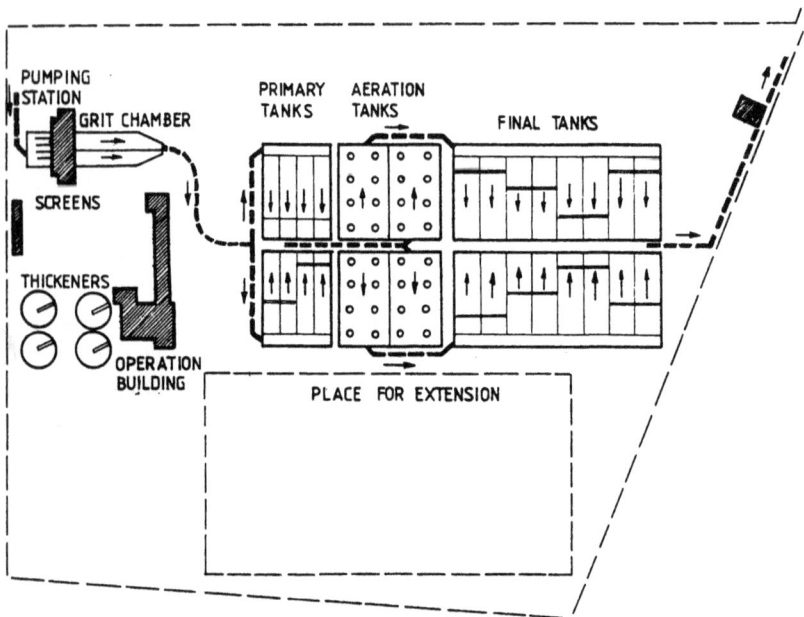

**Figure 3.13** Layout of main treatment plant Vienna-Simmering.

On occasion ferrous sulfate is added, mainly to control bulking. Because additionally the high organic loadings applied induce anaerobic zones favouring mobilization of phosphorus at the inlet ends of the aeration units, substantial removal of P can be achieved in the plant (see later section).

### 30:20 Effluent

Whereas activated sludge plants for producing effluent of this quality were quite common at inland sites, indeed almost the norm about two decades ago, the increasing tendency nowadays is to design for limitation of release of ammonia.

Many types of ASP have been used to produce 30:20 effluents as average values though probably the most common type comprises a compartmentalised aeration unit aerated either by a fine-bubble diffused air system or mechanical aeration. Loadings would normally be around 0.2–0.4 g/g day.

Among the exceptions is the world's largest deep shaft plant at Tilbury in the U.K. [26]. This treats a flow of some 35,000 m$^3$/day of a strong sewage (BOD 600) containing a considerable proportion of industrial wastewater before discharge to the Thames Estuary. It operates at a sludge loading of around 1 g/g day and although performance figures for a whole year do not yet appear to have been reported, during four months of the warmer part of the year the effluent complied with a 95-percentile 30:20 standard. It seems possible that such a standard might not be met at the above loadings in winter, in which case one might have to regard operation in this mode as rather more in the category of partial purification.

### Fully Nitrified Effluent (5 mg/l Limit for Ammonia or Better)

There are many fully nitrifying European plants. Up until the last decade the majority of these were simply lightly loaded conventional compartmentalized plants, with probably a slight preponderance in numbers of diffused air over mechanically aerated plants, and almost certainly a margin in favour of diffused air in terms of numbers of people served. In some cases the rate of aeration was automatically controlled according to the response of membrane electrode DO sensors and the oxygenation capacity was tapered. However, in many cases, particularly in diffused air plants, the rate of aeration toward the outlet end of the aeration units was often excessive, thus involving wastage of energy. Within the last decade it has become the practice to include an anoxic zone, usually as the first compartment in a chain, because of experience indicating that passage of mixed liquor through such a zone improves settleability of sludge and also reduces energy requirements by virtue of the fact that some of the oxygen injected into the aerobic zones and converted into nitrate is recovered. However, this is at the expense of having to provide somewhat more capacity (that of the anoxic zone) than would be necessary simply to produce a fully nitrified effluent meeting the same limit on ammonia. A bonus is that the nitrogen content of the effluent is significantly reduced; and the tendency of sludge to rise in final clarifiers in warm weather is probably also somewhat lessened. Another feature of modern practice stems from the recognition that reduction in longitudinal mixing improves sludge settleability. This effect is shown in Figure 3.14, in which SSVI for a large number of plants is plotted against dispersion number as reported for example by Chambers and Jones [8]. These authors have shown that dispersion number, a hydraulic parameter representative of the degree of longitudinal mixing, is related approximately but with adequate accuracy to the number of compartments in the aeration units by the expression:

$$\frac{D}{uL} = \frac{1}{2N} \tag{3.6}$$

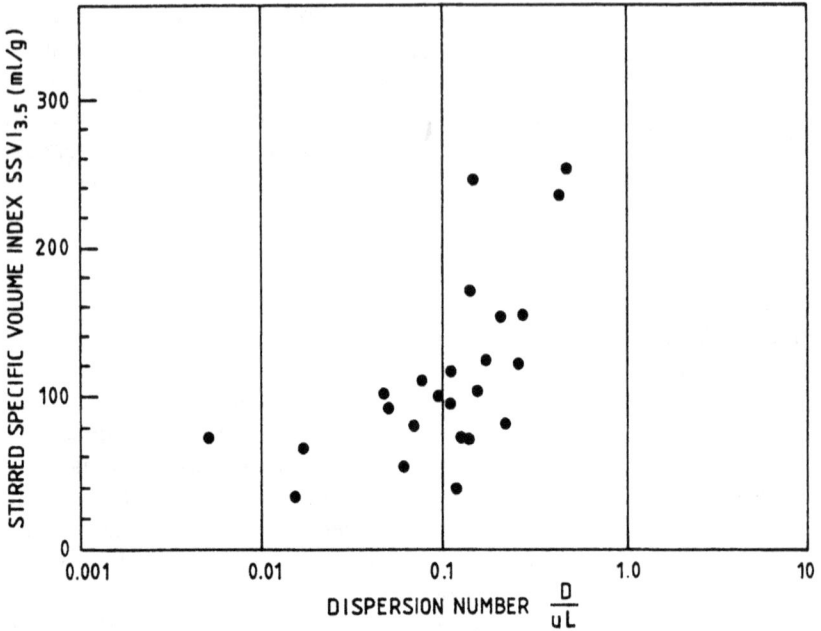

**Figure 3.14** Effect of longitudinal mixing on sludge settleability. (Reprinted from *Water Science & Technology, Vol 20*, figure 2, p. 129, with permission from the publishers, Pergamon Press, and the copyright holders, IAWPRC.)

where the group of terms on the left hand side is the dispersion number

$D$ = longitudinal dispersion coefficient
$u$ = average horizontal velocity of mixed liquor
$L$ = the length of the aeration unit
$N$ = the number of compartments into which the aeration units are divided

The implication is that if the compartments were perfectly mixed and in cascade (an arrangement adopted in at least one large European plant) then SSVI would approach a minimum if $N$ were 5 or more. In practice, to allow for longitudinal dispersion not being fully eliminated by baffles and to the desire in the interests of economy to match rate of aeration as closely as possible to oxygen demand, the number of compartments adopted in modern plants is usually between 5 and 16.

Using an early version of the WRc model, a design study of a typical situation—in which settled sewage having average SS, BOD, and ammonia of, respectively, 95, 120, and 30 mg/l and fluctuating diurnally in flow and composition between peak and minimum values around, respectively, 1.5 and 0.5 times the mean was treated in a diffused air plant—indicated that in an aeration unit having eight compartments, of which the first would be an anoxic zone, the average magnitudes of $k_La$ (per hour) required just to maintain about 1–2 mg/l DO in the aerobic units under steady average conditions were 6.5, 5.0, 5.0, 2.5, 2.5, 1.0, and 1.0. For this simulation, total sewage retention time was just under 10 hours, MLSS 3700 mg/l, temperature 12°C, and sludge age 9.3 days (i.e., sufficient for full nitrification).

In practice, the aeration units would have to be designed to cater for peak load conditions, and for these the corresponding values of $k_La$ were 14, 12, 10, 8, 6, 4, 2. However, the degree of tapering in the air flow and number of diffusers required to deliver the air would not be as great as that of the values of $k_La$ because of the influence of the α-factor, which was expected to vary along the length in the sequence 0.4, 0.4, 0.5, 0.6, 0.7, 0.8 and 0.9. As a result, the maximum air flow and the appropriate number of diffusers required in the first aerobic compartment, was calculated to be only about 1.7 times that for the last compartment, whereas the corresponding ratio for $k_La$ was 4 (Figure 3.15).

The incorporation of an anoxic zone in the first compartment of plug-flow nitrifying plants has now become almost a standard feature of modern design, the size of such zones usually being such as to provide retention times of between 0.5 and 1 hour and usually bringing about a 40–50 percent removal of N [27,28]. The average rate of denitrification declines with increasing retention and becomes too slow to make larger units economic. A recent U.K. design is described in a subsequent section concerned primarily with automation and control.

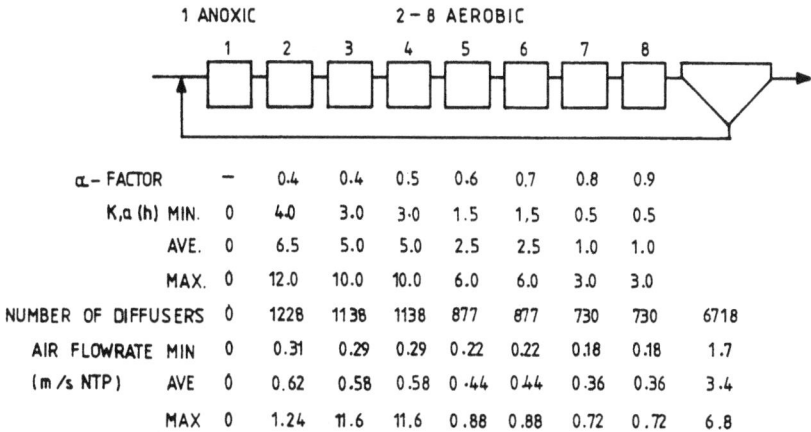

| | 1 ANOXIC | 2 | 3 | 4 | 5 | 6 | 7 | 8 | |
|---|---|---|---|---|---|---|---|---|---|
| α- FACTOR | — | 0.4 | 0.4 | 0.5 | 0.6 | 0.7 | 0.8 | 0.9 | |
| K,a (h) MIN. | 0 | 4.0 | 3.0 | 3.0 | 1.5 | 1.5 | 0.5 | 0.5 | |
| AVE. | 0 | 6.5 | 5.0 | 5.0 | 2.5 | 2.5 | 1.0 | 1.0 | |
| MAX. | 0 | 12.0 | 10.0 | 10.0 | 6.0 | 6.0 | 3.0 | 3.0 | |
| NUMBER OF DIFFUSERS | 0 | 1228 | 1138 | 1138 | 877 | 877 | 730 | 730 | 6718 |
| AIR FLOWRATE MIN | 0 | 0.31 | 0.29 | 0.29 | 0.22 | 0.22 | 0.18 | 0.18 | 1.7 |
| (m/s NTP) AVE | 0 | 0.62 | 0.58 | 0.58 | 0.44 | 0.44 | 0.36 | 0.36 | 3.4 |
| MAX | 0 | 1.24 | 11.6 | 11.6 | 0.88 | 0.88 | 0.72 | 0.72 | 6.8 |

**Figure 3.15** Diffuser layout and air flow rates for nitrifying activated sludge plant as determined by WRc model (Chambers, B. and G. L. Jones. 1985. "Energy Saving by Fine-Bubble Aeration." *Wat. Pollut. Control*, p. 82).

Another example is the plant recently installed at Valeton, southwest of Paris, in which the aerobic zones are concentric "plug-flow" circular channels surrounding a central anoxic zone [29]. This anoxic zone is covered and the contents are kept mixed by circulation of the liberated gas containing mainly $N_2$ and $CO_2$. DO in the aerated zones is automatically controlled to vary from about 1.5 mg/l at the inlet to 3.0 mg/l at the outlet.

## Nutrient Removal

The extent to which nutrient removal is practised in Europe still varies considerably from country to country but the EU Urban Waste Water Treatment Directive (1991) [1] has given the technology a large impetus. Examples of biological phosphorus removal can now be found in most of the EU countries certainly in Germany, Sweden, Denmark, France, UK, Holland and Italy. Research into relevant technology was first instigated in the early 1960s, particularly in countries such as Switzerland and those of Scandinavia, which had either many inland lakes or in the case of Norway semi-enclosed water bodies such as fjords that were already suffering or were thought likely to suffer from increasing eutrophication. In other countries already with a high degree of pollution control such as France, Germany, and the U.K. nutrient removal was not at the time considered necessary for various reasons including:

- Many lakes serving as sources for public supply were in upland areas not subject to significant pollution.
- Algal growth in rivers and coastal waters was not usually at a level then considered unacceptable.
- The technology of nutrient removal was still in its infancy and the most appropriate method in which to invest the substantial sums needed were far from clear.
- Nutrients were often derived from several sources and it was not obvious that eliminating them from sewage effluent alone would necessarily be beneficial in many cases.

It is fair to say that many of these reasons have now been challenged and in the last 10 years many more rivers, estuaries and the coast have been deemed "sensitive" under the EU's UWWTD and hence nitrogen and phosphorus removal processes are being widely installed as retrofits or new plants to meet the directive.

For the less wealthy countries, in which pollution control was much less advanced, remediation of other even more serious effects of pollution had greater priority.

While some of these features are still evident today, nutrient removal programs have been considerably extended in recent years and now embrace several more countries. Thus for example it has been reported that the number of nutrient removal plants in Scandinavia is to be increased from 50 to 500. In Germany new regulations have been introduced placing limits on the concentration of N in municipal plant effluent, and these limits will require many plants to be appropriately modified. Similar limits have also been imposed on individual plants in Austria, France, and Holland.

So far as the authors are aware, there are no European plants achieving N removal by other than biological processes, and very few make use of external sources of carbon to energise the denitrification processes. On the other hand, while biological processes for P removal are included in or planned for some plants, the majority depend at least partly on chemical precipitation.

When a large proportion of N must be removed to meet modern standards limiting inorganic, kjeldahl, or total nitrogen to 10 mg/l or less, then in larger works recirculation or spatial alternation processes are often employed with basic configurations as shown in Figure 3.16(a) and (b).

An alternative approach, popular in Austria, is to design for simultaneous nitrification-denitrification. Experiences in oxidation ditches (see later section), showing that in a nitrifying plant—if the oxygen introduced in

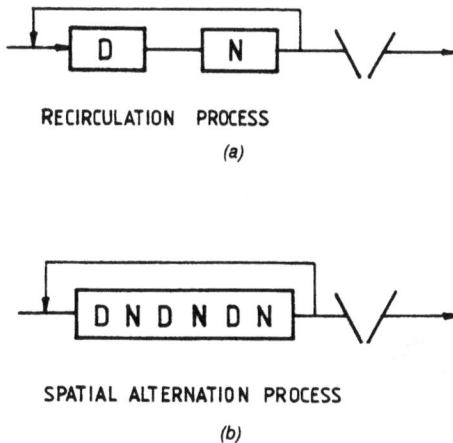

RECIRCULATION PROCESS

(a)

SPATIAL ALTERNATION PROCESS

(b)

LEGEND

D = DENITRIFICATION

N = NITRIFICATION

**Figure 3.16** Most common configurations used for biological denitrification in large plants.

MLSS as it passed through the region around aerators was not enough to prevent anoxic conditions developing before the MLSS once again passed into a zone of high aeration—both anoxic and aerobic zones could coexist in the same aeration unit. This principle has been made use of in comparatively large AS plants in Europe, notably that at Blumental, Vienna, where unsettled sewage from a population numbering more than 200,000 is treated in long baffled aeration units aerated by a series of Mammoth rotors (Figure 3.17) [30,31,32]. After some experimentation to decide the most favourable method of operation, the practice now adopted is to vary the aeration intensity according to the oxygen uptake rate of the sludge. This is determined by pumping mixed liquor to a small tank in which aeration is constant and recording DO continuously. Under these constant conditions, DO is inversely proportional to oxygen uptake rate (OUR) and so aeration intensity can be varied according to the DO recorded.

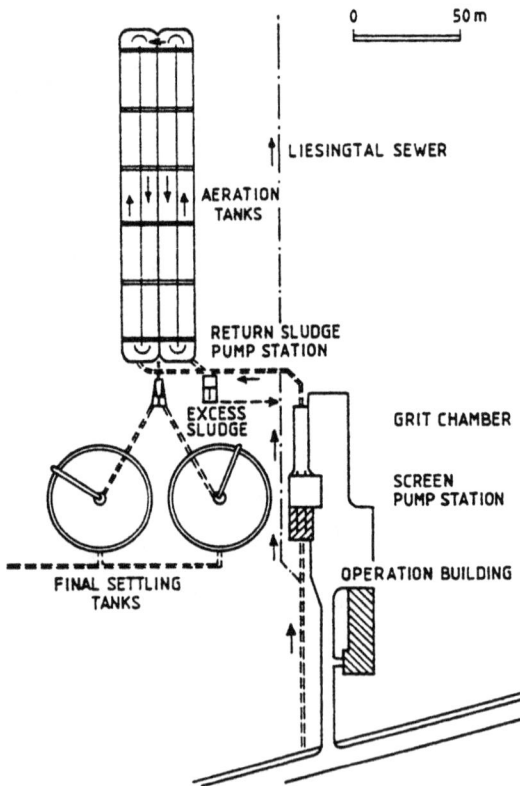

**Figure 3.17** Vienna Blumental treatment plant (Reprinted from *Progress in Water Technology, Vol. 8,* 1977, p. 629, with permission from Pergamon Press and IAWPRC, the copyright holders.)

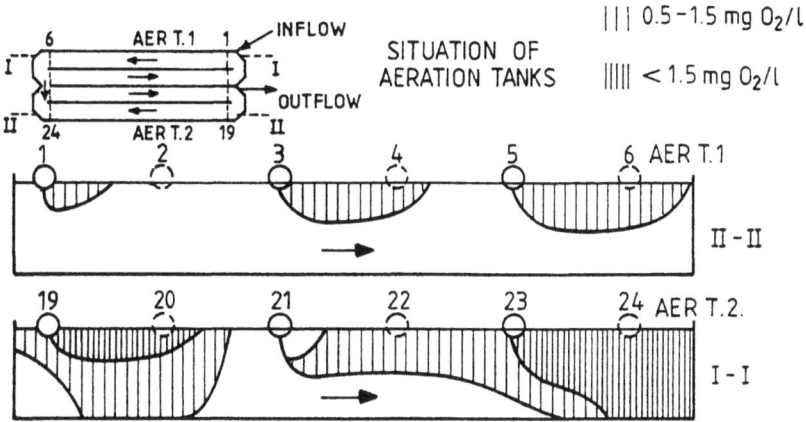

**Figure 3.18** Dissolved oxygen profiles in aeration tanks 1 and 2 of Vienna Blumental plant. (Reprinted from *Progress in Water Technology, Vol. 8,* 1977, p. 629, with permission from Pergamon Press and IAWPRC, the copyright holders.)

Performance depends on loadings, as expected, but at low loadings—around 0.12 g/g day—BOD removals of about 95 percent and TN of about 86 percent are achieved. The average proportion of the volume occupied by anoxic zones was not fully defined, but appears from published diagrams to be often between 30 and 40 percent, thus explaining the much greater percentage removal of nitrogen than is obtained in designs—such as those that have become popular in the U.K.—in which the anoxic zone occupies only 10–15 percent of total aeration tank volume (Figure 3.18).

In the case of the new plants to be built in Denmark and Scandinavia, the design has evolved from the joint researchers of six organisations collaborating in the socalled HYPRO project [33]. The main features of this design, shown in Figure 3.19, involve chemical precipitation of P in primary sedimentation tanks, hydrolysis of the sludge to release biodegradable organics, and feeding of these organics into a denitrification-nitrification reactor with recycle to provide "fuel" for denitrification.

The design of plants for the biological removal of phosphorus has tended to follow the practice pioneered in South Africa in which the sewage is first passed through an anaerobic zone to mobilise phosphorus so that it can be subsequently removed in the following aerobic stages by so-called "luxury uptake." A recent European version introduced in Denmark is the Bio-Denipho process shown in Figure 3.20.

An alternative approach has been adopted in several Austrian plants in the catchment areas of lakes, where new limits of 1 mg P/l in effluents are now mandatory. Although not originally designed to remove P, these plants have been modified to do so by addition of ferrous sulfate at the

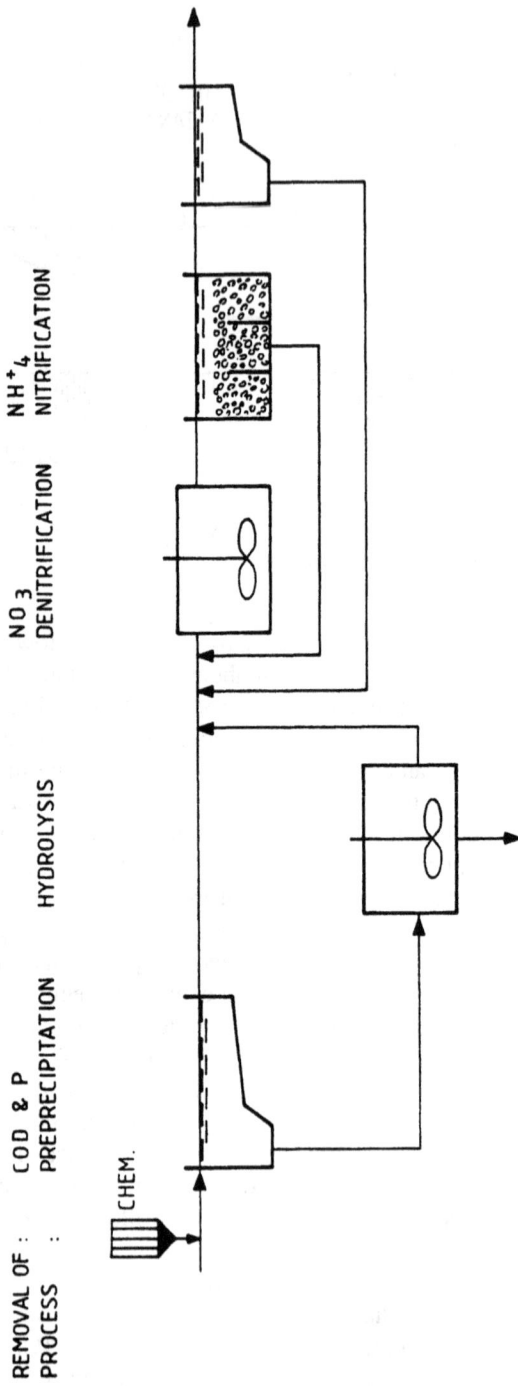

REMOVAL OF : COD & P              NH$^+_4$
PROCESS    : PREPRECIPITATION   HYDROLYSIS   DENITRIFICATION   NITRIFICATION
             NO$_3$

**Figure 3.19** HYPRO process for removal of N and P. (Reprinted with the permission of Springer-Verlag, Heidelberg.)

DN = DENITRIFICATION
 N = NITRIFICATION
AN = ANAEROBIC

PHASE A

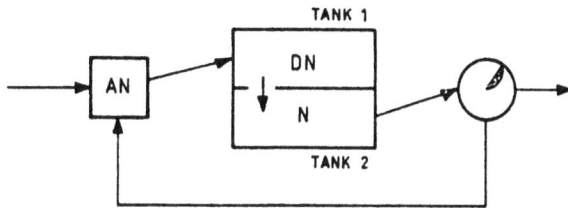

PHASE B

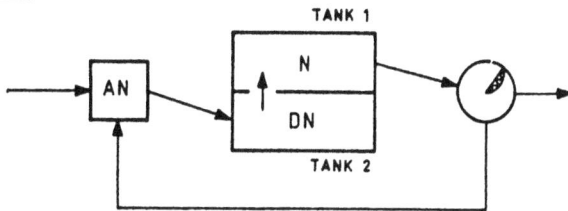

**Figure 3.20** Bio-Denipho process for N and P removal.

inlet end of the aeration units. This P removal can be maximised by adjusting aeration intensity to produce nearanaerobic conditions in the first of the two aeration tanks in series, thus fostering mobilisation of P in a form that can be both assimilated in luxury uptake and chemically precipitated [26,27,28]. It has been shown that in a plant at Wulkatal—when no iron is added if organic loadings are low, or when aeration intensity is too high so that nitrate appears in the effluent—concentration of P increases with an increase in concentration of nitrate. With nitrate-N up to more than 1 mg/l, 80 to 90 percent biological removal of P is obtained; however, with $NO_3$—N at 3 mg/l the biological removal reduces to 30 percent. When iron was added in dosages of 1–2.8 g/g P, the target limit of 1 mg P/l could be met [34].

In the main treatment plant of Vienna (Figure 3.13), in which the organic loading is too high to achieve nitrification, biological phosphorus removal in the absence of added iron was no more than about 50 percent, but it was substantially increased by addition of iron, reducing final effluent content to around 2–3 mg P/l. In this case no mandatory limit had to be met.

In several cases in which alteration of existing structures seemed likely to be uneconomic, addition of units simply to precipitate phosphorus from the secondary treated effluent has been preferred. An interesting recent design is that of the Crystalactor developed in Holland (Figure 3.21). In

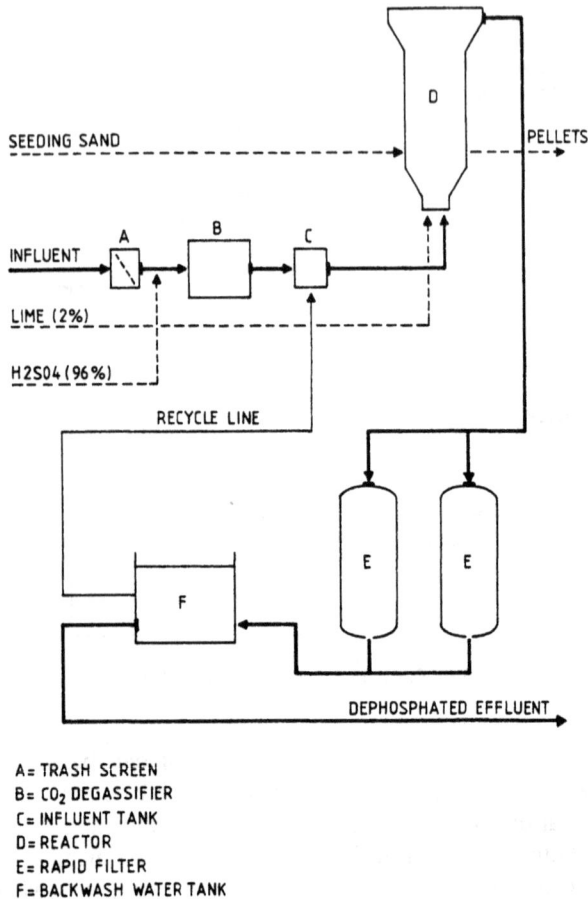

A= TRASH SCREEN
B= CO₂ DEGASSIFIER
C= INFLUENT TANK
D= REACTOR
E= RAPID FILTER
F= BACKWASH WATER TANK

**Figure 3.21** Process flow scheme for Crystalactor phosphate removal plant. (Reprinted from *Water Science & Technology, Vol. 23,* figure 1, p. 820, with permission from the publishers, Pergamon Press, and the copyright holders, IAWPRC.)

this device calcium phosphate is precipitated in granular form in a fluidized bed reactor [35]. Reported experience for the first year of operation indicates that effluent concentrations can be restricted to 0.5 mg/l or less.

In the UK there are a small number of biological phosphorus removal plants [36,37,38]. These are in the minority because most of the plants needing P removal are small to medium-sized plants (<50,000 pe) where it is more cost effective to use iron salts [39]. The cheapest route is to dose crystalline ferrous sulphate but it is not always convenient to fit the crystal dissolver on site and so in many small plants ferric sulphate dosing is applied.

The UK Biological phosphorus removal plants have also followed the practice established first in South Africa and later in the USA and Canada in the Bardenpho process (Figure 3.22), the UCT (University of Cape Town) processes (Figure 3.23) and the JHB Johannesburg processes (Figure 3.24). All of these processes have been called "main stream" biological phosphorus removal processes because the phosphorus is removed in the excess waste sludge under aerobic conditions as part of the normal process. The key to the processes is the release of the phosphorus from the sludge phase into the liquid phase in the anaerobic stage. The phosphorus in the liquid phase is then taken up again by the sludge during the aerobic stage but more is taken up in this stage than is usually the case, i.e. a higher concentration of phosphorus is achieved in the bacteria of the sludge of a biological phosphorus removal process. Essentially the bacteria capable of higher (excess or "luxury uptake") of phosphorus are selected by having an anaerobic selector stage. The keys to the process lie in (a) providing volatile fatty acids (VFAs) which are taken up in the anaerobic stage and trigger the release of phosphorus, and (b) maintaining a truly anaerobic condition in the first stage. The difference between the three main 'three tank' systems (Figures 3.22 to 3.24) relate to protection of the anaerobic stage from dissolved oxygen or nitrate in the return sludge. The Bardenpho system is the simplest and original system whereas the UCT and JHB systems pay more attention to removing recycled nitrate. These processes have been applied throughout Europe in the past 5 years. In the UK the Johannesburg process has been successfully tested on part of the Thames Water Beckton WwTP in East London [37,38] a range of Biological P removal processes have been tested at Severn Trent Water's Milcote plant at Stratford-on-Avon [36], and Anglian Water have also tested the Biological P removal options at their plants at Cambridge and Great Billing (Wellingborough).

The main plants in Berlin now use Biological P removal processes (with some chemical P removal back-up [40]. Biological P removal will also be used at the new treatment works at Rostock in Germany [41]. Denmark, Sweden, Norway and Switzerland are the countries with the highest percentage of wastewater to which phosphorus removal is applied [42]. It reaches up to 90% in Sweden and Switzerland. Both chemical and biological techniques of phosphorus removal are used but there is a trend towards biological phosphorus removal because of the lower sludge production and because the sludge has a fertiliser value. In Denmark all the mainstream biological phosphorus removal methods are used as well as Bio-denipho (Figure 3.20) [42].

In addition to the mainstream processes there is one major "side stream" process (or "sludge stream" process), this is the Phostrip process patented by the Biospherics Corporation in the USA. This process is shown in

144

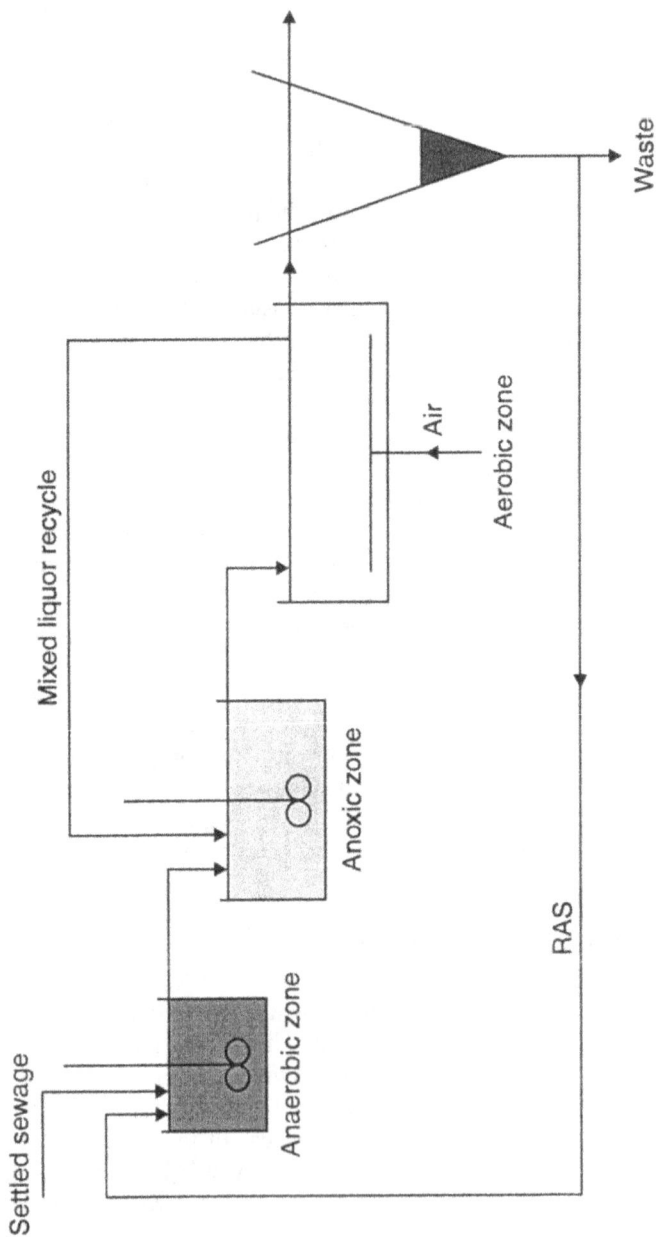

**Figure 3.22** Bardenpho arrangement.

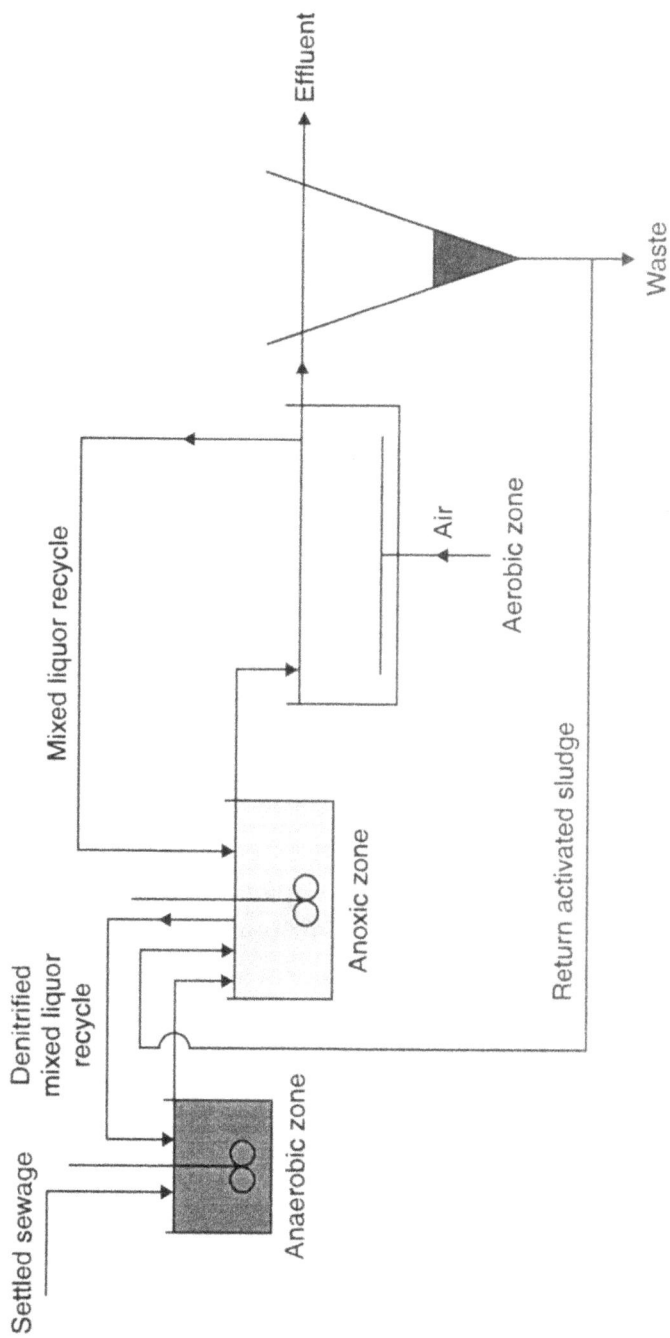

**Figure 3.23** Layout of the University of Cape Town/Virginia Initiative Process (UCT/VIP).

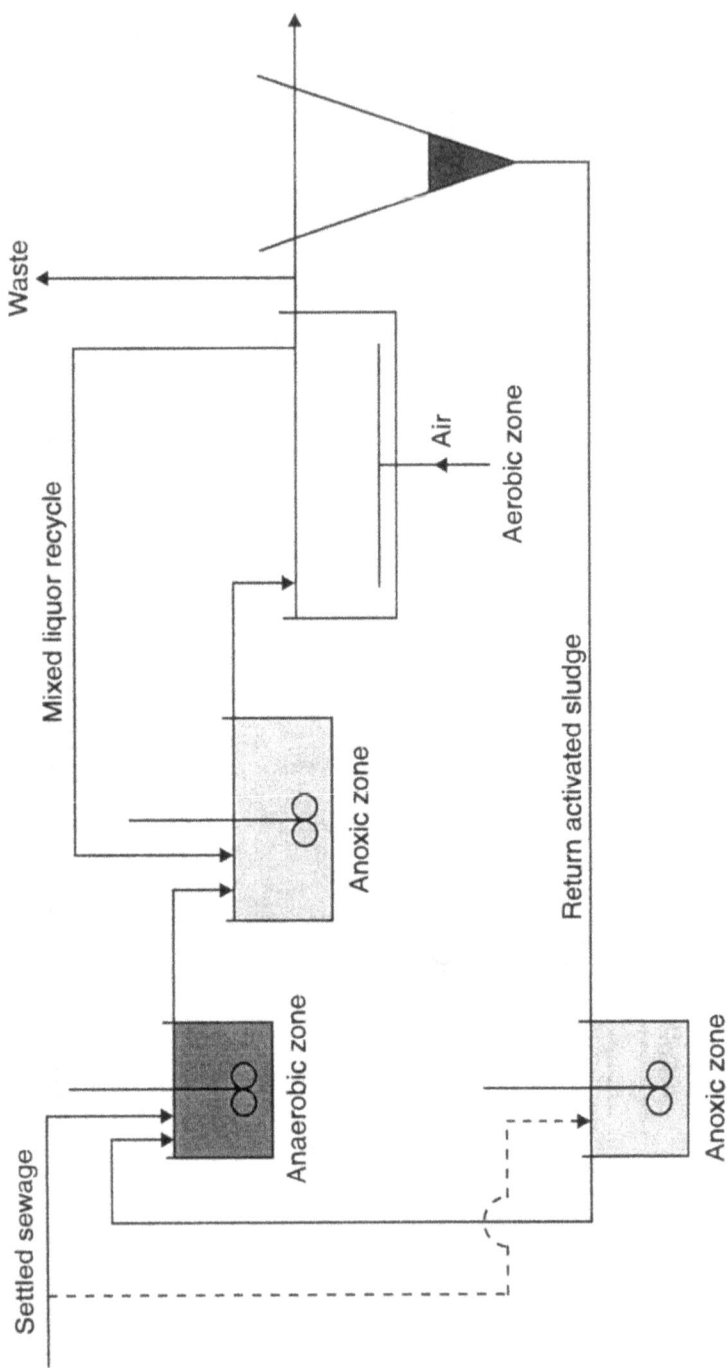

**Figure 3.24** The Johannesburg (JHB) process.

Figure 3.25. In the process the same biological principles are followed as with the main stream processes. However, the anaerobic stage is created in a separate side stream which takes a proportion of the recycled sludge. When the phosphorus is released the stripped sludge is recycled back into the main plant to allow it to pick up more phosphorus. The phosphorus-rich liquid in the side stream is precipitated with lime and the phosphorus-free liquid then returned to the main plant. Since the main plant is not changed in size and operation the Phostrip process may result in smaller plant. The Phostrip process is licensed to Severn Trent Water for Europe, with the exception of Italy.

Both the main stream and side stream processes have shown it is possible to get down to less than 2 mg total P/l and if carefully operated to less than 1 mg total P/l.

## SMALLER WORKS

Although many types of ASP are used for small populations, including various proprietary package plants, in terms of population served the most

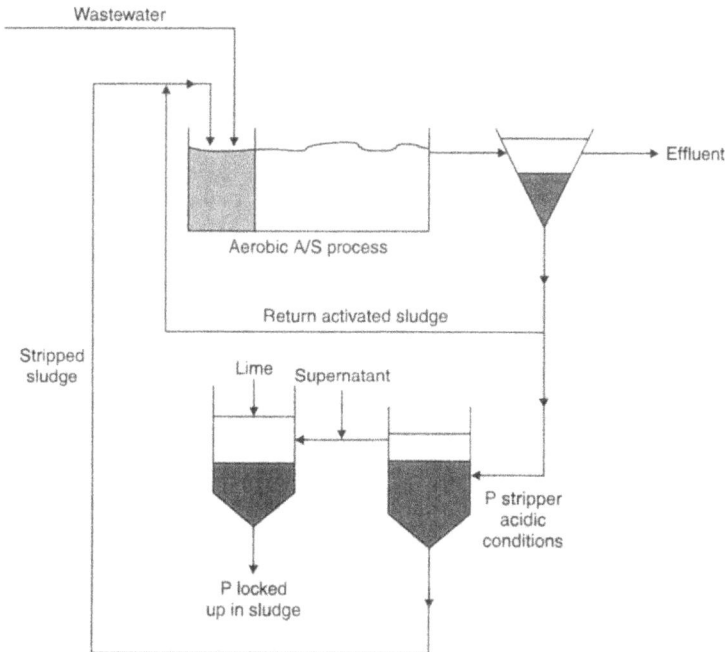

**Figure 3.25** Phostrip® side-stream P removal.

popular single type is the oxidation ditch. Possibly the most numerous though serving a much smaller total population is the Putox process, of which there are nearly one thousand in Austria alone.

### The Oxidation Ditch

Oxidation ditches were introduced into Europe by Pasveer (in Holland) in 1953 and though now widely used elsewhere they are still to be found in greater numbers and wider variety in Europe than in other parts of the world. Essentially, oxidation ditches are lightly loaded extended aeration AS plants in which the aeration units are endless channels around which mixed liquor is circulated and aerated, most commonly by mechanical agitation.

The first channel-type AS plants in Europe were in fact built in Sheffield in the U.K. in 1920, but the aerators used—rotating flat paddles—were inefficient. The ditches pioneered by Pasveer were equipped with much more efficient "brush" aerators in which the brushes—composed of metal combs attached to a horizontal rotating axis—were of a type originally devised by Kessener and later developed by Baars at the Dutch Research Institute TNO [43,44].

The original plants in Holland were designed to provide an inexpensive simple form of secondary treatment for small communities often located on land with a high water table. Accordingly, the ditches were quite shallow and often constructed by consolidating the excavation with simple linings rather than by use of reinforced concrete, which was employed only for the piers supporting the brushes.

Crude rather than settled sewage was treated, and in many of the early plants there was no separate secondary clarifier. The plants were either operated in a fill-anddraw mode or were provided with a simple baffled settlement zone within the channel.

The low sludge loadings adopted, commonly about 0.05 g/g day, conferred considerable performance stability, enabled high-quality nitrified effluents to be produced, minimised sludge production and rendered the sludge well-mineralized, thus facilitating its disposal on land.

Since those early days, the design and operating practice for ditches has undergone a series of developments. In the first phase of such developments the brush aerators were replaced by cage rotors developed at TNO. Then aeration rotors of much larger diameter (e.g., the so-called Mammoth rotors, see Figure 3.8) were introduced, providing greater oxygenation capacities and better vertical mixing, thus allowing channel depths to be increased to about 3 m. Then, in the next phase, vertical shaft cone aerators were introduced to create the "Carrousel system" enabling depths to be increased still further.

Such increases in OC and depth enabled plants to be designed for much larger populations, and with this extension in the range of application it was sometimes more appropriate to construct in reinforced concrete and to build formal secondary clarifiers. In some cases it was appropriate to include primary clarifiers also. Such developments enabled ditches to be built to cater for populations of 150,000 or more (and some built for industrial effluent treatment have catered for more than 300,000 PE). Generally, however, ditches have been employed for smaller populations—not often above 50,000, and usually considerably fewer.

It appears to be generally agreed that for this type of application, given that adequate land is available, the process is cheaper than other forms of ASP of the type used for large populations [45]. This economic advantage stems mainly from the cheaper form of construction and the ability to dispose of the well-stabilized sludge on rural land.

Figure 3.26 shows, in plan, four of the most common configurations of ditch used in the traditional applications.

More recently, further modifications adopted have included use of diffused air for aeration to induce circulation; the incorporation of in-channel settling units (floating in one proprietary design); and the operation of a combination of ditches sequentially in an automatically controlled fill-and-draw mode. Also in departure from the original channel configuration, plants have been built in which mixed liquor circulates through a block of four aeration compartments of rectangular cross section in plan and about 4 m deep (the Rotanox system).

Several of these recent designs of intermittent sequentially operating plants have emanated from Denmark [46]. Thus, for example, the so-

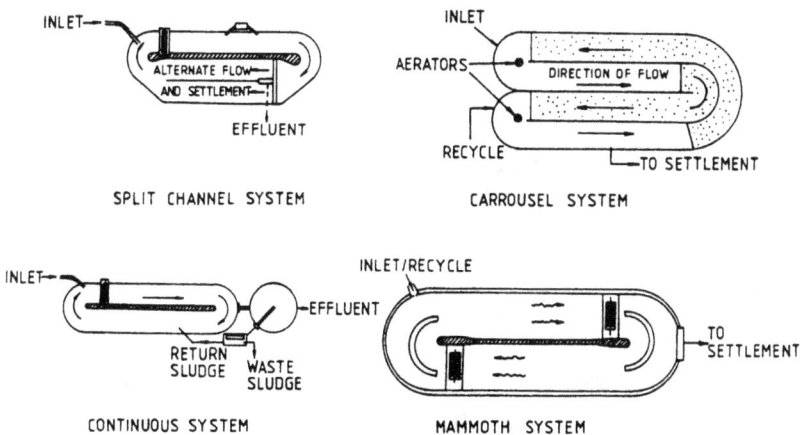

**Figure 3.26** Oxidation ditch configurations. (Reprinted by courtesy of *Effluent and Water Treatment Journal.*)

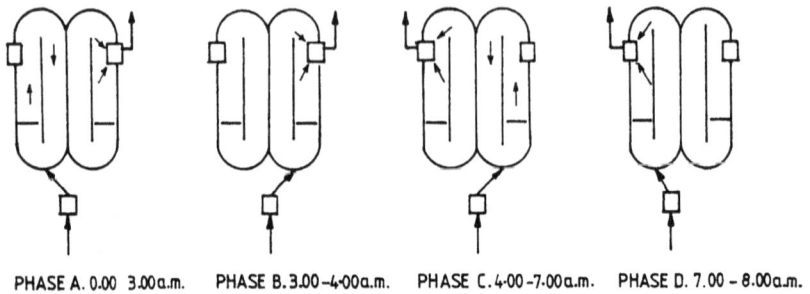

PHASE A. 0.00 3.00 a.m.     PHASE B. 3.00–4.00 a.m.     PHASE C. 4.00–7.00 a.m.     PHASE D. 7.00 – 8.00 a.m.

**Figure 3.27** Type D intermittent oxidation ditch system. (First published in the *Proceedings of the International Conference on Oxidation Ditch Technology, Amsterdam, 1982.*)

called Type D ditch operates in four phases as indicated in Figure 3.27. There is a continuous discharge, and both water level and aeration rotor immersion are constant. A disadvantage is that the degree of utilisation of the installed power and capacity of the rotors is rather low. To overcome this, two other designs have been evolved in Denmark, the Type VR and the Type T ditches.

The main features of the type VR, which is meant for PEs in the range 2000–6000, are shown in Figure 3.28. At each end of the partition wall there is a hinged flapgate that can open only towards the rotor. Depending on the direction of rotation of the reversible rotor and thus the flow of mixed liquor, one gate will open and the other shut. This enables one part of the ditch to be used for aeration while the other serves as a settlement zone. Treated effluent is drawn off continuously either from one or the other of the two side weirs, which are operated alternately automatically.

In the type T plant there are three ditches, two outer ones interlinked with a middle one by large-diameter pipes. There are six phases of automatically controlled operation involving the two outer ditches serving alternately

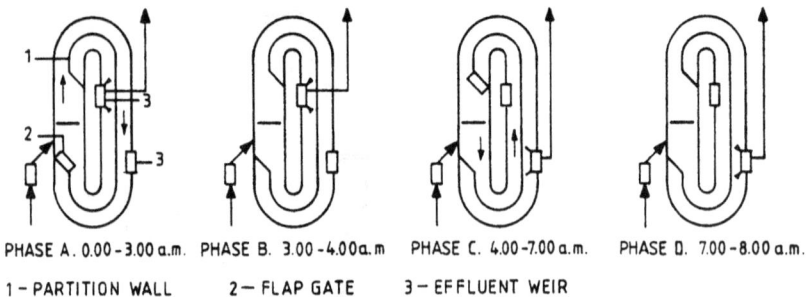

PHASE A. 0.00 –3.00 a.m.   PHASE B. 3.00 –4.00 a.m   PHASE C. 4.00 –7.00 a.m.   PHASE D. 7.00 –8.00 a.m.

1 – PARTITION WALL       2 – FLAP GATE       3 – EFFLUENT WEIR

**Figure 3.28** Type VR oxidation ditch system. (First published in the *Proceedings of the International Conference on Oxidation Ditch Technology, Amsterdam, 1982.*)

and intermittently as clarifiers as indicated in Figure 3.29. By this mode of operation the utilization of the capacity of the aeration rotors and their motors is improved, relative to that in the type D plants, by 50 percent.

Broadly, the relationships between loading and performance obtained in ditches designed to meet conventional standards for SS, BOD, and ammonia are consistent with those for other forms of ASP when account is taken of the oxygenation capacity provided.

Figure 3.30 shows compilations of performance data consistent with this view published in 1982. The data imply that either by original intent or because populations have increased some ditches were receiving sludge loads approaching 0.1 g/g day but the majority were around half this figure or even lower. In principle it would no doubt be possible to operate at loadings above 0.1 g/g day, but with increasing loading an increasing risk of prejudicing the various favourable features of the process (stability, good effluent quality, low sludge production).

In the most recent phase of development of the process, attention has been turned to its operation in a nitrification-denitrification mode by creat-

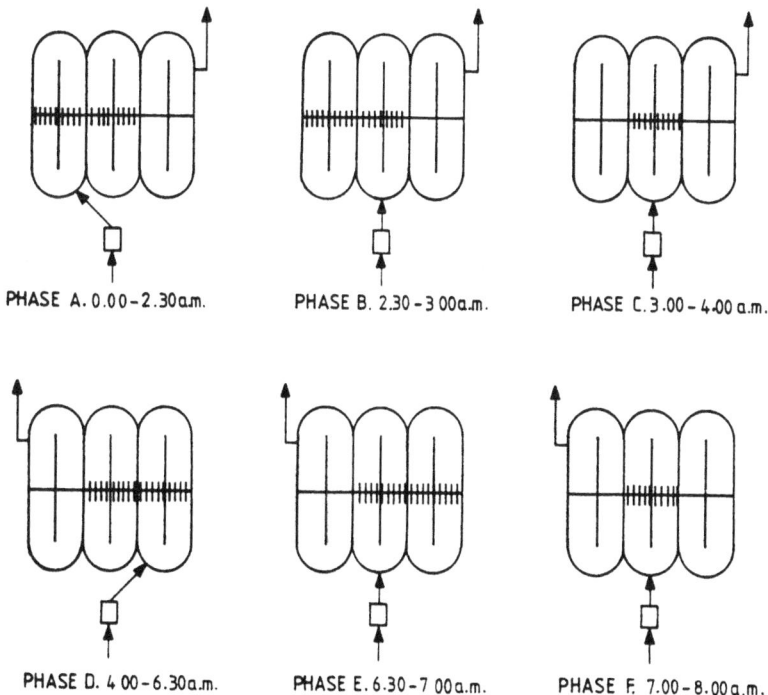

PHASE A. 0.00–2.30a.m.    PHASE B. 2.30–3 00a.m.    PHASE C.3.00–4.00 a.m.

PHASE D. 4 00–6.30a.m.    PHASE E. 6.30–7 00a.m.    PHASE F. 7.00–8.00a.m.

**Figure 3.29** Type T oxidation ditch system. (First published in the *Proceedings of the International Conference on Oxidation Ditch Technology, Amsterdam, 1982.*)

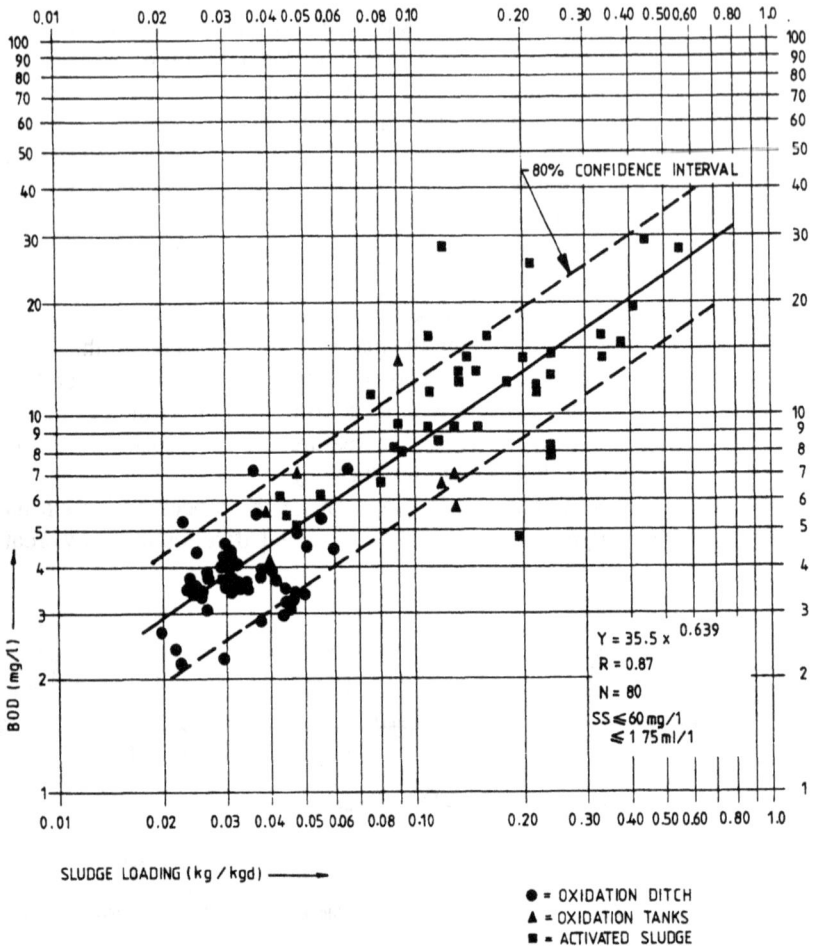

**Figure 3.30** Average yearly effluent BOD as a function of sludge loading (First published in the *Proceedings of the International Conference on Oxidation Ditch Technology, Amsterdam, 1982.*)

ing one or more anoxic zones and thus enabling sludge settleability to be improved, energy employed for nitrification to be recovered, and release of nitrogen to the environment to be reduced.

It can be shown that—if under winter conditions with minimum sewage temperatures of say 7°C (as in say France, Germany, Holland, and the U.K.) sludge were produced at the rate of 0.3 g/g BOD removed—then for a plant loaded at 0.05 g/g day containing 3000 mg/l MLSS and treating settled sewage having a BOD of 200 mg/l the retention period would be 1.33 days but the time required to nitrify all but a small residual of ammonia entering at

40 mg/l would be no more than about 0.25 days. Thus, even at peak flow, by reducing residence time to about 0.5 days a large proportion of the ditch could be made anoxic without risk of significant release of ammonia.

Studies of rates of denitrification of mixtures of sewage and activated sludge indicate that about 7 mg/l $NO_3$—N might be removed in the first 10 minutes of contact but the subsequent rate would fall to about 4.5 mg/l h [47]. Under these circumstances, to denitrify 40 mg/l $NO_3$—N would demand retention in the anoxic zone for about 7.5 h. At steady flow the residence time provided would probably be sufficient to ensure almost complete denitrification, but some leakage of nitrate could occur during periods of peak flow [48].

Dynamic models specific to oxidation ditches do not appear to have been produced to take account of the influence of diurnal fluctuations in flow and strength of sewage. In their absence, the preferred approach to design of ditches of the older style—in which all the contents were circulated continuously—has been to allocate a substantial proportion of the length of the ditch, up to about 40 percent, as potentially available for denitrification, to feed sewage directly to this section, to monitor DO immediately above this zone and control intensity of aeration so as to ensure that anoxic conditions are maintained.

In plants in which part of the capacity functions intermittently as a clarifier, the timing of the operation sequence can be modified to include a phase or phases in which power to the aeration rotors is reduced so as merely to provide mixing rather than aeration. This is the case in the type T plant shown in Figure 3.31, the method being known in Denmark as the Bio-denitro process. The main difference between the two sequences is that in the Bio-denitro process ditch 1 is operated anoxically during phase A and ditch 3 is operated anoxically in phase D. The duration of the phases can be varied to some extent depending on the amount of nitrogen to be removed, within the overall constraint of having to provide a long enough period of aerobic conditions at low temperatures to ensure that full nitrification is achieved if there is a low limit on ammonia in the effluent.

### The Putox Process

In the last 25 years more than 1000 small activated sludge plants for populations ranging from 10 to 500 have been built in Austria, mostly of the Putox design, basic features of which are shown in Figure 3.32 [49]. The treatment units comprising a two-compartment septic tank, a uniformly mixed aeration chamber, final clarifier, and pump well, are accommodated in excavations so that little is visible aboveground. The average residence time in the septic tanks is about 1 day, in the aeration unit about 0.5 day, and in the final clarifier around 7 h. Residence

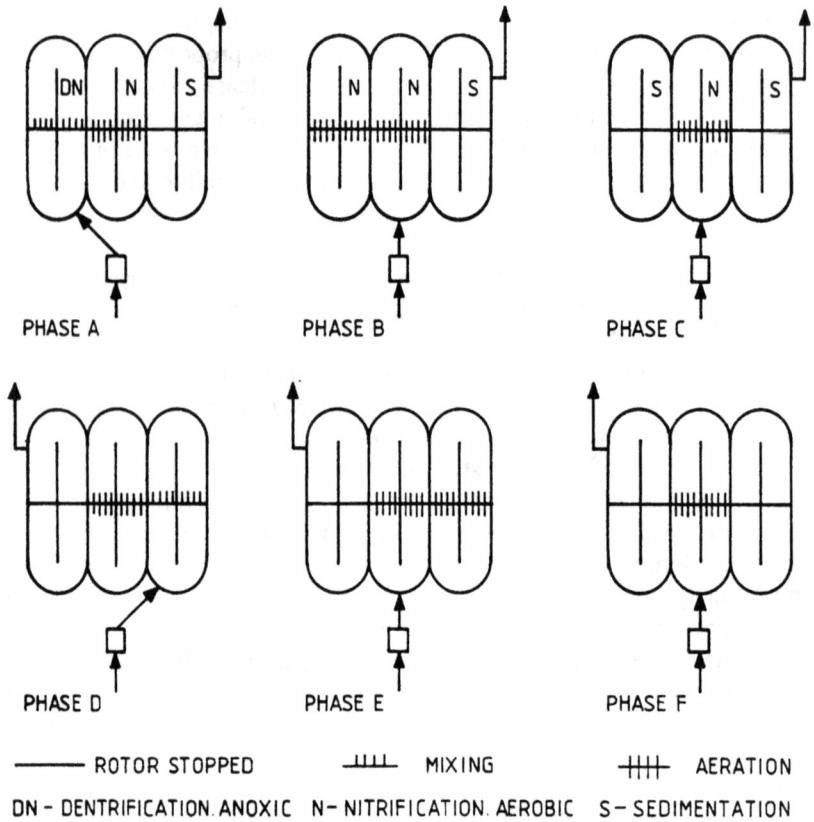

**Figure 3.31** Bio-denitro operation of Type T oxidation ditch. (First published in the *Proceedings of the International Conference on Oxidation Ditch Technology, Amsterdam, 1982.*)

time in the clarifier at peak flow is about 5.2 h. Oxygenation capacity provided is 1.5 g/g BOD load and aeration is by coarse bubbles over a 2-month period if observation in one plant during the winter of 1971 BOD of the effluent averaged 6.5 mg/l and ammonia 2 mg/l. Subsequently it was shown that by aerating intermittently, substantial removal of nitrogen could be achieved by denitrification during the periods when the aerators were switched off.

## INSTRUMENTATION, CONTROL, AND AUTOMATION (ICA)

The extent to which some degree of automatic control of AS plants is practised has increased considerably in the last decade, and at the

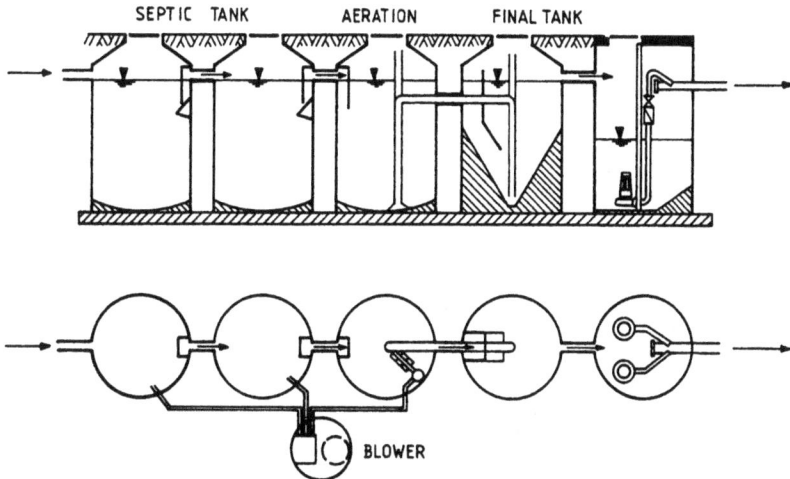

**Figure 3.32** Main features of Putox process.

very least control of aeration intensity according to DO concentration of the mixed liquor would now normally be provided in new plants. In many modern works the application of ICA is much more extensive, embracing most elements of the process sequence, though only about 2 percent of works in Europe, mainly in Germany, are fully computer controlled.

This feature could become more widely applied, particularly in plants designed for nutrient removal by biological means since experience in other parts of the world, notably South Africa, has shown that optimisation of performance is difficult to achieve in the face of substantial diurnal variations in flow and composition.

A good example of modern design is afforded by the contact-stabilization diffused air AS plant at Holdenhurst near Bournemouth in the U.K. (Figure 3.33) [50]. This works treats a DW flow of some 55 ml/d to produce effluent with an average BOD of 8 mg/l and ammonia content of 0.6 mg N/l. A feature of the works is that flow is partially balanced by retaining sewage in the main interceptor sewer feeding the plant by operation of an automatically activated penstock. Also controlled are the operation of inlet works, primary sedimentation tanks, the AS plant including final clarifiers and sludge return, effluent recirculation pumps, and sludge treatment. This is achieved by the action of six "intelligent outstations" serially linked to a host computer and a further outstation linked via a modem as shown schematically in Figure 3.34.

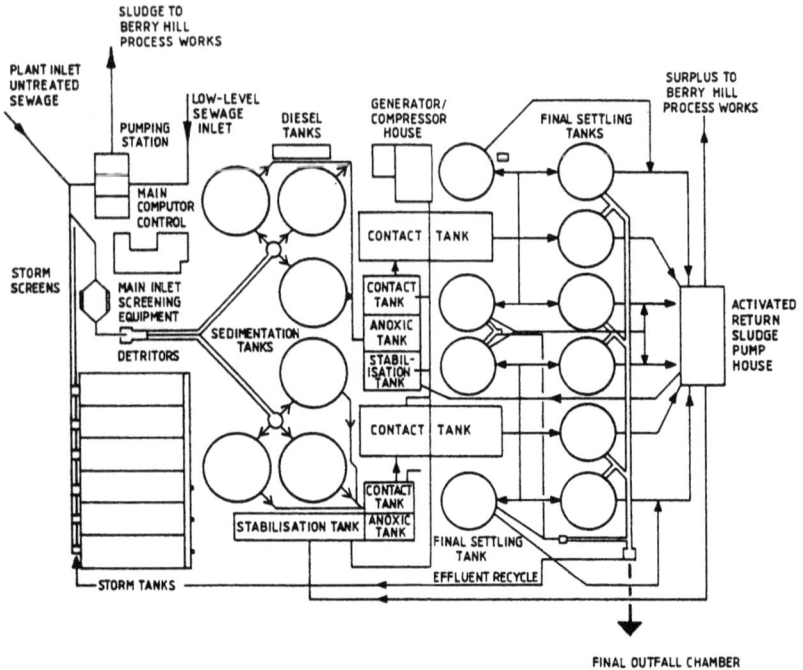

**Figure 3.33** Layout of Holdenhurst sewage treatment works (Robinson, M. S. 1990. "Operating Experiences of Instrumentation, Control and Automation at Holdenhurst STW, Bournemouth." *J. Instn. Wat. & Envir. Mangt.*, 4:560, 564, 565)

In addition to the supervision of outstations the functions of the host computer include:

- data logging
- alarm monitoring
- data display, including printing and graph plotting

The complete system includes about 400 instruments and some 25 km of connecting cables. Analogue and digital inputs are monitored respectively every 10 and 6 seconds.

The aeration units are divided into two trains, each subdivided into stabilisation and anoxic contact zones, the anoxic zone being incorporated in accordance with modern U.K. practice referred to earlier for control of sludge settleability and for energy recovery. In the aerated contact and stabilization zones, DO is controlled to a target value according to the output from a DO sensor. A cascade PID control loop generates a difference signal that is used to generate movements of the values controlling the air flow to the porous disc diffusers. Operation of these valves alters the pressure in the air main. The changes in pressure are used to alter the vane angle on the blowers and thus the blower output. The reference pressure is automatically adjusted if the opening of the most open of the valves exceeds a dead band of 65 percent of fully open. A second or third blower is started when the vane angle has resulted in a blower delivering maximum output for 1 h. Blower selection is based on accumulated running hours so as to even out usage according to the arrangements shown in Figures 3.35 and 3.36. Each air valve has a mechanical stop that serves to maintain a minimum air flow to prevent external fouling of the discs.

MLSS concentrations are monitored continuously according to concentration but not controlled. Instead, surplus activated sludge is drawn off at a rate preset by the works management.

The ASP effluent and final works effluent are monitored for SS, ammonia, and temperature. If the effluent quality contravenes pre-set limits, the ASP effluent is returned to the storm tanks by screw pump.

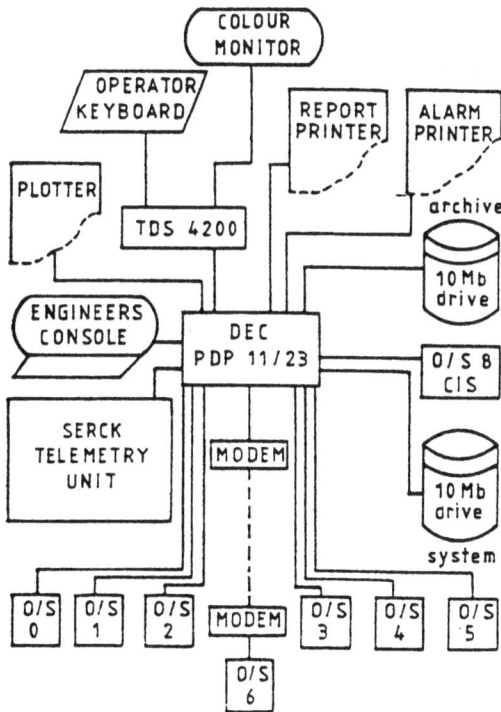

**Figure 3.34** Arrangement of computer, outstation and peripherals (Robinson, M. S. 1990. "Operating Experiences of Instrumentation. Control and Automation at Holdenhurst STW, Bournemouth." *J. Instn. Wat. & Envir. Mangt.*, 4:560, 564, 565)

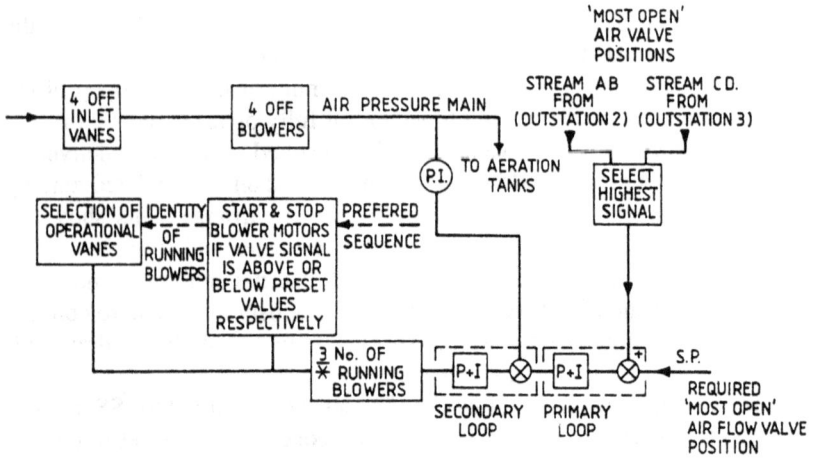

**Figure 3.35** Outstation 0—air pressure control system (Robinson, M. S. 1990. "Operating Experiences of Instrumentation, Control and Automation at Holdenhurst STW, Bournemouth." *J. Instn. Wat & Envir. Mangt.*, 4:560, 564, 565).

## SOME OTHER PROCESS VARIANTS

### SEQUENCING BATCH REACTORS

Sequencing Batch Reactors (SBRs) have attracted a good deal of attention in Europe in the past 5 years. The original activated sludge processes in Manchester and Salford, UK in 1914, were "fill and draw plants" [51,52,53] but they were gradually replaced by continuous flow systems which possibly needed less manual attention. The interest in the SBR processes started in the mid 1970's by Irvine and colleagues in the USA [54,55] and was taken up enthusiastically by Goronszy in Australia, USA

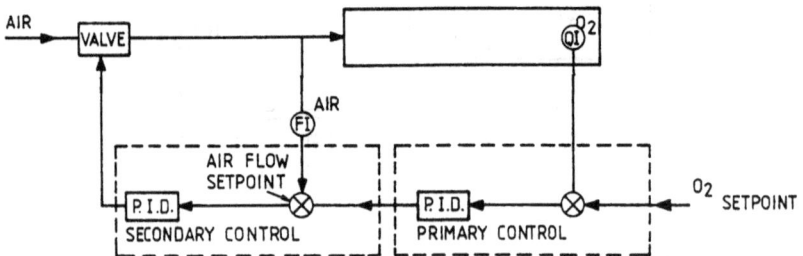

**Figure 3.36** Outstation 3—aeration control system (Robinson, M. S. 1990. "Operating Experiences of Instrumentation, Control and Automation at Holdenhurst STW, Bournemouth." *J. Instn. Wat. & Envir. Mangt.*, 4:560, 564, 565).

and more lately in Europe [56]. A conference reviewing the developments has recently been held [57,58]. Over the last four years they have started to be applied in Europe. The first new SBR type system in the UK has been built by Yorkshire Water at Wath-upon-Dearne and Dwyr Cymru Welsh Water are intending to put SBRs in at sites in Swansea (South Wales) and Ganol (North Wales). In Germany they are being applied to the treatment of flows from small communities [59]. They are also starting to be more readily applied outside the municipal sewage treatment field. In France they have been applied to treating winery wastewaters [60] and in the Netherlands they are being applied for the treatment of the wastewater from cleaning up road and railway vehicles [61].

The recent interest in SBR processes probably derives from the fact that the size of the SBR system can be smaller (and hence cheaper), than the overall size of the conventional systems (aeration tanks plus settlement tanks). The control of the fill and draw (drain) cycles can be achieved by the more sophisticated electrical and computer control packages now available. This allows very precise control of a whole range of cycle options. It has been shown all the processes possible in continuous-flow system BOD removal, nitrification, denitrification and biological phosphorus removal can be achieved in SBR type systems [56,57].

Another benefit of the SBR system is that they do not suffer from sludge building and hence have less problems with solid liquid separation. They can still produce stable foam (often called ''chocolate mousse'') but this need not cause a problem because the withdrawal of the liquid effluent may be done from below the liquid foam interface.

There are now also a number of systems which combine fixed-film processes with SBRs and some with sequenced fixed-film processes [57].

It seems likely that the potential for SBR systems will continue to be investigated and developed in the next few years. Originally SBRs were only considered for small systems but they are now being exploited for larger populations.

## FLUIDIZED BEDS

Although not strictly AS plants, fluidized beds are sufficiently similar in principle to justify inclusion in this chapter. They may in fact be described as suspended fixed-film processes in that the particles on which the bacteria grow (usually sand) are in suspension in the up-flowing liquid. It has been shown that the rate of BOD removal and nitrification is very similar in mass loading rate terms (kg BOD/kg MLSS·d) for Fluidized Beds as for Activated Sludge processes [62]. Where they differ is the volumetric loading rate (kg BOD/m$^3$ of reactor·d) for the same removal. This is because the Fluidized Bed Reactors can achieve biomass concentrations of 5 to 10 times ($10 \to 20$ gMLSS/l) higher than the Activated Sludge Process which is limited to about 5 g MLSS/l by the solid liquid separation stage. This inclusion is merely for the purpose of recording that although fluidised beds have been the object of substantial investigation in Europe, there appear to be no full-scale European plants for sewage treatment. Probably the reason is that, as research studies have shown, the advantages of savings in space do not necessarily outweigh the disadvantages of greater complexity, including a requirement to use oxygen rather than air, and uncertainty, owing to lack of extensive "track records," about the possibility of operational difficulties—for example in sludge-sand separation. Moreover, the evolution of the aerated filter (which uses air rather than oxygen) has probably appeared to offer a more attractive means of reducing land requirements and rendering plants less obtrusive.

## AERATED FILTERS

Though aerated filters can hardly be described as variants of ASP they depend on aeration of biomass which, though not fully fluidized is nevertheless not wholly static in the manner that it would be in, say, a trickling filter. It thus has some of the characteristics of conventional ASP, and so for completeness can appropriately be included in this section.

Such aerated filters are becoming increasingly popular, particularly in France where the most widely used version, the Biocarbone process, has been pioneered by Compagnie Generale des Eaux [63]. In this process sewage is passed through a drowned bed of small shale particles into which air is injected about 1 m above the bottom of the bed. This rising flow of air though insufficient to fluidise the shale bed, helps to maintain hydraulic permeability and supplies the necessary oxygen to meet the respiratory demand of the biomass that attaches to the particles and metabolises the biodegradable components of the sewage. The section below the point of air injection acts as a purely physical filter, and penetration of SS is so small that secondary clarifiers are not required (Figure 3.37).

Biomasses per unit volume are typically equivalent to about 15,000–20,000 mg/l so that, for a given effluent standard, loadings per unit volume can be applied that are several times those treatable in conventional AS plants.

Depending on choice of loadings, aerated filters can be operated simply to meet limits on BOD and SS or to produce nitrified effluents in a single stage or to function as second-stage high-rate nitrifying units. Excess sludge, which otherwise would accumulate in the bed, is removed by backwashing intermittently.

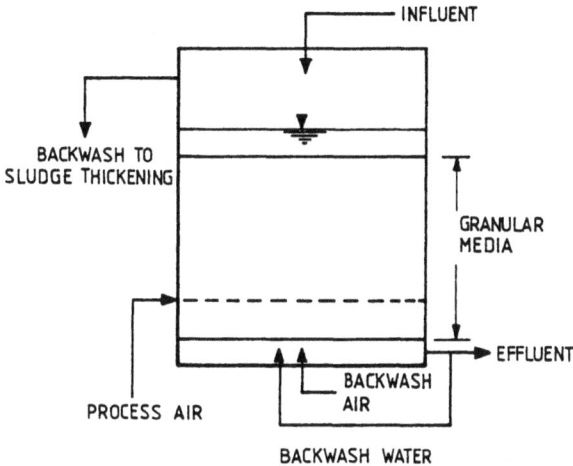

**Figure 3.37** Biological aerated filter system (Biocarbone).

## ROPE-TYPE BIOFILM REACTORS

A novel form of reactor has recently been introduced at Geiselbullach in Germany as a means of uprating a conventional plant serving a population numbering about 0.25 million to meet new standards including a limit of 10 mg/l ammonia-N from May to October. The method adopted consisted of suspending media in the aeration units on which biofilms could develop, thus in effect creating a hybrid of the AS and aerated filter processes [64]. Various types of media were examined in preliminary trials, that selected being a so-called Ring-Lace material consisting of flexible ropes of modified PVC incorporating a large number of woven rings. The ropes were suspended vertically in the aeration units and although attached in cages were able to move in response to liquid turbulence. Although the results so far published cover a period of less than a year, it appears that the biomass in the aeration units (expressed as equivalent MLSS) could be increased substantially by at least 1700 mg/l when no chemicals were added to precipitate P, and by 5000–8000 mg/l when P-precipitation was in progress. These increases had a consequential benefit on nitrification.

## FUTURE TRENDS

For various reasons, ASP can be expected to continue. These reasons are that:

- AS plants can be expected to last at least 50 years.
- There is a large investment in such plants.

- A great many plants in Europe are less than 20 years old.
- Most such plants can be uprated without major difficulty to meet such new standards as can be foreseen.
- There is no reason to suppose that biological treatment per se will be superseded by better and cheaper methods in the foreseeable future.

It can be expected that ASP will continue to be the dominant method of sewage treatment in Europe well into the next century. At the same time it seems clear that demands for reducing odours and more generally rendering sewage treatment works less obtrusive are likely to result in a trend in new plants for larger populations towards use of processes that are less demanding on space and that can if desired be totally enclosed above or below ground. Although there may be ways of reducing the size of AS plants, involving for example the use of membrane processes to separate and recycle biomass and thus to increase MLSS, it seems more probable that designers will prefer to specify aerated filters. For small- to moderate-sized towns, these are likely to be small enough to enclose; and when used in conjunction with cross-flow membrane filters (a combination already shown to be feasible for removal of carbonaceous BOD to produce effluent that would comply with, say, a hypothetical 3:3 standard) can be made very small indeed, for example with residence times of less than 10 minutes. Almost certainly the other major factor that will influence design in the next two decades will be requirements to limit release of the nutrients N and P in a very much wider range of circumstances than hitherto considered necessary. This seems most likely to be accomplished by internal modification of existing plants to facilitate biological removal of N and P; and by addition of supplementary units, including aerated filters and anoxic filters for N removal by nitrification-denitrification, and of chemical precipitation processes for P removal.

For smaller works, the oxidation ditch seems sure to remain the most popular AS process. Conceivably, at some sites where ditches might otherwise be used, reed beds will be preferred if the promising early performances of existing units are continued in the longer term.

## REFERENCES

1 Council of European Communities. 1991. Directive concerning urban waste water treatment. (91/271/EEC) *Official Journal of the European Communities.* L135/40, 30 May 1991.

2 U.S.EPA, 1975. *A Guide to the Selection of Cost-Effective Wastewater Treatment Systems,* EPA-430-9-75/002.

3 Spearing, B. W., ed. 1987. ''Sewage Treatment Optimization Model—User Manual

and Description,'' *WRc Technical Report 144,* Second Edition, Medmenham, U.K.: Water Research Centre.

4 Downing, A. L., H. A. Painter and G. Knowles. 1964. "Nitrification in the Activated Sludge Process," *J. Inst. Sew. Purif.,* (2):130.

5 Downing, A. L. and A. P. Hopwood. 1964. "Some Observations on the Kinetics of Nitrifying Activated Sludge Plants," *Schweitz Z. Hydrol.,* 26:271.

6 Gujer, W. 1977. "Design of Nitrifying Activated-Sludge Process with the Aid of Dynamic Simulation," *Proj. Wat. Tech.,* 9:323–336.

7 Jones, G. L. 1978. In: *Mathematical Models In Water Pollution Control.* A. James, ed., John Wiley and Sons.

8 Chambers, B. and G. L. Jones. 1988. *Wat. Sci. Tech,* 20:121.

9 Chambers, B. and G. L. Jones. 1985. *Wat. Pollut. Control,* 84:70.

10 Smith, M., Cooper, P. F., McMurchie, J., Stevenson, D., Mann, B., Stocker, D, Bayes, C. and Clark, D. 1996. Nitrification trials at Dunnswood STW, Cumbernauld and parallel process modelling. Paper presented to CIWEM Branch Meeting, Glasgow, UK, January 1996.

11 IAWPRC. 1987. "Activated Sludge Model No. 1." Scientific and Technical Reports No. 1.

12 Gujer, W. and M. Henze. 1991. "Activated Sludge Modelling and Simulation," *Wat. Sci. Tech.,* 23:1011–1024.

13 IWEM. 1987. *Manual of British Practice in Water Pollution Control—Activated Sludge.* London: The Institution of Water and Environmental Management.

14 Thomas, V. K., B. Chambers and W. Dunn. 1989. *Wat. Sci. Tech.,* 21:1403.

15 Haines, D., M. Bailey, J. Ousby and F. Roseler. 1975. "Deep Shaft Aeration Process for Effluent Treatment," *Inst. Chem. Engrs. Symposium No. 41.*

16 British Oxygen Company Environmental. 1991. *The Vitox System.* London.

17 Matt, K. 1990. *Abwassertechnik,* 41:62.

18 Dick, R. I. 1972. "Gravity Thickening of Waste Sludges," *Filtrn. and Separn.,* 9:177–183.

19 Dick, R. I. and K. W. Young. 1972. "Analysis of Thickening Performance of Final Settling Tanks," *27th Ind. Waste Conf.,* Purdue University, 34 pp.

20 Eckenfelder, W. W. and D. L. Ford. 1967. "Effect of Process Variables on Sludge Floc Formation and Settling Characteristics," *J. Wat. Pollut. Control Fed.,* 39(11):1850–1859.

21 White, M. J. D. 1976. *Wat. Pollut. Control,* 75:459.

22 Von der Emde, W. 1982. "Design and Operation Interaction—An Example: Main Treatment Plant Vienna," *Wat. Sci. Tech.,* 14:494.

23 Von der Emde, W. 1972. "Design Considerations for Large Treatment Plants," *Wat. Res.* 6:5–567.

24 Marais, G. V. R. and G. Ekama. 1976. "The Activated Sludge Process Part 1," Steady-State Behaviour Water SA, 2/4 163–200.

25 Hopwood, A. P. and A. L. Downing. 1965. "Factors Affecting the Rate of Production and Properties of Activated Sludge in Plants Treating Domestic Sewage," *J. Inst. Sew. Purif.,* (5):435.

26 Irwin, R. A., W. J. Brignal and M. A. Biss. 1989. *J. IWEM,* 3:280.

27 Cooper, P. F., Drew, E. A., Bailey, D. A. and Thomas, E. V. 1977. Recent advances in sewage effluent denitrification Part I. *Wat. Pollut. Control,* Vol. 76. No. 3, p. 287.

**28** Cooper, P. F., Collinson, B. and Green, M. K. 1977. "Recent advances in sewage effluent denitrification Part II." *Wat. Pollut. Control,* Vol. 76, No.4, p. 389.

**29** Gonsailles, M., J. M. Rovel and R. Nicol. 1991. *Wat. Sci. Tech.,* 23:773.

**30** Matsche, N. F. and G. Spatzierer. 1977. *Proj. Wat. Tech.,* 8:501.

**31** Matsche, N. F. 1977. *Prog. Wat. Tech.,* 8:625.

**32** Matsche, N. F. 1980. *Prog. Wat. Tech.,* 12:551.

**33** Henze, M. and P. Harremoes. 1990. *Chemical Water and Wastewater Treatment,* H. H. Hahn and R. Klute, eds., Berlin: Springer-Verlag, p. 499.

**34** Spatzierer, G., C. Ludwig and N. F. Matsche. 1985. *Wat. Sci. Tech.,* 17:163.

**35** Eggers, E., A. H. Dirkzwager and H. Honing. 1991. *Wat. Sci. Tech.,* 23:819.

**36** Upton, J. E., 1994. "A full scale comparison of mainstream and side stream EBPR." Paper presented at Water Biotreatment Club, Cranfield University, Bedford, UK. October, 1994.

**37** Williams, S. and Wilson, A. W., 1994. Beckton demonstration Biological Nutrient Removal Plant. *Journal of the Institution of Water and Environmental Management,* Vol. 8, No.6, December, pp. 664–670.

**38** Williams, S. 1994. "EBPR operation at Beckton—an update." Paper presented at Water Biotreatment Club, Cranfield University, Bedford, UK. October 1994.

**39** Cooper, P. F., Upton, J. E., Smith, M. and Churchley, J. 1995. "Biological nutrient removal: design snags, operational problems and costs." *Journal of the Institution of Water and Environmental Management,* Vol. 9, No.1, pp. 7–18, February.

**40** Peter-Frohlich, A. 1996. "Biological phosphorus elimination in Berlin's WWTPs." *World Water and Environmental Engineering.* November, pp. 12–13.

**41** Strohmeir, A., 1996. "Upgraded WwTP confirms Rostock's clean image." *World Water and Environmental Engineering,* September, pp. 14–15.

**42** Henze, M., 1996. "Biological Phosphorus Removal from Wastewater: process and technology." *Water Quality International,* July/August, pp. 32–36.

**43** Pasveer, A. 1959. "A Contribution to the Development in Activated Sludge Treatment," *J. Proc. Inst. Sew. Purif.,* (4):436.

**44** 1982. *Oxidation Ditch Technology.* Norwich: CEP Consultants Ltd.

**45** Bender, J. H. 1979. *EPA Environmental Research Brief, The Oxidation Ditch Process.* Cincinnati: EPA.

**46** Bungaard, E. and G. H. Kristensen. 1982. "The Operation of Oxidation Ditches Including the Bio-denitro System in Denmark," *Oxidation Ditch Technology.* Norwich: CEP Consultants Ltd., pp. 117–128.

**47** Forster, C. F. and D. W. M. Johnstone. 1984. *Eff. Wat. Treat.,* p. 258.

**48** Van Haandel, A. C., G. A. Ekama and G. V. R. Marais. 1981. *Water Res.,* 15:1135.

**49** Begert, A. 1978. *Österreichische Abwasser-Rundschau,* p. 59.

**50** Robinson, M. S. 1990. *J. IWEM,* 4:559.

**51** Ardern, E. and Lockett, W. T. 1914. "Experiments on the oxidation of sewage without the aid of filters." *J. Soc. Chem. Ind.* 33, p523.

**52** Duckworth, W. H. 1914. Aeration experiments with activated sludge. *Proc. Annual Meeting.* Manchester District Branch of the Association Managers of Sewage Disposal Works. p. 50.

**53** Melling, S. E. 1914. "Purification of Salford sewage along the line of the Manchester experiments." *J. Soc. Chem. Ind.* Vol. 33 p. 1124.

**54** Irvine, R. L. and Bush, A. W. 1979. "Sequencing batch biological reactors—an overview." *J. Wat. Pollut. Cont. Fed.* Vol. 51 p. 235.

**55** Irvine, R. L., Wilderer, P. A. and Flemming H-C. 1997. "Controlled unsteady state processes and technologies—an overview." *Wat. Sci. Tech.* Vol. 35, No.1 , pp. 1–10.

**56** Demoulin, G., Goronszy, M. C., Wutscher, K. and Forsthuber, E. 1997. "Co-current nitrification/denitrification and biological P-removal in cyclic activated sludge plants by redox controlled cycle operation." *Wat. Sci. Tech.* Vol. 35, No.1, pp. 215–224.

**57** Wilderer, P. A., Irvine, R. L. and Doellerer, J. (Editors) 1997. "Sequencing batch reactor technology." *Wat. Sci. Tech.* Vol. 35, No.1, pp. 278.

**58** Morgenroth, E. and Wilderer, P. A. 1997. "Sequencing batch reactor technology—concepts, design, experiences." Presented to the *Chartered Institution of Water and Environmental Management Conference, Activated Sludge into the 21st Century.* Manchester, UK, September.

**59** Schleypen, P., Michel, I., and Seiwert, H. E. 1997. "Sequencing batch reactors with continuous inflow for small communities in rural areas in Bavaria." *Wat. Sci. Tech.* Vol. 35, No.1 pp. 269–276.

**60** Torrijos, M. and Moletta, R. 1997. "Winery wastewater depollution by sequencing batch reactor." *Wat. Sci. Tech.* Vol. 35, No.1, pp. 249–257.

**61** Zilverentant, A. G. 1997. "Pilot-testing, design and full-scale experience of a sequencing batch reactor system for the treatment of the potentially toxic waste water from a road and rail car cleaning site." *Wat. Sci. Tech.* Vol. 35, No.1, pp. 259–267.

**62** Cooper, P. F. and Wheeldon, D. H. V. 1982. "Complete treatment in a two-stage fluidised bed system Part 1." *Wat. Pollut. Control.* Vol. 81, No.4, pp. 447–462.

**63** Rogalla, F., M. Payraudeau, G. Bacquet, M. Bourbigot, J. Sibony and P. Gilles. 1990. *J. Wat. Pollut. Control. Fed.,* 62:169.

**64** Lessel, T. H. 1991. *Wat. Sci. Tech.,* 23:825.

# CHAPTER 4

# Activated Sludge Treatment of Industrial Waters

## INTRODUCTION

NEW regulations in the United States have imposed limitations on a wide variety of organic and inorganic pollutants. This in turn requires modifications to the conventional design approach for activated sludge treatment of industrial wastewaters. Depending on the wastewater and the applicable permit limitations, many different process design and/or operational control parameters must be considered. These are shown in Figure 4.1 and are discussed in this chapter.

## PRETREATMENT OF INDUSTRIAL WASTEWATER

Efficient operation of the activated sludge process requires control of toxic substances, incompatible pollutants such as oil and grease, and load fluctuation through a variety of pretreatment technologies. These are shown in Figure 4.2. Guidelines for the limiting influent concentration of pollutants to the activated sludge process are shown in Table 4.1.

### EQUALIZATION

Since most industrial wastewaters fluctuate in flow rate and/or pollutant

W. Wesley Eckenfelder, Eckenfelder, Inc. Nashville, TN and Jack L. Musterman, Musterman & Associates, Nashville, TN.

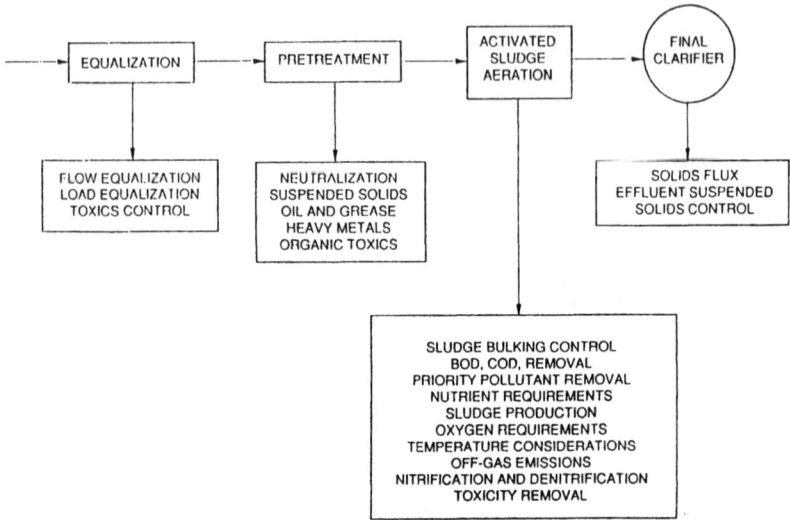

**Figure 4.1** Design variables for the activated sludge treatment of industrial wastewaters.

concentration with time, equalization is required to dampen these fluctuations and maintain stable process operation. The equalization basin (EQB) should be completely mixed and can be operated in either a constant volume (variable outflow) or variable volume (constant outflow) mode. If the wastewater flow rate remains reasonably constant with time, a constant volume basin will provide adequate load equalization at lower

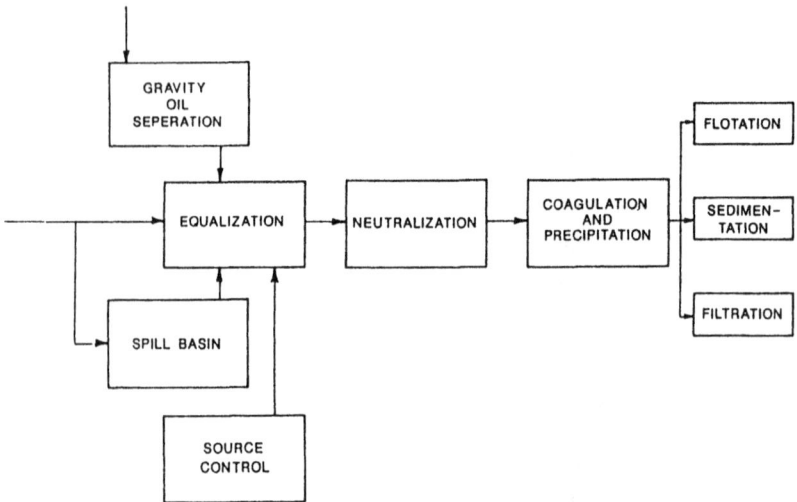

**Figure 4.2** Pretreatment technologies.

TABLE 4.1. Concentration of Pollutants That Make Prebiological Treatment Desirable.

| Pollutant or System Condition | Limiting Concentration | Kind of Pretreatment |
|---|---|---|
| Suspended solids | < 50 to 125 mg/L | Sedimentation, flotation, lagooning |
| Oil or grease | 30 mg/L | Skimming tank or separator or DAF |
| Toxic ions | | Precipitation or ion exchange |
| Pb | ≤0.1 mg/L | |
| Cu + Ni + CN | ≤1 mg/L | |
| $Cr^{+6}$ + Zn | ≤3 mg/L | |
| $Cr^{+3}$ | ≤10 mg/L | |
| pH | 6 to 9 | Neutralization |
| Alkalinity | 0.5 lb alkalinity as $CaCO_3$/lb BOD removed | Neutralization for excessive alkalinity |
| Acidity | Free mineral acidity | Neutralization |
| Organic load variation | 1:2 | Equalization |
| Sulfides | < 100 mg/L | Precipitation or stripping with recovery |
| Phenols[1] | < 70 to 300 mg/L | Extraction, adsorption, internal dilution |
| Ammonia | < 500 mg/L (as N) | Dilution, -ion exchange, pH adjustment and stripping |
| Dissolved salts[2] | < 10 to 16 g/L | Dilution, ion exchange |
| Temperature | 10 to 38°C in reactor | Cooling, steam addition |

[1]Can treat higher concentrations in complete mix reactor.
[2]Can acclimate process to 7% TDS.

cost. By contrast, a plant employing batch production processes with rapid changes in both flow and load should employ a variable volume basin with a constant volumetric withdrawal rate. As a general rule, the basin should be designed to provide an effluent load peaking factor of 1.2 (maximum mass discharge/average mass discharge). The sustained variation in basin effluent load should also be considered by determining the ratio of the standard deviation of the effluent loading rate ($SD_{eff}$) to the average effluent load. Sufficient basin volume should be provided such that this ratio is less than 0.2. Equalization basin volume requirements for wastewater from a pulp mill are shown in Figures 4.3a and 4.3b. In this case a variable volume basin provided greater equalization performance than equivalent tankage operated in the constant volume mode. A 3.5 MG basin was selected and produced a ratio of $SD_{eff}$ to mean effluent BOD load of approximately 0.15. In cases of readily degradable wastewaters, aeration should be provided in order to avoid septic conditions.

In industries subject to spill events and/or periodic shock loads, a spill

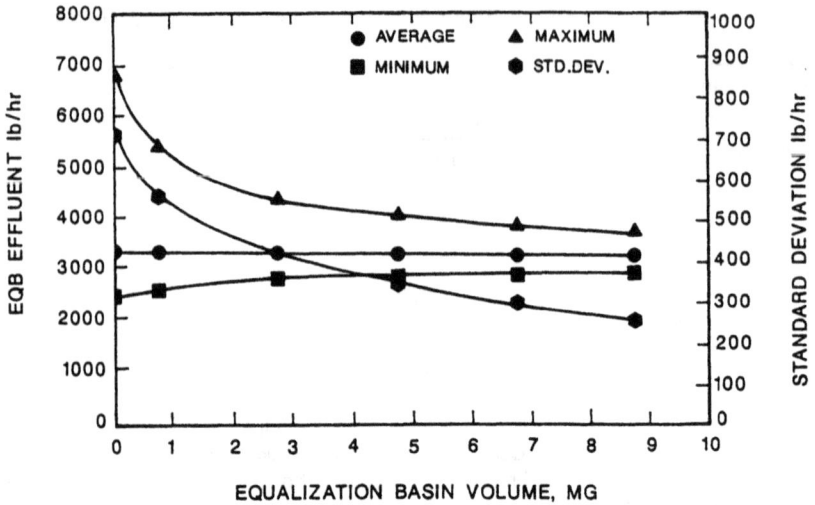

**Figure 4.3a** Effect of variable volume equalization on EQB effluent BOD loading.

basin should be provided to divert the influent wastewater flow when the concentration exceeds a predetermined value. Parameters requiring spill basin control are wastewater strength (usually defined as TOC or TOD), TDS, temperature, and toxic organics and metals. A flow schematic for an off-line spill basin is shown in Figure 4.4.

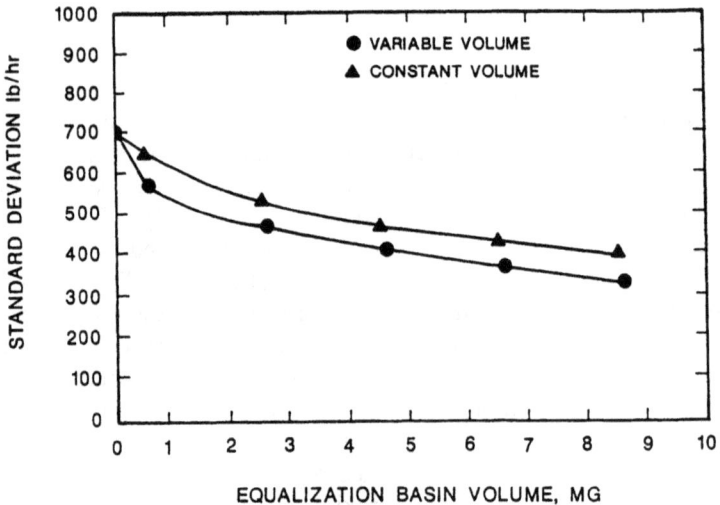

**Figure 4.3b** Comparison of variable volume and constant volume equalization on EQB effluent BOD loading.

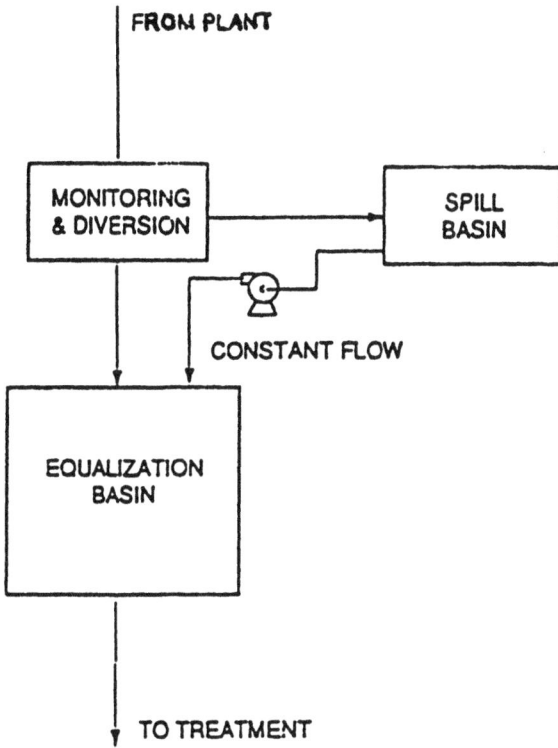

**Figure 4.4** Flow schematic for off-line control of spills and shock loads.

High concentrations of oil such as those found in petroleum refinery wastewater should be removed by gravity separation in an API or corrugated plate separator prior to activated sludge treatment. In many cases this is followed by a dissolved air flotation unit (DAF) in order to achieve levels compatible with the biological treatment process and effluent permit requirements.

## HEAVY METALS

The presence of heavy metals poses several problems. They contribute to effluent aquatic toxicity and generally have very restrictive permit limitations. Metals removal through the activated sludge process has been observed by many investigators and has been summarized in Table 4.2 [1]. The variability of metals removal through the process is related to the following removal mechanisms.

TABLE 4.2.  Metal Removal Efficiencies during Activated
Sludge Treatment [1].

| Metal/Plant | Activated Sludge Effluent ($\mu$g/l) | Range of Removal Efficiency (%) | Average Removal Efficiency (%) |
|---|---|---|---|
| Aluminum | | | |
| FS | 500–1,750 | 70–98 | 92 |
| PP | 250–350 | | 51 |
| Bismuth | | | |
| LS | 3[a] | 57–79[b] | 70 |
| Cadmium | | | |
| 4-FS | 12–10 | 7–84 | 51 |
| 3-FS | 10–30 | 11–64 | 36 |
| 2-FS | 30–120 | 0–92 | 49 |
| Chromium | | | |
| 4-FS | 10–60 | | 57 |
| FS | 202[a] | 0–98 | 46 |
| FS | 30–800 | | 54 |
| FS | 300–315 | 80–84 | 82 |
| FS | 4,400–38,000 | 75–99 | 88 |
| Cobalt | | | |
| LS | 17[a] | <0–9 | 1 |
| LS | 79[a] | 0–19[b] | 3 |
| Copper | | | |
| 10-FS | 10–170 | | 66 |
| FS | 40–660 | | 60 |
| 3-PP | 20–100 | | 70 |
| 3-PP | 160–920 | | 48 |
| PP | 9,000 | | 75 |
| Iron | | | |
| FS | 1,047 | | 72 |
| FS | 457–700 | 87–88 | 88 |
| FS | 1,000–2,950 | 95–98 | 97 |
| Lead | | | |
| 7-FS | 20–230 | | 63 |
| FS | 980–1,100 | 43–63 | 53 |
| FS | 10–490 | | 79 |
| FS | | | 54 |
| 2-PP | 5–15 | | 85 |
| 3-PP | 55–95 | | 51 |
| LS | 35–85 | | 28 |
| LS | 2,100–25,500 | 97–99 | 98 |

TABLE 4.2. **(continued).**

| Metal/Plant | Activated Sludge Effluent ($\mu$g/l) | Range of Removal Efficiency (%) | Average Removal Efficiency (%) |
|---|---|---|---|
| Manganese | | | |
| FS | 20–100 | | 6 |
| FS | 32–38 | 25–31 | 28 |
| PP | 67 | | 25 |
| 2-LS | 89[a]–101[a] | 0–17[b] | 5 |
| Mercury | | | |
| 3-FS | 0.2–1.4 | | 62 |
| FS | 1–9 | | 62 |
| PP | 0.51 | | 69 |
| Molybdenum | | | |
| LS | 11[a] | <0–18[b] | 2 |
| LS | 15[a] | 39–93[b] | 63 |
| Nickel | | | |
| 3-FS | 20–50 | | 25 |
| 4-FS | 60–250 | | 35 |
| FS | 30–1,600 | | 1 |
| PP | 70–200 | | 21 |
| 3-PP | 700–10,000 | | 20 |
| Silver | | | |
| PP | <5 | | 44 |
| LS | 10[a] | 64–94[b] | 82 |
| Thallium | | | |
| LS | 4[a] | 0–28[b] | 12 |
| Zinc | | | |
| 7-FS | 180–510 | 91–97 | 53 |
| FS | 310–600 | 67–90 | 78 |
| FS | 1,527[a] | | 57 |
| FS | 240–8,940 | | 50 |
| 6-PP | 190–530 | | 54 |
| Cyanide | | | |
| 20-FS | >10 | | 54 |

[a] Mean effluent concentration.
[b] Range of mean removals observed at 6 SRTs.
Note: LS = laboratory simulations, PS = pilot plant and FS = full-scale works. Number of plants indicated by prefix.

- Entrapment of precipitated metals in the sludge floc matrix. Parameters that affect floc size and character will also affect metals removal.
- Bacterial extracellular polymer binding of soluble metals. Extracellular polymer production is a function of sludge age and organic loading rate.
- Accumulation of soluble metals by the cell.
- Limited volatilization of selected metals.

The bio-concentration effects can result in high levels of regulated metals in the stabilized sludge and thereby limit the options for ultimate solids disposal. Metals removal in the activated sludge process from a petroleum refinery wastewater is shown in Table 4.3. Recent data has shown that when metals are present in low concentrations, high removals are usually achieved. Data from one activated sludge plant and modeled according to Patterson et al. [2] are shown in Figure 4.5. As the solids retention time (SRT) is increased, metal accumulation on the sludge will also increase, as shown in Figure 4.6 for accumulation of copper.

Metals should be removed by source control or chemical precipitation prior to the activated sludge process.

## pH NEUTRALIZATION

The activated sludge process operates effectively over a pH range of 6.5 to 8.5 and thus neutralization may be required for wastewaters that are outside of this pH range. There are, however, exceptions to this in which highly alkaline or acidic wastewaters do not require pH adjustment for effective treatment by activated sludge. If a complete mix activated sludge process is used, hydroxyl ions ($OH^-$) will react with the carbon dioxide ($CO_2$) produced by microbial respiration yielding bicarbonate ion ($HCO_3^-$), which will tend to buffer the system at a pH near 8.0. As a general rule, 0.5 lbs of hydroxide alkalinity (as calcium carbonate) will be neutralized by 1 lb of BOD removed in the process. High-strength textile mill wastewaters having a pH of 10 or greater entering a complete

TABLE 4.3. Heavy Metal Removal in the Activated Sludge Process Treating Petroleum Refinery Wastewater.

| Heavy Metal | Activated Sludge Plant | |
| --- | --- | --- |
| | Influent (mg/l) | Effluent (mg/l) |
| Cr | 2.2 | 0.9 |
| Cu | 0.5 | 0.1 |
| Zn | 0.7 | 0.4 |

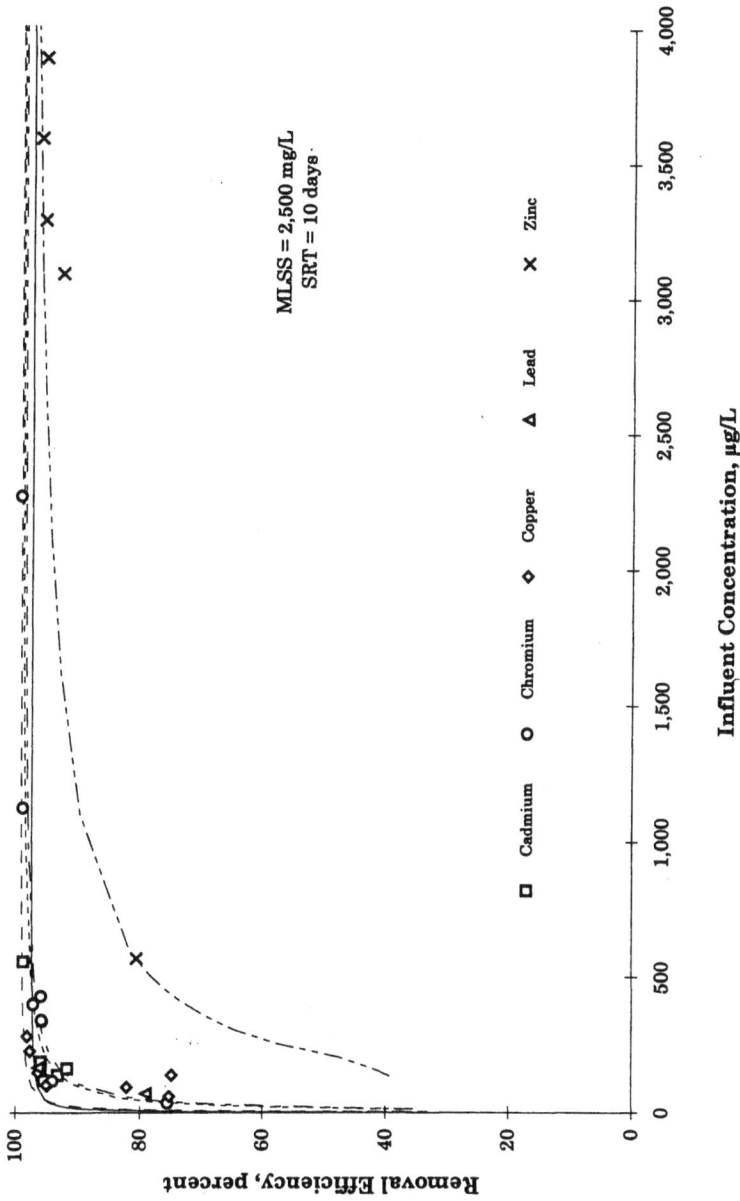

**Figure 4.5** Heavy metals removal in the activated sludge process.

MLSS = 2,500 mg/L
SRT = 10 days

□ Cadmium   ○ Chromium   ◇ Copper   ▲ Lead   ✕ Zinc

Influent Concentration, µg/L

Removal Efficiency, percent

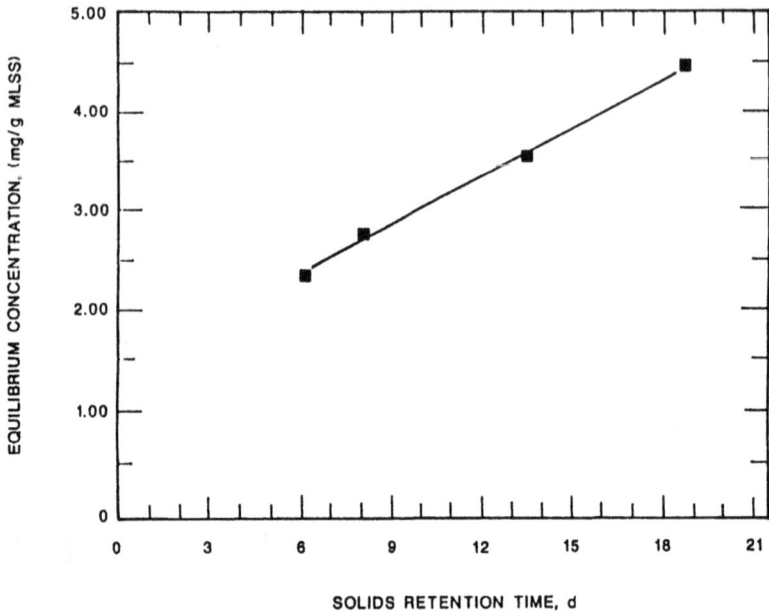

**Figure 4.6** Copper accumulation in the activated sludge process as a function of solids retention time.

mix biological process were neutralized to a pH near 8.0 by microbial respiration and $CO_2$ production. Similar biological neutralization effects occur in wastewaters that have been lime treated to pH 10 or 11 for phosphorus and/or metals removal. It should be noted however, that the amount of hydroxide alkalinity rather than the actual pH of the influent wastewater determines the degree of neutralization required.

In like manner, organic acids will be biologically oxidized to $CO_2$ and water. The $CO_2$ will be air-stripped from the process to its equilibrium concentration, thus reducing the acidity. Wastewaters from the synthetic fibers and organic chemicals industries containing acetic acid and having a pH less than 4.0 have been successfully treated in a complete mix activated sludge process without pre-neutralization.

There are also cases in which the biological reactions generate acidity or alkalinity. Nitrification generates acidity and destroys alkalinity. In many cases supplemental alkalinity must be added to offset the effect of nitrification and maintain an optimal pH in the treatment process. Biological oxidation of sulfonates yields sulfuric acid ($H_2SO_4$). The wastewater from one plant containing sulfonates required pH adjustment to 11.5 prior to entering the aeration basin in order to maintain the recommended mixed liquor pH. Wastewaters containing high concentrations of salts of weak organic acids, e.g., sodium acetate, will produce alkalinity in the form of carbonates and cause the mixed liquor pH to increase to a range of 8.5 to 9.2. These operating conditions exceed the recommended mixed liquor pH range and may require acid addition directly to the aeration basin.

Where neutralization is required, acidic wastewaters can be neutralized using hydrated lime ($Ca(OH)_2$), caustic soda (NaOH) or magnesium hydroxide ($Mg(OH)_2$). $Mg(OH)_2$ frequently has the advantage of producing less, more concentrated sludge. Since pH is a logrithmic function, highly acidic wastewater requires a two stage neutralization process to maintain a stable effluent pH as shown in Figure 4.7. In cases where the pH is fairly constant a limestone bed can be employed Alkaline wastewaters are neutralized using sulfuric acid ($H_2SO_4$) or flue gas ($CO_2$) if weakly buffered.

## TOXIC SUBSTANCES AND OFF-GAS CONTROL

Toxic organic compounds in wastewaters should be modified or removed by pretreatment prior to the activated sludge process. This includes

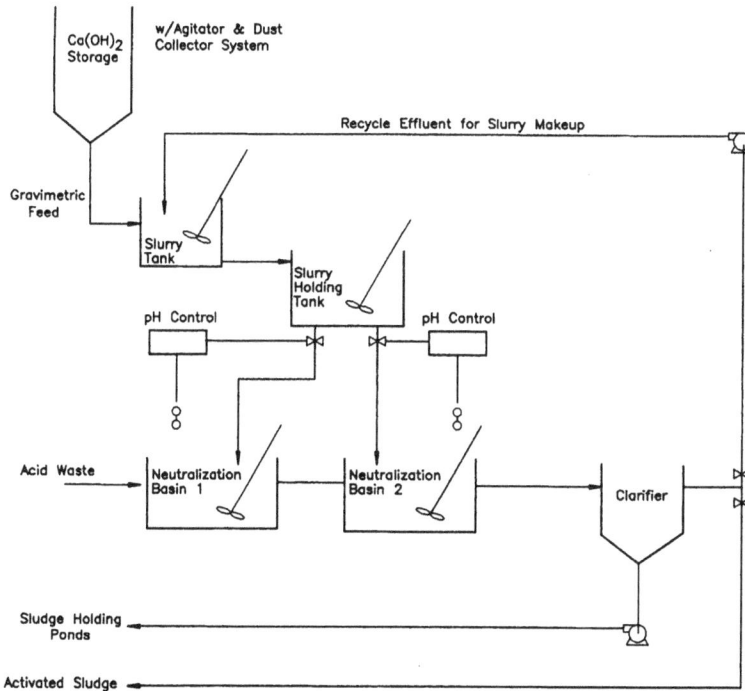

**Figure 4.7** Schematic diagram of two-stage neutralization system.

organics that are toxic to aquatic life as measured by bioassay methods as well as organics that are inhibitory to the biomass in the activated sludge process. Applicable technologies for pretreatment of toxic wastewaters are shown in Figure 4.8 [3]. It should be noted that in many cases, the primary objective of the pretreatment process is detoxification and enhanced biodegradability.

Restrictions on air emissions may require prestripping and off-gas capture prior to activated sludge treatment of wastewaters containing high concentrations of volatile organic and inorganic substances (VOS). If the VOS are not controlled and treated at the stripper they will be volatilized at the aeration basin where off-gas capture and control is more difficult and costly.

## CHARACTERIZATION OF INDUSTRIAL WASTEWATER

COD or TOC (rather than BOD) may also be an effluent permit

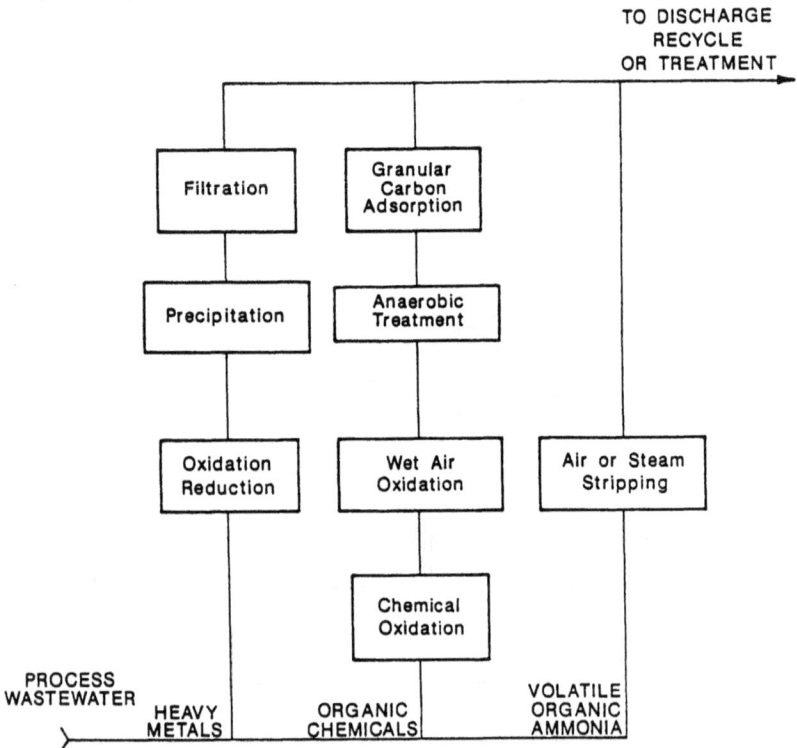

**Figure 4.8** Applicable technologies for treatment of toxic wastewaters [2].

TABLE 4.4. Evaluation of COD and BOD with Respect to Theoretical Oxygen Demand of Selected Organic Compounds [4].

| Chemical Group | ThOD (mg/mg) | Measure COD (mg/mg) | $\frac{COD}{ThOD}$ (%) | Measure $\frac{COD}{ThOD}$ (%) | $\frac{BOD_5}{BOD_5}$ (mg/mg) |
|---|---|---|---|---|---|
| Aliphatics | | | | | |
| Methanol | 1.50 | 1.05 | 70 | 1.12 | 75 |
| Ethanol | 2.08 | 2.11 | 100 | 1.58 | 76 |
| Ethylene glycol | 1.26 | 1.21 | 96 | 0.39 | 29 |
| Isopropanol | 2.39 | 2.12 | 89 | 0.16 | 7 |
| Maleic acid | 0.83 | 0.80 | 96 | 0.64 | 77 |
| Acetone | 2.20 | 2.07 | 94 | 0.81 | 37 |
| Methyl ethyl ketone | 2.44 | 2.20 | 90 | 1.81 | 74 |
| Ethyl acetate | 1.82 | 1.54 | 85 | 1.24 | 68 |
| Oxalic acid | 0.18 | 0.18 | 100 | 0.16 | 89 |
| Group average | | | 91 | | 56 |
| Aromatics | | | | | |
| Toluene | 3.13 | 1.41 | 45 | 0.86 | 28 |
| Benzaldehyde | 2.42 | 1.98 | 80 | 1.62 | 67 |
| Benzoic acid | 1.96 | 1.95 | 100 | 1.45 | 74 |
| Hydroquinone | 1.89 | 1.83 | 100 | 1.00 | 53 |
| o-Cresol | 2.52 | 2.38 | 95 | 1.76 | 70 |
| Group average | | | 84 | | 58 |
| Nitrogenous organics | | | | | |
| Monoethanolamine | 2.49 | 1.27 | 51 | 0.83 | 34 |
| Acrylonitrile | 3.17 | 1.39 | 44 | nil | 0 |
| Aniline | 3.18 | 2.34 | 74 | 1.42 | 44 |
| Group average | | | 58 | | 26 |
| Refractory | | | | | |
| Tertiary butanol | 2.59 | 2.18 | 84 | 0 | 0 |
| Diethylene glycol | 1.51 | 1.06 | 70 | 0.15 | 10 |
| Pyridine | 3.13 | 0.05 | 2 | 0.06 | 2 |
| Group average | | | 52 | | 4 |

parameter. The COD and TOC both measure degradable plus nondegradable organics and hence adjustments must be made to define those organics that are removable in the activated sludge process. The relationship between COD and $BOD_5$ for a variety of pure organic compounds is shown in Table 4.4 [4]. The relationships between $BOD_5$, COD, and TOC for several untreated industrial wastewaters and process waste streams are shown in Table 4.5 [4]. Several conclusions can be drawn from these data:

- Degradable aromatics exhibit higher $BOD_5$/COD ratios than degradable aliphatics. The degradable aromatics listed generally exhibit higher $BOD_5$/COD ratios than the domestic sewage

TABLE 4.5.  **Wastewater Characterization Results [4].**

| Wastewater | $BOD_5/COD$ | COD/TOC |
|---|---|---|
| Acrylonitrile | 0.19 | 2.0 |
| Butadiene-styrene | 0.05 | 3.8 |
| Cumene | 0.12 | 5.6 |
| EDC-direct | 0.49 | 1.6 |
| EDC-oxyhydrochlorination | 0.64 | 1.8 |
| Ethylene oxide | 0.35 | 17.0* |
| Olefins | 0.25 | 3.4 |
| Polystyrene | 0.44 | 3.3 |
| Polyvinylchloride | 0.10 | 1.9 |
| Propylene oxide | 0.45 | 5.0 |
| Propylene glycol | 0.48 | 4.9 |
| Propylene tetramer | 0.34 | 0.7 |
| Sewage | 0.37 | 3.4 |
| Synthetic rubber | 0.51 | 3.9 |
| Urea | 0.79 | 0.8 |
| Vinyl chloride | 0.04 | 0.9 |

*TDS > 20,000 mg/l.

$BOD_5/COD$ ratio, whereas the degradable aliphatics listed exhibit lower $BOD_5/COD$ ratios than domestic sewage.

- The COD to TOC ratio is highly variable depending on the specific organics in the raw wastewater and no correlation should be anticipated between TOC and $BOD_5$ or COD. Biologically treated effluents however, will usually show a more consistent correlation of parameters. This correlation however, cannot be predicted from the influent characteristics and must be empirically determined.

Since only biodegradable organics are removed in the activated sludge process, the soluble COD remaining in the effluent [$(SCOD)_{eff}$]will consist of nondegradable organics present in the influent wastewater [$(OCD_{nd})_o$] plus residual degradable organics (as defined by the soluble $BOD_5$) plus the soluble microbial products (SMP) generated in the process. The SMP are not biodegradable (designated as $SMP_{nd}$) and thus exert a soluble COD (or TOC) but no BOD. COD relationships during biological treatment are shown in Figure 4.9. The effluent COD can, therefore be calculated as follows:

$$(COD_{nd})_o = COD - COD_{deg}$$
$$COD_{deg} = \frac{BOD_5}{f \cdot 0.92}$$

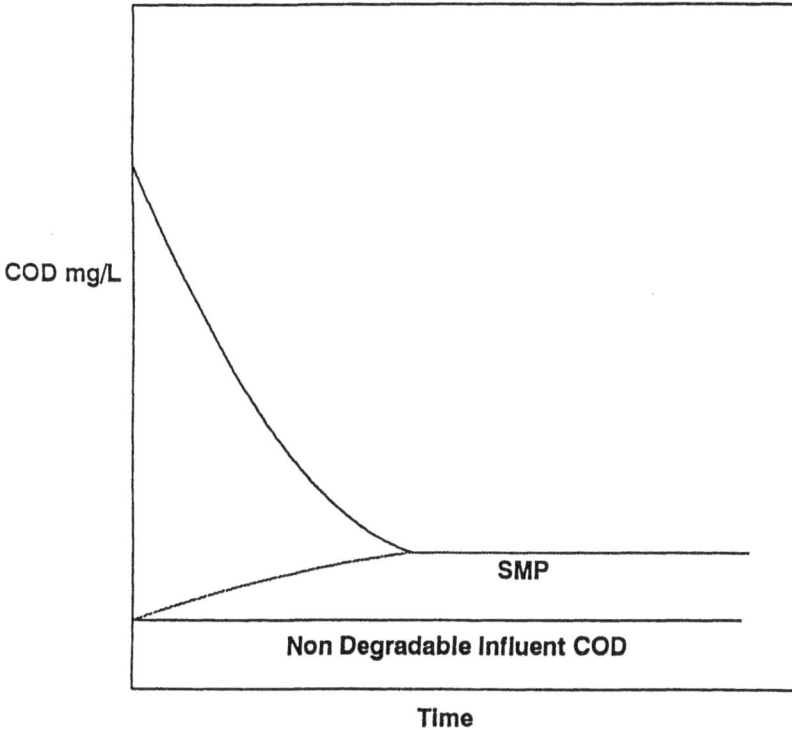

**Figure 4.9** COD relationships during biological oxidation.

where $f = \dfrac{BOD_5}{BOD_{ult}}$

$$SCOD_{eff} = (COD_{nd})_o + \frac{(SBOD_5)_{eff}}{0.65} + SMP_{nd}$$

The $SMP_{nd}$ may vary from 2 to 10 percent of the degradable influent COD. Effluent characteristics for a variety of industrial wastewaters following activated sludge treatment are shown in Table 4.6.

Example 4.1 illustrates the calculation of the effluent COD after biological treatment.

*Example 4.1*

Influent COD = 300 mg/l
Influent BOD₅ = 192 mg/l

Effluent $BOD_5$ = 10 mg/l (soluble)
Effluent COD = Influent nondegradable COD + degradable COD
remaining + $SMP_{nd}$

$$\text{Influent nondegradable COD} = 300 - \left(\frac{192}{0.8} \cdot \frac{1}{0.92}\right) = 39 \text{ mg/L}$$

In the above calculation it was assumed that the influent $BOD_5$ was 80 percent of the ultimate BOD ($BOD_u$) and that the $BOD_u$ was 92 percent of the COD.

$$\text{Effluent degradable SCOD} = \frac{10}{0.65} \cdot \frac{1}{0.92} = 17 \text{ mg/L}$$

In the above calculation, it was assumed that the effluent soluble $BOD_5$ was 65 percent of the $BOD_u$ and that the $BOD_u$ was 92 percent of the SCOD.

Assuming that the $SMP_{nd}$ is 8 percent of the influent degradable COD, then

$$SMP_{nd} = (300 - 39)\,0.08 = 21 \text{ mg/L}$$

Therefore, the total effluent SCOD = 39 + 17 + 21 = 77 mg/L. The effluent of total COD[$(TCOD)_{eff}$] includes the contribution of "particulate COD" (PCOD) due to the effluent of TSS.

## PRINCIPLES OF BIOLOGICAL OXIDATION

### ORGANIC REMOVAL MECHANISMS

#### Partitioning

Removal of some soluble organics occurs by adsorption in the biological solids. This removal mechanism is termed "partitioning" and has been related to the octanol-water partition coefficient (log $p$) of the organic.

$$K_{sw} = \ln K_{ow}^n \tag{4.1}$$

where
$K_{sw}$ = biosolids accumulation factor, ratio of organic sorbed and in solution, mg/mg/mg/l
$K_{ow}$ = octonal water partition coefficient

TABLE 4.6. COD, BOD and SMP Relationships for Industrial Wastewaters.

| Wastewater | Influent | | Effluent | | SMP$_{nd}$[a] (mg/L) | (COD$_{nd}$)$_e$[b] (mg/L) | (BOD$_5$)$_e$/ (COD$_{deg}$)$_i$[c] |
| --- | --- | --- | --- | --- | --- | --- | --- |
| | BOD (mg/L) | COD (mg/L) | BOD (mg/L) | COD (mg/L) | | | |
| Pharmaceutical | 3,290 | 5,780 | 23 | 561 | 261 | 265 | 0.60 |
| Diversified chemical | 725 | 1,487 | 6 | 257 | 62 | 186 | 0.56 |
| Cellulose | 1,250 | 3,455 | 58 | 1,015 | 122 | 804 | 0.47 |
| Tannery | 1,160 | 4,360 | 54 | 561 | 190 | 288 | 0.28 |
| Alkylamine | 893 | 1,289 | 12 | 47 | 62 | — | 0.69 |
| Alkyl benzene sulfonate | 1,070 | 4,560 | 68 | 510 | 202 | 204 | 0.25 |
| Viscose rayon | 478 | 904 | 36 | 215 | 35 | 125 | 0.61 |
| Polyester fibers | 208 | 559 | 4 | 71 | 24 | 41 | 0.40 |
| Protein process | 3,178 | 5,355 | 5 | 245 | 256 | — | 0.59 |
| Tobacco | 2,420 | 4,270 | 139 | 546 | 186 | 146 | 0.59 |
| Propylene oxide | 532 | 1,124 | 49 | 289 | 42 | 172 | 0.56 |
| Paper mill | 380 | 686 | 7 | 75 | 31 | 33 | 0.58 |
| Vegetable oil | 3,474 | 6,302 | 76 | 332 | 298 | — | 0.55 |
| Vegetable tannery | 2,396 | 11,663 | 92 | 1,578 | 504 | 933 | 0.22 |
| Hardboard | 3,725 | 5,827 | 58 | 643 | 259 | 295 | 0.67 |
| Saline organic chemical | 3,171 | 8,597 | 82 | 3,311 | 264 | 2,921 | 0.56 |
| Coke | 1,618 | 2,291 | 52 | 434 | 93 | 261 | 0.79 |
| Coal liquid | 2,070 | 3,160 | 12 | 378 | 139 | 221 | 0.70 |
| Textile dye | 393 | 951 | 20 | 261 | 35 | 196 | 0.53 |
| Kraft paper mill | 308 | 1,153 | 7 | 575 | 29 | 535 | 0.50 |

[a] 0.05 (COD$_{deg}$)$_i$.
[b] (COD$_{nd}$)$_e$ = (SCOD)$_e$ − [(BOD$_5$)$_e$/0.65] − SMP$_{nd}$.
[c] (COD$_{deg}$)$_i$ = (COD)$_i$ − (COD$_{nd}$)$_e$ + SMP$_{nd}$.

$K_{sw}$ = has been reported to vary from $1.38 \times 10^{-5}$ to 4.3
$\times 10^{-7}$ and n from 0.58 to 1.0 (5)(6)(7)(8)

In most industrial wastewaters partitioning provides negligible SCOD removal but may be a method of bio accumulation of certain lipid-soluble organic compounds.

## Stripping

Volatile organic carbon (VOC) will strip in biological treatment plants. The general relationship between the volatilization and biodegradation at certain organic compounds is shown in Figure 4.10 [40] and is discussed in detail on page *00*.

## Biooxidation

In the activated sludge process, when organic matter is removed from solution by biological metabolism, oxygen is consumed by the organisms and new cell mass is synthesized. The organisms also undergo progressive auto-oxidation (endogenous decay) of their cellular mass. These phenomena are shown by Equations (4.2) and (4.3).

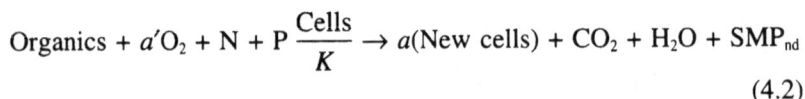

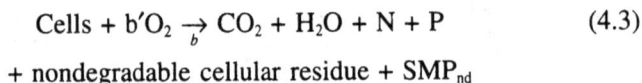

$$\text{Organics} + a'O_2 + N + P \frac{\text{Cells}}{K} \rightarrow a(\text{New cells}) + CO_2 + H_2O + SMP_{nd}$$

$$(4.2)$$

$$\text{Cells} + b'O_2 \xrightarrow{b} CO_2 + H_2O + N + P \qquad (4.3)$$
$$+ \text{ nondegradable cellular residue} + SMP_{nd}$$

In Equation (4.2), $a'$ is the oxygen equivalent of the fraction of organic matter removed that is oxidized to end products and $a$ is the fraction of organic matter removed that is synthesized into biomass. Term $K$ is a reaction rate coefficient that is related to the biodegradability of the specific wastewater. In Equation (4.3), the coefficient $b$ is the fraction per day of degradable biomass that is endogenously oxidized and "Cells + $b'O_2 \xrightarrow{b}$" is the oxygen required to support the endogenous decay.

A small portion of the organics removed for synthesis [Equation (4.2)] and endogenous metabolism [Equation (4.3)] remains as nondegradable cellular residue and $SMP_{nd}$. The nondegradable cell residue is 20 percent of the volatile suspended solids generated in Equation (4.2). The $SMP_{nd}$

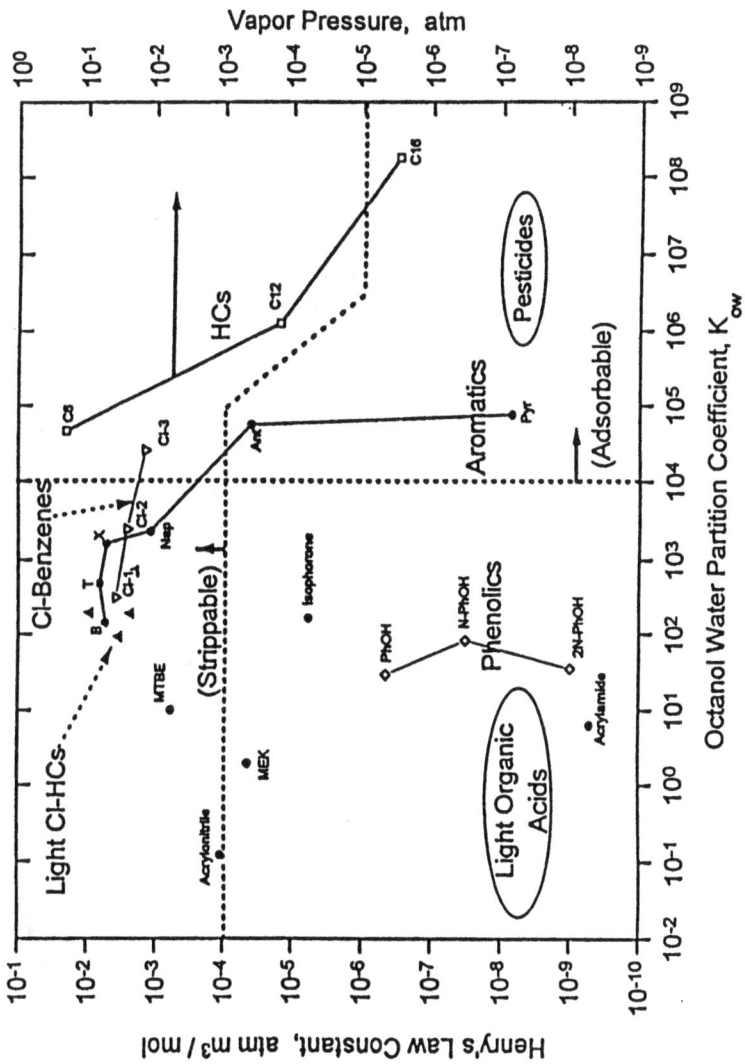

**Figure 4.10** Categorization of organic compounds—strippability vs. sorbability [39].

varies with the organic composition of the raw wastewater and sludge age, and varies from 2 to 10 percent of the influent degradable COD. These reactions are shown schematically in Figure 4.11.

Since all but a small portion of the organics removed are either oxidized to end products (i.e., $CO_2$ and $H_2O$), or synthesized to biomass, terms $a$ and $a'$ can be combined on a total oxygen demand basis using either COD or the adjusted ultimate BOD value ($BOD_u/0.92$). Therefore, on a COD basis

$$a_{COD} + a'_{COD} = 1 \qquad (4.4a)$$

The actual sum will always be less than unity due to the formation of $SMP_{nd}$. Since the biomass is usually expressed as volatile suspended solids (VSS) it is convenient to express the term $a$ in VSS units. On the average, it takes 1.4 g $O_2$ to oxidize 1 g of biomass expressed as VSS. Therefore:

$$1.4a_{VSS} + a'_{COD} \sim 1 \qquad (4.4b)$$

where
$a$ = mg VSS produced per mg of COD (or $BOD_u/0.92$) removed
$a'$ = mg $O_2$ consumed per mg of COD (or $BOD_u/0.92$) removed

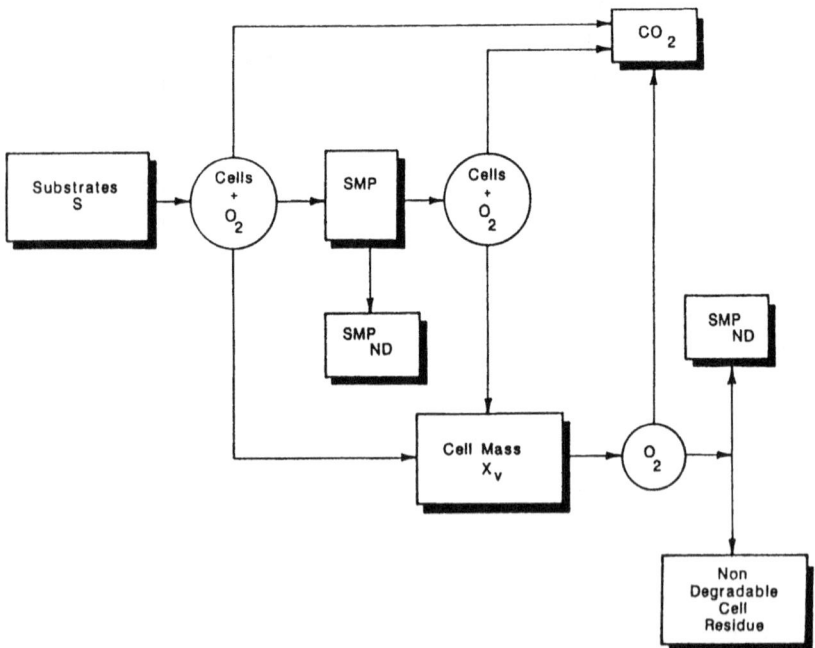

**Figure 4.11** Mechanism of biodegradation.

In order to design an activated sludge process, $a$, $a'$, $b$ and $K$ must be determined for the specific wastewater. This permits material balances to be developed around the process for substrate removal, oxygen requirements, and biomass production. Values of these parameters are available for the more common industrial wastewaters such as those from petroleum refining, food processing, and pulp and paper manufacturing. Wastewaters from the organic chemical and pharmaceutical industries, however, require laboratory or pilot plant studies to define these parameters. Test procedures for this purpose are discussed in this chapter.

The design basis and operating criteria for an activated sludge plant are usually expressed in terms of the organic loading rate on the process ($F/M$) and the sludge age ($\theta_c$) or solids retention time. The organic loading rate is expressed as

$$F/M = \frac{S_o}{X_v t} \tag{4.5}$$

where

$F/M$ = food to biomass ratio, mg/mg-d
$S_o$ = BOD or degradable COD in the influent, mg/L
$X_v$ = MLVSS, mg/L
$t$ = hydraulic retention time, d

The sludge age or SRT is expressed as

$$\text{SRT} = \theta_c = \frac{X_v t}{\Delta X_v} \tag{4.6}$$

where

$\theta_c$ = sludge age, d
$\Delta X_v$ = volatile suspended solids wasted per day, mg/L
$X_v t$ = volatile suspended solids in reactor, mg-d/L

The biomass generated in Equation (4.2) is about 80 percent biodegradable. As the SRT is increased, the degradable portion of the biomass will be endogenously oxidized and the biodegradable fraction of the remaining volatile biomass (designated as $X_d$) will decrease. The degradable fraction of the volatile biomass can be estimated by Equation (4.7). This relationship for the treatment of a food processing wastewater is shown in Figure 4.12.

$$X_d = \frac{0.8}{1 + 0.2b\theta_c} \tag{4.7}$$

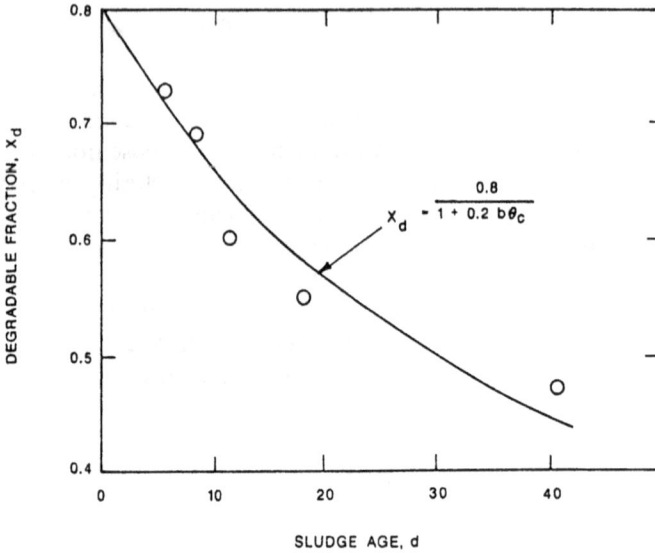

**Figure 4.12** Relationship between degradable fraction and SRT for a food processing wastewater.

## SLUDGE PRODUCTION

Using the above relationships, the net sludge yield for a wastewater that contains primarily soluble organic substrates can be expressed by Equation (4.8). This relationship is shown in Figure 4.13 for a soluble wastewater from pharmaceutical manufacturing.

$$\Delta X_v = aS_r - bX_dX_vt \tag{4.8}$$

where $S_r$ = organic removal, mg/L.

If the influent contains VSS, such as in a pulp and paper mill wastewater, Equation (4.8) is modified to include this contribution.

$$\Delta X_v = a[S_r + f_df_xX_i] - bX_dX_{vb}t + (1 - f_d)f_xX_i + (1 - f_x)X_i \tag{4.9a}$$

in which

$X_i$ = influent VSS, mg/L
$X_{nb}$ = biological VSS, mg/L
$f_x$ = fraction of influent VSS which is degradable
$K_p$ = degradation rate coefficient of influent VSS
$f_d$ = fraction of degradable influent VSS degraded
$f_b$ = fraction of the biological VSS

The degradation of the influent VSS is a function of the SRT

$$(1 - f_d) = e^{-K_p\theta_c} \tag{4.9b}$$

It is assumed that 1 mg VSS solubilized generates 1 mg COD and

$$f_b = \frac{a[S_r + f_d f_x X_i] - bX_d X_{vb}t}{a[S_r + f_d f_x X_i] - bX_d X_{vb}t + (1 - f_d)f_x X_i + (1 - f_x)X_i} \tag{4.9c}$$

Most pulp and fiber in pulp and paper mill wastewaters are essentially nondegradable, and hence $(1 - f_d)$ is approximately 1. In food processing wastewaters, however, $(1 - f_d)$ may be less than 0.2. If the influent contains high levels of VSS, the value $(1 - f_d)$ of must be experimentally determined in

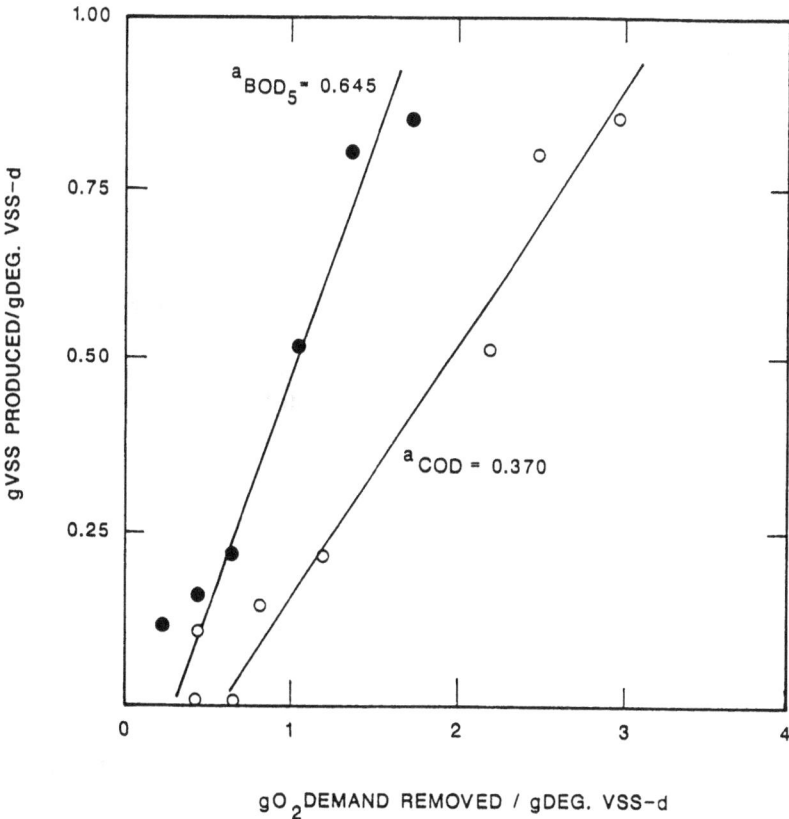

**Figure 4.13** Cell synthesis relationship for a soluble pharmaceutical wastewater.

order to accurately predict the volatile sludge production rate and true biomass yield.

The impact of influent non-volatile suspended solids on mixed liquor characteristics and sludge production rate can be significant. The quantity of inert material (measured as non-volatile suspended solids) generated is related to the SRT, the fraction of influent non-volatile suspended solids that is nondegradable and the formation of nonbiomass particulates in the activated sludge process. This relationship may be expressed as follows.

$$\Delta NVSS = a*S_r f_{ibnd} + f_{oi} X_{oi} + \text{Inert Production} \qquad (4.10)$$

where

$\Delta NVSS$ = inert suspended solids produced, mg/L
$X_{oi}$ = influent inert solids, mg/l
$f_{ibnd}$ = fraction of inert biomass not degraded
$f_{oi}$ = fraction of influent inert solids not degraded
$a*$ = biomass produced per unit of substrate removed, mg TSS/mg BOD (or COD)

Inert material may also be produced in the activated sludge system through precipitation reactions of the wastewater. This later accumulation term is indicated but not characterized in Equation (4.10) however, since it is difficult to quantify unless the influent suspended solids concentration is negligible and the inert accumulation is significant.

Total sludge production can be calculated by summing the results of Equations (4.9) and (4.10). The impact of the total suspended solids production on operating aeration basin MLSS concentrations is a direct function of SRT. As SRT increases, the aeration basin MLSS will increase. Consideration must be given in secondary clarifier solids loading rate design to accommodate the inert and volatile suspended solids generation while maintaining the required SRT for substrate removal.

## OXYGEN REQUIREMENTS

The biological oxygen requirements can be computed by Equation (4.11):

$$O_2 = a'S_r + (1.4b)X_d X_v t \qquad (4.11)$$

The lumped coefficient "1.4b" is frequently referred to as $b'$. Equation (4.11) is used to determine $a'$ and $b'$ and the system oxygen requirements. This relationship is shown in Figure 4.14 for a food processing wastewater.

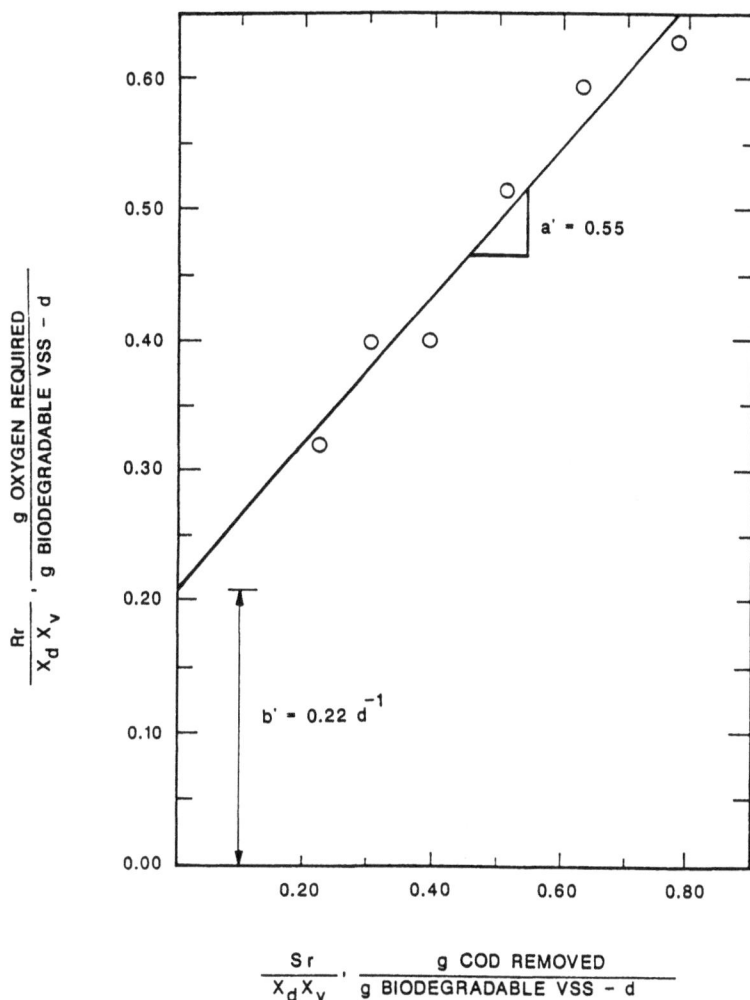

**Figure 4.14** Development of oxygen utilization coefficients for a food processing wastewater at 25°C.

The oxygen utilization coefficients $a'$ and $b'$ and the sludge yield coefficient $a$ can be determined from wastewater treatment plant operating data using the above relationships. These calculations are illustrated in Example 4.2.

*Example 4.2*

Operating data from a high purity oxygen activated sludge plant treating an organic chemicals wastewater were analyzed. The first stage of the oxygen process served as a sludge reaeration basin and the oxygen uptake rate was assumed to be the endogenous demand. Most of the organic removal (expressed as TOD) occurred in the second stage of the process.

First-stage operating characteristics

OUR = 27 mg/L-hr
VSS = 9,000 mg/L
SOUR = 3 mgO$_2$/gVSS-hr = $k_e$

Second-stage operating characteristics

OUR = 104 mg/L-hr
VSS = 6,000 mg/L
SOUR = 17.3 mgO$_2$/gVSS-hr = $k_r$
TOD removed = 12,000 lb/d
V = 0.26 MG

$$b' = \frac{3 \text{ mgO}_2/\text{gVSS} - \text{hr}}{1,000} \cdot 24 = 0.07/\text{d}$$

$$a' = \frac{8.34V(k_r - k_e)X_v \cdot 24}{S_r}$$

$$a' = \frac{8.34 \cdot 0.26(17.3 - 3) \cdot 24}{12,000}$$

$$a' = 0.37 \text{ mgO}_2/\text{mg TOD removed}$$

$$a = \frac{1.0 - 0.37}{1.4} = 0.45 \text{ mg VSS/mg TOD removed}$$

## NUTRIENT REQUIREMENTS

The biomass requires nitrogen and phosphorus in order to affect metabolism and removal of organics in the process. In addition to this, trace levels of other micro nutrients are required to assure good floc formation. These are usually present in the carriage water of most process waste streams, except in cases where deionized water is used in production and thus constitutes the process wastewater. Addition of small amounts of iron (and sometimes other nutrients) will usually solve the problem. Trace nutrient requirements are shown in Table 4.7 [9].

When insufficient nitrogen is present, the amount of cellular material

TABLE 4.7. Trace Nutrient Requirements for Activated Sludge [9].

| Micronutrient | Requirement (mg/mg BOD) |
|---|---|
| Mn | $10 \times 10^{-5}$ |
| Cu | $15 \times 10^{-5}$ |
| Zn | $16 \times 10^{-5}$ |
| Mo | $43 \times 10^{-5}$ |
| Se | $14 \times 10^{-10}$ |
| Mg | $30 \times 10^{-4}$ |
| Co | $13 \times 10^{-5}$ |
| Ca | $62 \times 10^{-4}$ |
| Na | $5 \times 10^{-5}$ |
| K | $45 \times 10^{-4}$ |
| Fe | $12 \times 10^{-3}$ |

synthesized per unit of organic matter removed increases due to an accumulation of polysaccharide. At some point, nitrogen-limiting conditions restrict the rate of BOD removal. Nutrient-limiting conditions will also stimulate filamentous growth.

Nitrogen is available to the biomass in the form of ammonium ($NH_4^+$) and/or nitrate ($NO_3^-$). Organic nitrogen, present in the wastewater as protein or amino acids, must first be biologically hydrolyzed to release ammonium in order to be available to the biomass. Therefore, in wastewaters containing organic nitrogen as the primary nitrogen source, experiments must be conducted to determine the availability of the organic nitrogen for biomass utilization.

The organic removal rate with ammonia as the nitrogen source is substantially higher than with nitrate [10]. This is due to the fact that with ammonia as the nitrogen source, the maximum substrate removal rate is determined by the rate of production of energy or carbon building units. When nitrate is the nitrogen source, however, the maximum removal rate depends on the rate of nitrate reduction to ammonia. When nitrate serves as the nitrogen source, the percentage of oxidation is higher than with ammonia since the formation of ammonia requires more energy.

Phosphorus must be in the form of soluble orthophosphate ($PO_4$) in order to be assimilated by the biomass. Therefore, complex inorganic and organically bound phosphorus must first be bio-hydrolyzed to orthophosphate in order to be available for biosynthesis.

The "rule of thumb" to assure adequate nitrogen and phosphorus for BOD removal is to provide a nutrient mass ratio of 100 : 5 : 1 (BOD : N : P). A higher ratio (e.g., 150 : 5 : 1) will reduce the rate of

BOD removal and promote filamentous growth. In a continuous flow process, however, the actual nitrogen and phosphorus requirement will depend on the net biomass synthesis (i.e., nitrogen assimilation due to growth and nitrogen release through endogenous respiration). This balance will be related to the nitrogen and phosphorus content of the wasted sludge as determined by the wastewater characteristics and the SRT. The nitrogen content of the biological sludge as generated in the process is approximately 12.3 percent (by weight) based on the VSS. However, the nitrogen content of the sludge declines in the endogenous phase and when nitrogen is limiting growth. The nitrogen content of the nondegradable cellular mass has been shown to average 7 percent. This relationship is shown in Figure 4.15. In like manner, the phosphorus content of sludge at generation has been found to average 2.6 percent, with the nondegradable cellular mass having a phosphorus content of approximately 1 percent.

The nitrogen and phosphorus requirements can be calculated considering the nitrogen and phosphorus content of the sludge wasted from the process.

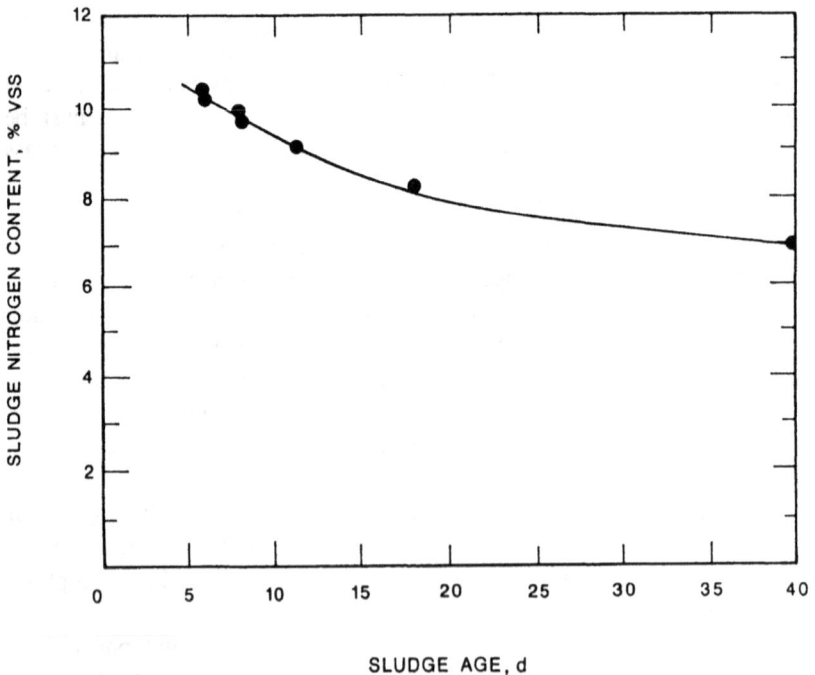

Figure 4.15 Nitrogen content of activated sludge as related to sludge age.

$$N = 0.123 \frac{X_d}{0.8} \Delta X_{vb} + 0.07 \frac{0.8 - X_d}{0.8} \Delta X_{vb} \quad (4.12)$$

$$P = 0.026 \frac{X_d}{0.8} \Delta X_{vb} + 0.01 \frac{0.8 - X_d}{0.8} \Delta X_{vb} \quad (4.13)$$

In Equations (4.12) and (4.13) when $\Delta X_{vb}$ is expressed as lbs/d, $N$ and $P$ are the nutrient requirements in lbs/d expressed as nitrogen and phosphorus, respectively. Example 4.3 illustrates the calculation of nitrogen and phosphorus requirements.

*Example 4.3*

An activated sludge plant treating an industrial wastewater operated under the following conditions:

Flow = 1.6 mgd
$S_o$ = 560 mg/l (BOD basis)
$S_e$ = 20 mg/l
$X_v$ = 3000 mg/l
$a$ = 0.55
$b$ = 0.1/day
$NH_3$–$N$ = 5 mg/l
$P$ = 3 mg/l
$F/M$ = 0.4/day
$\theta_c$ = 7 days

Compute the $N$ and $P$ which must be added to the process.

$$t = \frac{S_o}{X_v \, F/M} = \frac{560}{3000 \cdot 0.4} = 0.47 \text{ days}$$

$$\Delta X_v = aS_r - bX_dX_vt$$

$$X_d = \frac{0.8}{1 + (0.2 \cdot 0.1 \cdot 7)} = 0.7$$

$$= 0.55 \, (540) - 0.1 \cdot 0.7 \cdot 3000 \cdot 0.47$$

$$= 198 \text{ mg/l or } 2642 \text{ lbs/day}$$

$$N = 0.123 \frac{0.7}{0.8} \cdot 2642 + 0.07 \frac{0.8 - 0.7}{0.8} \cdot 2642$$

$$= 284 + 23 = 307 \text{ lbs/day}$$

$$N_{INFLUENT} = 5 \cdot 1.6 \cdot 8.34 = 67 \text{ lbs/d}$$

$$N_{ADDED} = 307 - 67 = 240 \text{ lbs/day}$$

$$P = 0.026 \cdot \frac{0.7}{0.8} \cdot 2642 + 0.01 \cdot \frac{0.8 - 0.7}{0.8} \, 2642$$

$$= 60 + 3.3$$
$$= 63.3 \text{ lbs/day}$$
$$P_{\text{INFLUENT}} = 3 \cdot 1.6 \cdot 8.34 = 40 \text{ lbs/day}$$
$$P_{\text{ADDED}} = 63.3 - 40 = 23.3 \text{ lbs/day}$$

## ACCLIMATION OF BIOLOGICAL SLUDGES

When treating industrial wastewaters, particularly for removal of specific organics, it is necessary to acclimate the biomass to the wastewater. The source of the biomass (i.e., municipal or an industrial sludge treating a similar wastewater), the operating temperature and the sludge age will determine the time required for acclimation. Typical acclimation periods are from several days to 5 to 6 weeks or acclimation may not occur at all. Acclimation results obtained by Tabak et al. [11] for several organic compounds when starting with a municipal biomass as seed are shown in Figure 4.16. Acclimation of a municipal sludge to benzidine is shown in Figure 4.17 [12].

If the wastewater is readily degradable and susceptible to filamentous bulking, acclimation of the biomass is best achieved on a batch (fill and

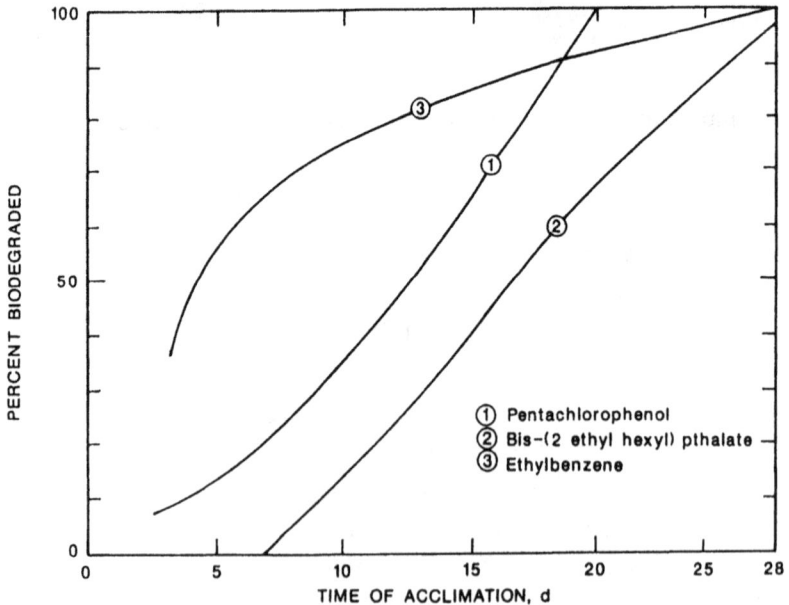

Figure 4.16 Acclimation of activated sludge to specific organics.

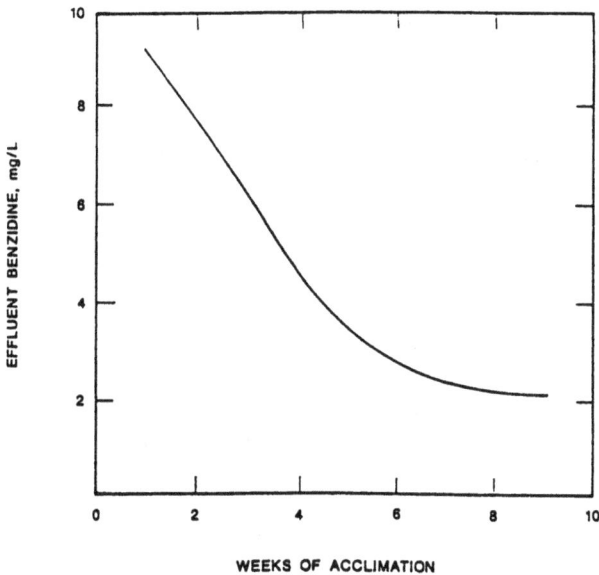

**Figure 4.17** Acclimation of activated sludge to benzidine.

draw) operating basis. After acclimation is established, the operation can be converted to a continuous flow basis using either a plug-flow mixing regime or a biological selector to control bulking. A wastewater containing biorefractory organics can be acclimated in a continuous flow system. If the wastewater (or specific organics contained in the wastewater) are bioinhibitory, acclimation must start at concentrations well below the inhibition threshold concentration. This concentration can be defined using the fed-batch reactor (FBR) or other similar test procedures as subsequently discussed in this chapter. As acclimation proceeds and biodegradation commences, the concentration of the target wastewater can be gradually increased as long as the concentration in the reactor does not exceed the inhibition threshold. Acclimation of the biomass to a specific wastewater composition is assumed to be complete when the specific oxygen uptake rate (SOUR) or the residual organic concentration reaches a steady state condition.

It has been found in some cases that the acclimated biomass possesses a "genetic memory" [13]. This capacity allows the mixed liquor to retain its ability to degrade specific organics even after several SRTs (exceeding theoretical washout) without the presence of the organic in the wastewater stream. This is of great importance for those industries with intermittent production of specific products.

## KINETICS OF ORGANIC REMOVAL

Most industrial wastewaters contain multi-component mixtures of various types of organics (carbohydrates, protein, fatty acids, organic complexes, etc.). These organics can be broadly classified as "readily degradable," (e.g., dairy, food processing, pulp and paper) or "bio-refractory" (e.g., tannery, agricultural chemicals, and certain pharmaceuticals). The removal characteristics of the wastewaters by activated sludge can be depicted as shown in Figure 4.18.

In the case of a readily degradable wastewater, there is a rapid "sorption" of organics by the biological floc immediately on contact of the wastewater with the return sludge. For refractory wastewaters, however, there may be little or no sorption depending on the organic composition. For comparative purposes, the activated sludge removal characteristics of typical domestic sewage is also shown in Figure 4.18. Domestic sewage contains approximately one-third suspended organics, one-third colloidal organics, and one-third soluble organics. On contact with activated sludge, the suspended organics are removed by being enmeshed in the biological floc. The colloidal organics are partially adsorbed and entrapped by the floc and a portion of the soluble organics are "sorbed" by the floc. The

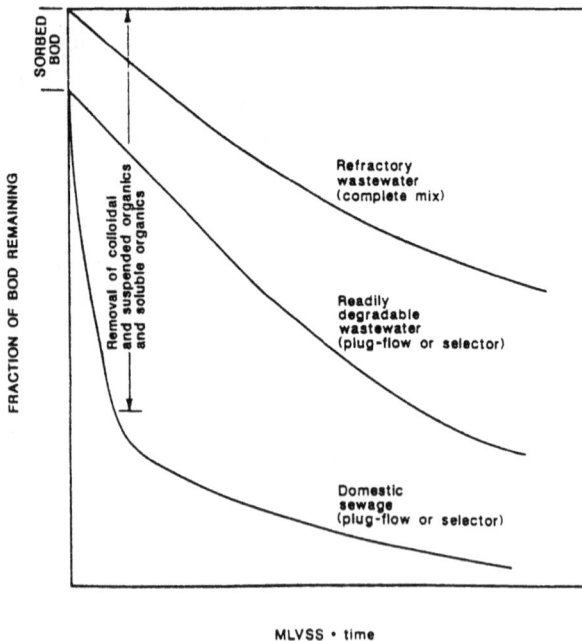

**Figure 4.18** Organic removal characteristics by activated sludge.

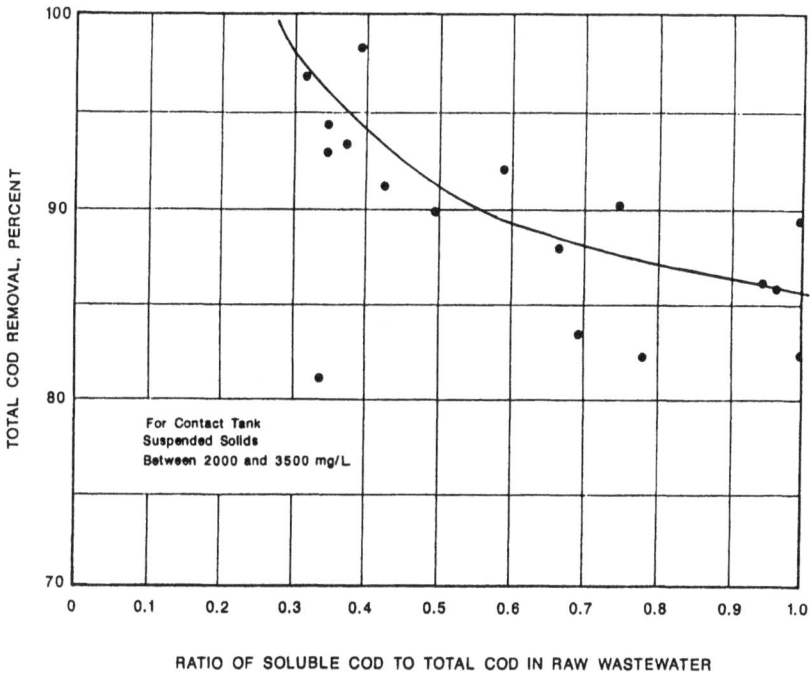

**Figure 4.19** The relationship between SCOD/TCOD and TCOD removal.

result is up to 85 percent removal of the total COD after 10 to 15 minutes of wastewater-sludge contact. This phenomenon was the basis for development of the contact-stabilization process. The effect of the SCOD/TCOD ratio of the influent wastewater on the removal of TCOD in the contact zone of the contact-stabilization process is shown in Figure 4.19.

The phenomenon of rapid biosorption of readily degradable organics occurs on contact of wastewater with activated sludge. The extent of the substrate (expressed as BOD or COD) biosorption can be related to the "floc load." Floc load is defined as the mg COD or BOD applied/g VSS. Sorption vs floc load relationships obtained for several wastewaters are shown in Figure 4.20. The sorption efficiency is related to the active mass of the sludge, which decreases as the *F/M* decreases. A lower *F/M* (or higher SRT) implies a lower degradable fraction of biomass and hence a lower active mass. These relationships for a dairy wastewater are shown in Figure 4.21. The sorption phenomenon provides the basis for selector design such that sufficient organics are sorbed for subsequent utilization and growth by the floc-forming organisms in preference to the growth of filamentous organisms.

It has been shown that a complete mix activated sludge (CMAS) process

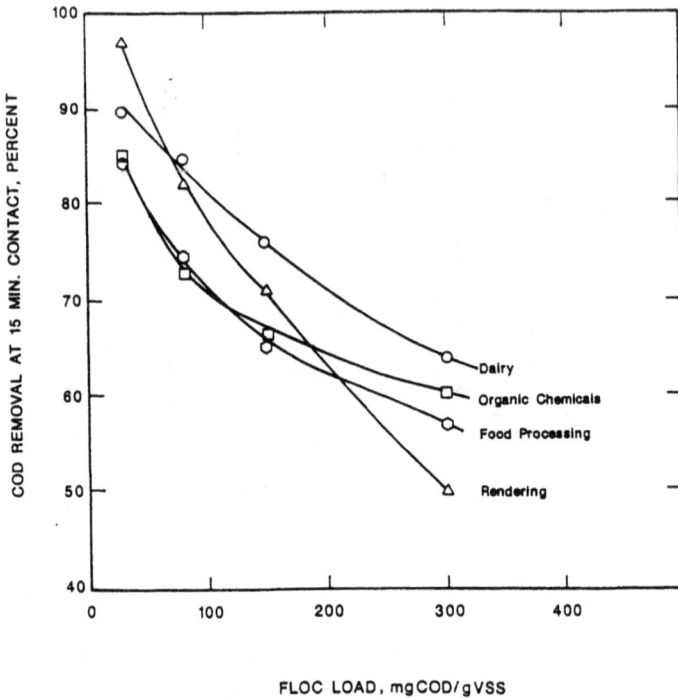

**Figure 4.20** COD removal by biosorption vs floc load for several industrial wastewaters.

treating readily degradable wastewater results in filamentous bulking. In these cases, a plug-flow regime or a biological selector located ahead of the complete mix basin should be used. Wastewaters that contain more slowly degraded substrates and adequate nutrients do not stimulate filamentous growth and can be treated in a complete mix flow configuration. These alternatives are shown schematically in Figure 4.22.

## PLUG-FLOW KINETICS

When developing kinetic relationships it is important to consider the active biomass. In the case of a soluble wastewater the active biomass is related to the SRT and can be estimated by Equation (4.14).

$$X_{va} = X_v \cdot \frac{X_d}{0.8} \tag{4.14}$$

Depending on the nature of the organics present in the wastewater and considering multiple zero order kinetics, the overall organic removal in a batch or plug flow system can be estimated by pseudo first or second order kinetics

$$\frac{S_t}{S_i} = e^{-k_b X_{va} t / S_i} \tag{4.15}$$

$$S_t = \frac{S_i^2}{S_i + K_b X_{va} t} \tag{4.16}$$

in which $S_i$ is the initial concentration after biosorption.

Batch oxidation with and without biosorption is shown in Figure 4.23. A first order relationship for a kraft pulp and paper mill wastewater is shown in Figure 4.24.

## COMPLETE MIX KINETICS

In a complete mix reactor the overall removal rate decreases as the concentration of organics remaining in the reactor decreases. This is because the more readily degradable organics are removed first, resulting in a mixture of substrates and metabolites that are progressively

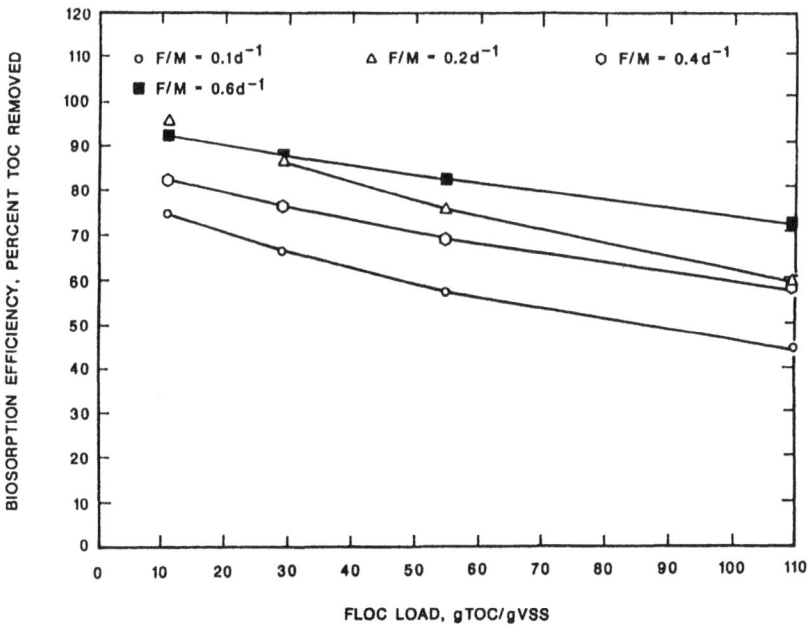

**Figure 4.21** Relationship between biosorption efficiency and floc load at different *F/M* levels for a dairy wastewater.

more difficult to degrade. The relationship defining substrate removal in the complete mix activated sludge process is presented as Equation (4.17).

$$\frac{S_o - S_e}{X_{va}} = K \frac{S_e}{S_o} \qquad (4.17)$$

where
$S_e$ = effluent soluble substrate, mg/L
$S_o$ = total influent BOD, mg/L
$K$ = complete mix reaction rate coefficient, $d^{-1}$

This relationship is shown in Figure 4.25 for a multi-product organic chemicals wastewater. The effect of SRT on the reaction rate coefficient, $K$, is shown in Figure 4.26. The application of this correction for active mass is shown in Example 4.4.

*Example 4.4*

An activated sludge plant with an influent BOD of 800 mg/l and a $K$ of 6/day is operating at a SRT of 10 days. What is the effluent quality

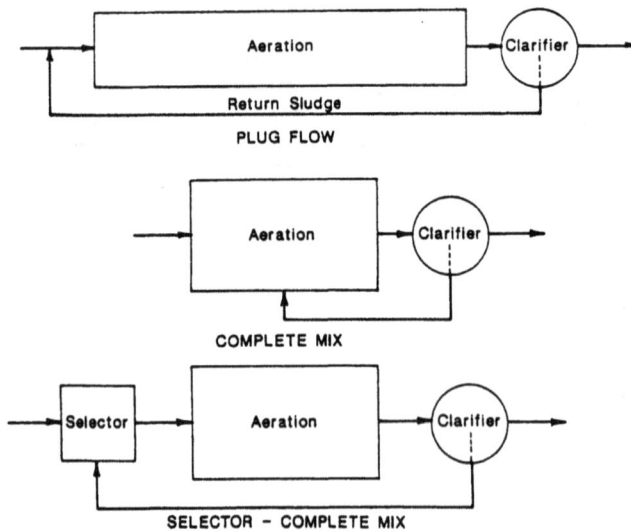

**Figure 4.22** Activated sludge process configurations.

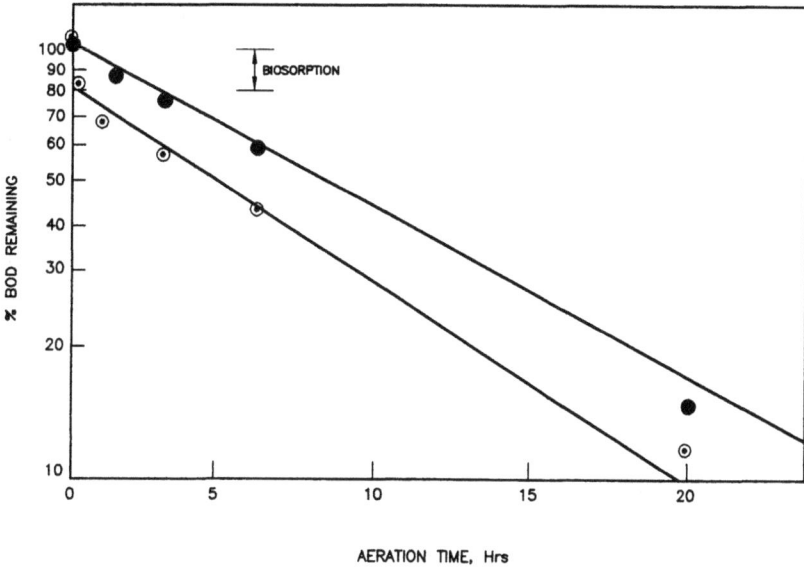

**Figure 4.23** Batch activated sludge with and without biosorption for a pulp and paper mill wastewater.

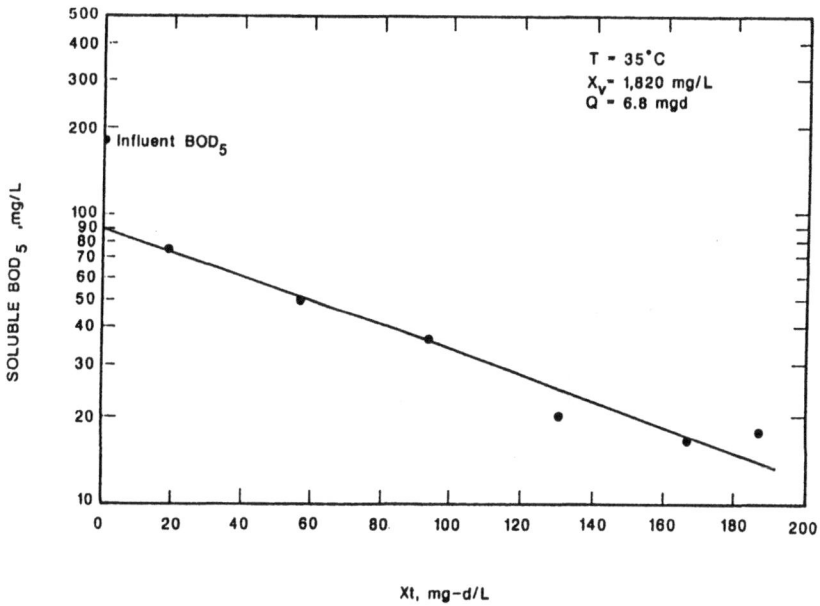

**Figure 4.24** Plug-flow BOD removal kinetics for a bleached kraft pulp and paper wastewater.

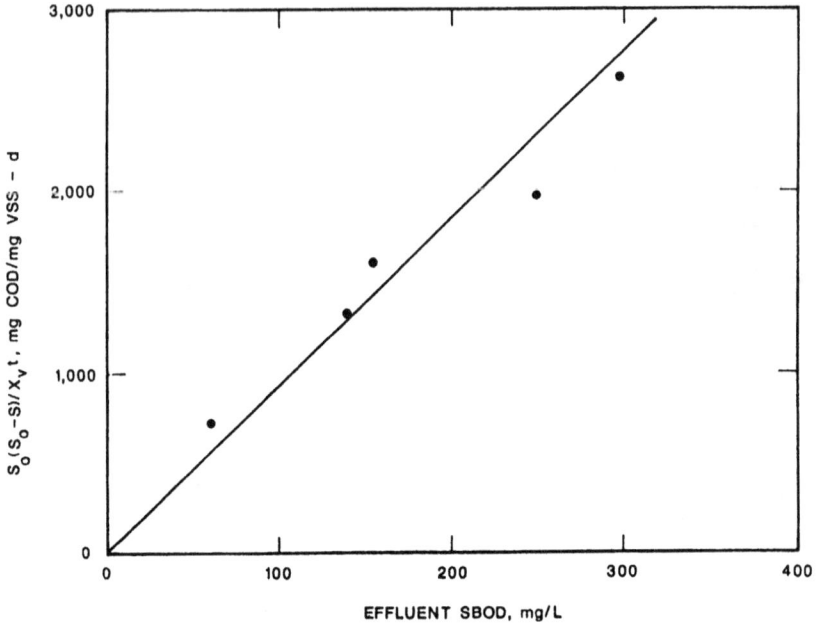

**Figure 4.25** Complete mix BOD removal kinetics for an organic chemicals wastewater.

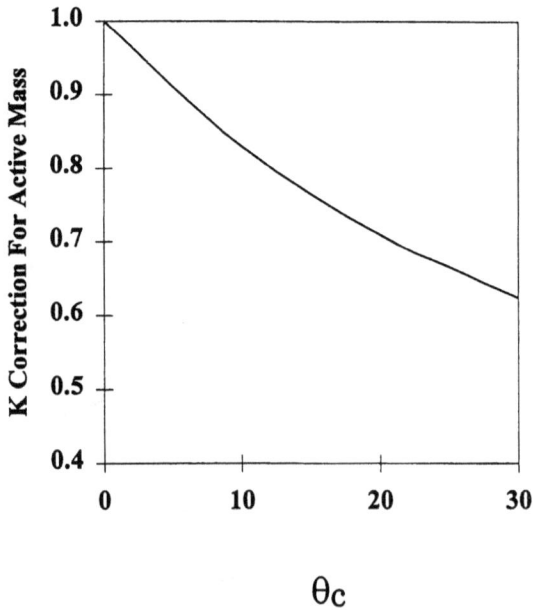

$\theta_c$

**Figure 4.26** Relationship between $K$ and $\theta_c$ considering biological active mass.

204

and what will the effluent quality be if the SRT is increased to 30 days. $a$ is 0.5 and $b$ is 0.1/day.

$$\frac{1}{\theta_c} = aK \frac{S_e}{S_o} - bX_d$$

At $\theta_c = 10$ days

$$\frac{1}{10} = 0.5 \cdot 6 \cdot \frac{S_e}{S_o} - 0.1 \cdot \left(\frac{0.8}{1 + 0.2 \cdot 0.1 \cdot 10}\right)$$

$$\frac{S_e}{S_o} = 0.056$$

and $S_e = 0.056 \cdot 800 = 45$ mg/l

At $\theta_c = 30$ days, the $K$ can be adjusted by Figure 4.26.

$$K = 6 \cdot \frac{0.625}{0.83} = 4.5$$

$$\frac{1}{30} = 0.5 \cdot 4.5 \frac{S_e}{S_o} - 0.1 \cdot \left(\frac{0.8}{1 + 0.2 \cdot 0.1 \cdot 30}\right)$$

$$\frac{S_e}{S_o} = 0.0368 \text{ and}$$

$$S_e = 0.0368 \cdot 800 = 30 \text{ mg/l}$$

If $X_v$ is 2500 mg/l and $t = 0.9$ days at $\theta_c = 10$ days, the $X_v t$ at 30 days SRT is

$$X_v t = \frac{\theta_c a S_r}{1 + \theta_c b X_d}$$

$$= \frac{30 \cdot 0.5 \cdot 770}{1 + 30 \cdot 0.1 \cdot 0.5}$$

$$= 5133 \text{ mg/l}$$

As previously described the presence of influent non-degradable volatile suspended solids will reduce the active biomass under aeration. This effect is shown in Example 4.5.

*Example 4.5*

Given

$S_o = 700$ mg/l

$S_e = 30$ mg/l
$K = 10$/day
$X_v = 3000$ mg/l
$a = 0.4$

Rearranging Equation (4.17) the hydraulic retention time, $t$, to produce $S_e = 30$ mg/L is

$$t = \frac{S_o S_r}{KX_v S_e}$$

$$t = \frac{700 \cdot 670}{10 \cdot 3000 \cdot 30} = 0.52 \text{ days}$$

and

$$X_v t = 1560 \text{ mg}^{-d}/L$$

The SRT for these conditions is

$$\theta_c = \frac{X_v t}{aS_r - bX_d X_v t}$$

Assume $X_d = 0.7$

$$\theta_c = \frac{1560}{0.4 \cdot 670 - 0.1 \cdot 0.7 \cdot 1560}$$

$$= \frac{1560}{159} = 9.8 \text{ days say 10 days}$$

Check assumption for $X_d = 0.7$ of SRT $= 10$ days

$$X_d = \frac{0.8}{1 + 0.2 \cdot 0.1 \cdot 10}$$

$$X_d = 0.67$$

Determine the required hydraulic retention time to produce the same effluent $S_e = 30$ mg/L if the influent non-degradable VSS = 50 mg/L
  Assume $t = 0.69$ days (by trial and error) and calculate the accumulation of the influent VSS in the mixed liquor (MLVSS$_i$)

$$\text{MLVSS}_i = \frac{50 \cdot 10}{0.69} = 725 \text{ mg/l}$$

The residual biomass VSS in the mixed liquor

$$X_{vb} = 3000 - 725 = 2275 \text{ mg/l}$$

Calculate required $t$

$$t = \frac{X_{vb}t}{X_{vb}} = \frac{1560}{2275} = 0.69 \text{ day}$$
$$f_b = 2275/3000 = 0.76$$

Check sludge age

$$\theta_c = \frac{X_v t}{(aS_r - bX_{vb}X_d t) + X_i}$$
$$= \frac{3000 \cdot 0.69}{(268 - 0.1 \cdot 2275 \cdot 0.7 \cdot 0.69) + 50}$$
$$= \frac{2070}{208} = 9.9 \text{ days}$$

The greater hydraulic retention time is required to produce the same effluent quality since the nondegradable influent VSS accumulated in the constant MLVSS concentration of 3000 mg/L. This is shown in Figure 4.27.

The performance of CMAS treating several industrial wastewaters is shown in Table 4.8. The reaction rate coefficient [for Equation (4.17)] for a variety of wastewaters is summarized in Table 4.9 and for several pulp and paper mill wastewaters in Table 4.10. Treatment parameters and kinetic coefficients for activated sludge treatment of petroleum refinery wastewaters are shown in Table 4.11 and for a variety of wastewaters in Table 4.12.

## TWO-STAGE ACTIVATED SLUDGE

In some cases a two-stage activated sludge process has been employed for the treatment of industrial wastewaters. Since the more readily degradable organic components will be removed in the first stage, the rate coefficient $K$ will decrease in the second stage. If there is no intermediate clarifier and zero order kinetics apply the reaction rate $K$ in the second stage can be estimated

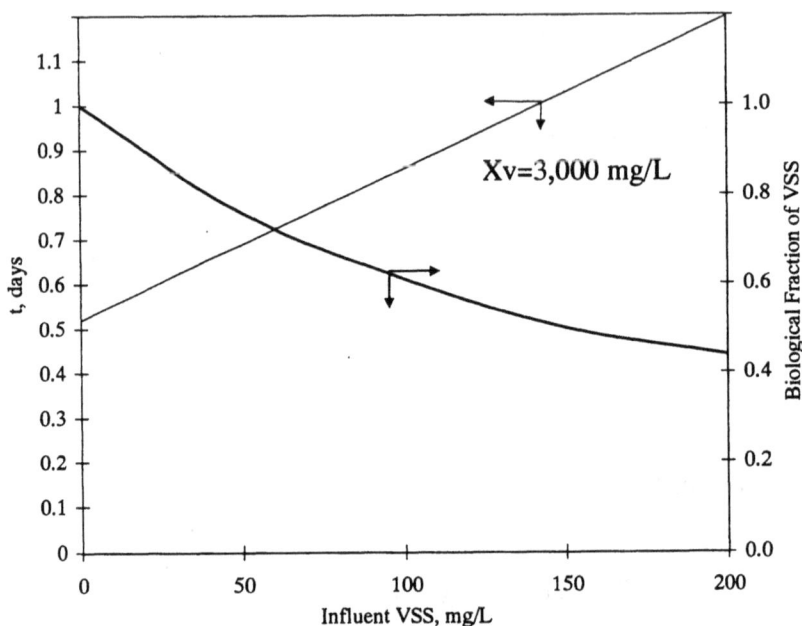

**Figure 4.27** The effect of influent VSS on detention time and biological fraction of biomass for Example 4-5.

$$K_2 = K_1 \left(\frac{S_i}{S_2}\right)^2 \tag{4.18}$$

in which $S_i$ and $S_2$ are the influent concentrations to stage 1 and stage 2 respectively.

If there is an intermediate clarifier the biomass in stage 2 will reflect only the residual organics entering stage 2 and the reaction rate $K$ can be estimated

$$K_2 = K_1 \left(\frac{S_i}{S_2}\right) \tag{4.19}$$

The difference in rate coefficients is shown in Figure 4.28 for a synthetic fiber wastewater and in Figure 4.29 for an organic chemical monomers wastewater.

If the only concern is total organics (i.e., BOD or COD) in the effluent then there is little advantage to a two-stage process. However, if specific refractory constituents (e.g., phenol) must be removed to low effluent

TABLE 4.8. CMAS Treatment Performance for Selected Industrial Wastewaters.

| | Influent | | Effluent | | | F/M | | | | | | |
| Wastewater | BOD (mg/l) | COD (mg/l) | BOD (mg/l) | COD (mg/l) | T (°C) | BOD (d⁻¹) | COD (d⁻¹) | SRT (d) | MLVSS (mg/l) | HRT (d) | SVI (ml/g) | ZSV (ft/hr) |
|---|---|---|---|---|---|---|---|---|---|---|---|---|
| Chemical intermediates plant | 621 | 978 | 18 | 103 | | 0.55 | 0.79 | | 2,228 | 0.51 | 97 | 8.4 |
| Pharmaceutical | 2,950 | 5,840 | 65 | 712 | 10.4 | 0.11 | 0.19 | | 4,970 | 5.4 | | |
| | 3,290 | 5,780 | 23 | 561 | 20.8 | 0.11 | 0.18 | | 5,540 | 5.4 | | |
| Coke and by-products chemical plant | 1,880 | 1,950 | 65 | 263 | | 0.21 | 0.18 | | 2,430 | 4.1 | 42.4 | 26 |
| Diversified chemical industry | 725 | 1,487 | 6 | 257 | 21 | 0.41 | 0.71 | | 2,874 | 0.61 | 11.9 | 4.54 |
| Cellulose | 1,250 | 3,455 | 58 | 1,015 | 19.5 | 0.51 | 1.03 | | 3,280 | 0.75 | | |
| Custom pharmaceutical manufacture | 2,328 | 4,724 | 15 | 507 | 8 | 0.120 | 0.23 | 91 | 2,449 | 7.65 | 21 | 12.7 |
| | 2,170 | 3,993 | 8 | 563 | 21.5 | 0.081 | 0.13 | 126 | 2,768 | 9.7 | 44 | 15.9 |
| Tannery | 1,020 | 2,720 | 31 | 213 | 21 | 0.18 | 0.45 | 16 | 1,900 | 3 | | |
| | 1,160 | 4,360 | 54 | 561 | 21 | 0.15 | 0.49 | 20 | 2,650 | 3 | | |
| Alkylamine manufacturing | 893 | 1,289 | 12 | 47 | | 0.146 | 0.21 | | 1,977 | 3.1 | 133 | 4.2 |
| ABS | 1,070 | 4,560 | 68 | 510 | 33.5 | 0.24 | 0.94 | 6 | 2,930 | 1.5 | 23 | 28.7 |

(continued)

209

TABLE 4.8. (continued).

| Wastewater | Influent BOD (mg/l) | Influent COD (mg/l) | Effluent BOD (mg/l) | Effluent COD (mg/l) | T (°C) | F/M BOD (d⁻¹) | F/M COD (d⁻¹) | SRT (d) | MLVSS (mg/l) | HRT (d) | SVI (ml/g) | ZSV (ft/hr) |
|---|---|---|---|---|---|---|---|---|---|---|---|---|
| Complex organic | 1,630 | 1,660 | 111 | 415 |  |  |  |  |  | 2.58 | 98.5 | 7.1 |
| Viscose rayon | 478 | 904 | 36 | 215 |  | 0.30 | 0.47 |  | 2,759 | 0.57 | 117 | 4.7 |
| Polyester and nylon fibers | 207 | 543 | 10 | 107 | 13.1 | 0.18 | 0.40 |  | 1,689 | 0.664 | 116 | 7.9 |
| Protein processing | 208 | 559 | 4 | 71 | 22.4 | 0.20 | 0.48 |  | 1,433 | 0.712 | 144 | 8.6 |
|  | 3,178 | 5,355 | 10 | 362 | 10 | 0.054 | 0.08 |  | 2,818 | 21 | 180 | 2.9 |
| Tobacco processing | 3,178 | 5,355 | 5.3 | 245 | 26.2 | 0.100 | 0.16 |  | 2,451 | 12.7 | 215 | 2.7 |
|  | 2,420 | 4,270 | 139 | 546 | 29.4 |  |  |  | 3,840 |  | 61 | 1 |
| Emulsified chemical | 1,990 |  | 42 |  |  | 0.20 |  |  | 1,280 | 7.6 |  |  |
| Propylene oxide | 532 | 1,124 | 49 | 289 | 20 | 0.20 | 0.31 |  | 2,969 | 1 | 51 | 12.5 |
|  | 645 | 1,085 | 99 | 346 | 37 | 0.19 | 0.25 |  | 2,491 | 1.4 | 32 | 3.7 |
| Paper mill | 375 | 692 | 8 | 79 | 9.3 | 0.111 | 0.19 |  | 1,414 | 2.38 | 63 | 22 |
|  | 380 | 686 | 7 | 75 | 23.3 | 0.277 | 0.45 | 18.9 | 748 | 1.83 | 504 | 10 |
| Synthetic fuel | 990 |  | 15 |  | 15 | 0.32 |  |  | 3,100 | 1 |  |  |
|  | 790 |  | 18 |  | 23 | 0.21 |  | 5.2 | 2,500 | 1.5 |  |  |
| Combined industrial | 669 | 1,548 | 16 | 293 | 12 | 0.28 | 0.53 |  | 1,819 | 1.32 | 107 | 4.1 |
| Vegetable oil | 669 | 1,548 | 26 | 325 | 23.6 | 0.25 | 0.47 |  | 1,738 | 1.53 | 91 | 4.2 |
|  | 3,474 | 6,302 | 76 | 332 |  | 0.57 | 1.00 |  | 1,740 | 3.5 | 49.2 | 30 |

TABLE 4.8. (continued).

| Wastewater | Influent BOD (mg/l) | Influent COD (mg/l) | Effluent BOD (mg/l) | Effluent COD (mg/l) | T (°C) | F/M BOD (d⁻¹) | F/M COD (d⁻¹) | SRT (d) | MLVSS (mg/l) | HRT (d) | SVI (ml/g) | ZSV (ft/hr) |
|---|---|---|---|---|---|---|---|---|---|---|---|---|
| High strength chemical | 3,110 | 3,910 | 174 | 450 | | | | | | 2.25 | 176 | 2.2 |
| Tetraethyl lead | 41 | 92 | 1.2 | 26 | | | | | | 0.045 | 36 | 8.8 |
| Organic chemicals | 453 | 1,097 | 3 | 178 | 20.3 | 0.10 | 0.21 | | 2,160 | 2.02 | 111 | 6.9 |
| Viscose cellulose | 224 | 367 | 7 | 60 | | 0.35 | 0.50 | | 1,984 | 0.32 | 20 | 26 |
| Synthetic fibers | 269 | 343 | 16.5 | 89 | 11.1 | 0.35 | 0.36 | 5.6 | 2,730 | 0.28 | 43 | 8.7 |
| | 272 | 365 | 11.4 | 47 | 19.2 | 0.26 | 0.32 | 8.1 | 3,340 | 0.31 | 71 | 6.5 |
| Vegetable tannery | 2,396 | 11,663 | 171 | 1,637 | 8 | 0.06 | 0.27 | | 8,454 | 4.5 | 83 | |
| Air products and chemicals | 2,396 | 11,663 | 92 | 1,578 | 22 | 0.10 | 0.43 | | 9,582 | 2.5 | 74 | |
| | 2,690 | 3,935 | 28 | 400 | | 0.29 | 0.38 | | 3,068 | 3.05 | 68 | 11 |
| Hardboard mill | 3,725 | 5,827 | 58 | 643 | 22.2 | 0.30 | 0.42 | | 2,793 | 4.45 | 250 | 2 |
| High strength saline organic chemicals | 3,030 | 8,750 | 830 | 6,125 | 6 | 0.75 | 0.95 | | 233 | 17.4 | | |
| | 3,171 | 8,597 | 82 | 3,311 | 24 | 0.24 | 0.41 | | 328 | 40 | | |
| Coke | 1,618 | 2,291 | 52 | 434 | | 0.23 | 0.27 | 15.5 | 3,060 | 2.3 | 152 | 3.1 |
| Chemical | 634 | | 52 | | 8.5 | 0.12 | | | 2,813 | 1.85 | 52 | 10.6 |
| | 423 | | 21 | | 15.2 | 0.22 | | | 2,780 | 0.7 | 50 | 13.5 |
| | 888 | | 16 | | 23.9 | 0.18 | | | 2,854 | 1.78 | 47 | 11.9 |

(continued)

211

TABLE 4.8. (continued).

| Wastewater | Influent BOD (mg/l) | Influent COD (mg/l) | Effluent BOD (mg/l) | Effluent COD (mg/l) | T (°C) | F/M BOD (d$^{-1}$) | F/M COD (d$^{-1}$) | SRT (d) | MLVSS (mg/l) | HRT (d) | SVI (ml/g) | ZSV (ft/hr) |
|---|---|---|---|---|---|---|---|---|---|---|---|---|
| H coal liquification | 2,030 | 3,090 | 34 | 405 | 13.4 | | | | 2,610 | | 70 | 9.3 |
| | 2,070 | 3,160 | 12 | 378 | 23.8 | | | | 2,770 | | 71 | 11.5 |
| Textile dye | 393 | 951 | 20 | 261 | 10.5 | 0.12 | 0.23 | 18.8 | 2,620 | 1.21 | 185 | 6 |
| | 393 | 951 | 14 | 173 | 22 | 0.22 | 0.44 | 13.4 | 2,440 | 0.75 | 350 | 2.2 |
| Kraft and semi-chemical pulp and paper | 308 | 1,153 | 7 | 575 | | 0.28 | 0.36 | | 2,800 | 0.58 | 83 | 9.7 |

TABLE 4.9. Reaction Rate Coefficient for Selected Wastewaters.

| Wastewater Source | $K$ (d$^{-1}$) | Temperature (°C) |
|---|---|---|
| Vegetable tannery | 1.2 | 20 |
| Cellulose acetate | 2.6 | 20 |
| Peptone | 4.0 | 22 |
| Organic phosphates | 5.0 | 21 |
| Vinyl acetate monomer | 5.3 | 20 |
| Organic intermediates | 5.8 | 8 |
| | 20.6 | 26 |
| Viscose rayon and nylon | 6.7 | 11 |
| | 8.2 | 19 |
| Domestic sewage (solubles) | 8.0 | 20 |
| Polyester fiber | 14.0 | 21 |
| Formaldehyde, propanol, methanol | 19.0 | 20 |
| High-nitrogen organics | 22.2 | 22 |
| Potato processing | 36.0 | 20 |

concentrations, a second stage may be considerable advantage. In the second stage, the slower growing "substrate specific" biomass can be concentrated by operating at a higher SRT than would be possible in a single- or first-stage reactor due to MLSS buildup. Furthermore, if the specific substrate removal rate is first-order, the effluent concentration will be lower in a two- or multiple-stage reactor than in a single reactor system.

## REMOVAL OF SPECIFIC ORGANICS IN THE ACTIVATED SLUDGE PROCESS

The removal rate of "total organics" (i.e., BOD or COD) by the activated sludge process can be expressed by Equations (4.15), (4.16) and (4.17). However, the removal of specific organics (i.e., phenol) is best expressed using the Monod relationship

TABLE 4.10. Reaction Rate Coefficient for Pulp and Paper Mills.

| Type of Mill | $K$ (d$^{-1}$) | Temperature (°C) |
|---|---|---|
| Oxygen bleached kraft | 13.5 | 35 |
| Virgin pulp and wastepaper | 13.6 | 23 |
| Unbleached kraft | 4.5 | 38 |
| Sulfite | 5.0 | 18 |
| Bleached sulfite | 6.2 | — |
| Bleached kraft | 5.2 | — |
| Bleached kraft | 4.4 | 34 |

TABLE 4.11. Biological Treatment Coefficients for Petroleum Refinery Wastewaters.

| Influent | | Organic Removal Rate, $K^a$ | | Sludge Growth Coefficients | | | | Oxygen Requirements Coefficients[b] | | Residual COD |
| | | | | BOD Basis | | COD Basis | | | | |
| BOD (mg/l) | COD (mg/l) | BOD (d⁻¹) | COD (d⁻¹) | $a$ | $bX_d$ | $a$ | $bX_d$ | $a'$ | $b'X_d$ | (mg/l) |
|---|---|---|---|---|---|---|---|---|---|---|
| 244 | 509 | 4.15 | 2.74 | — | — | — | — | 0.57 | 0.1 | 106 |
| 575 | 981 | — | 7.97 | — | — | 0.5 | 0.06 | 0.60 | 0.11 | 53 |
| 396 | 782 | — | 5.86 | — | — | 0.5 | 0.06 | 0.34 | 0.06 | 100 |
| 153 | 428 | — | 2.92 | 0.5 | 0.08 | 0.44 | 0.1 | 0.35 | 0.08 | 22 |
| 170 | 600 | — | 5.0 | — | — | 0.26 | 0.03 | 0.46 | 0.05 | 100[c] |
| 248 | 563 | 4.11 | 7.79 | — | — | 0.2 | 0.08 | 0.40 | 0.01 | 76 |
| 345 | 806 | — | 7.24 | — | — | 0.43 | 0.10 | 0.52 | 0.14 | 82 |
| 196 | 310 | 4.70 | — | 0.6 | 0.05 | — | — | 0.46 | 0.14 | 50 |
| 138 | 275 | — | — | 0.58 | — | 0.25 | — | 0.60 | 0.09 | 42 |

[a] At 24°C.
[b] COD basis.
[c] TOD.

214

TABLE 4.12. Activated Sludge Kinetic Coefficients for Industrial Wastewaters.

| Wastewater | $T$ (°C) | BOD Basis | | | | COD Basis | | | | | |
| | | $K$ (d$^{-1}$) | $a$ | $a'$ | $K$ (d$^{-1}$) | $a$ | $a'$ | $bX_d$ (d$^{-1}$) | $b'X_d$ (d$^{-1}$) |
|---|---|---|---|---|---|---|---|---|---|
| Chemical intermediates plant | | | | | 8.9 | 0.2 | | 0.06 | |
| Pharmaceutical | 10.4 | | | | 2.43 | | | | |
| | 20.8 | | | | 4.2 | | | | |
| Coke and by-products chemical plant | | | | | 6.44 | | 0.6 | | 0.02 |
| Diversified chemical industry | 21 | | 0.46 | 0.5 | 8.6 | 0.27 | 0.22 | 0.03 | 0.07 |
| Cellulose | 19.5 | 21.3 | 0.56 | 0.83 | | | | 0.08 | 0.2 |
| Custom pharmaceutical manufacture | 8 | 15.1 | 0.225 | | | | | 0.08 | |
| Tannery | 21.5 | 21.7 | | | 8.98 | | | | |
| | 21 | 5.61 | | | 4.36 | | | | |
| | | 3 | | | | | | | |
| Alkylamine manufacturing | | 4.2 | 0.28 | 0.58 | | | | 0.025 | 0.012 |
| ABS | 33.5 | 4.07 | 0.86 | 1.28 | | | | 0.01 | 0.03 |
| Complex organic waste | | 6.8 | 0.318 | 0.333 | | | | 0.125 | 0.153 |

*(continued)*

TABLE 4.12. (continued).

| Wastewater | T (°C) | BOD Basis | | | COD Basis | | | | |
|---|---|---|---|---|---|---|---|---|---|
| | | $K$ (d⁻¹) | $a$ | $a'$ | $K$ (d⁻¹) | $a$ | $a'$ | $bX_d$ (d⁻¹) | $b'X_d$ (d⁻¹) |
| Viscose rayon | | 3.6 | | 1.1 | | | | | 0.1 |
| Polyester and nylon fibers | 13.1 | 0.99 | 0.77 | 0.73 | 0.67 | 0.46 | 0.77 | 0.1 | 0.12 |
| | 22.4 | 4 | | | 1.31 | | | | 0.18 |
| Protein processing | 10 | 14.6 | 0.37 | 1.25 | | 0.24 | 0.6 | 0.013 | 0.1 |
| | 26.2 | 60.4 | | | | | | | |
| Tobacco processing | 15 | 10.6 | 0.743 | 0.446 | | | | 0.07 | 0.05 |
| | 21.5 | 14 | | | | | | | |
| Emulsified chemical | | | 0.65 | 0.54 | | | | 0.02 | 0.15 |
| Propylene oxide | 20–23 | 0.77 | | 1.3 | | | 3 | | 0.05 |
| | 34–37 | 3.71 | | | | | | | |
| Paper mill | 9 | 5.2 | 0.78 | 0.68 | 3.3 | 0.52 | 0.42 | 0.02 | 0.12 |
| | 23 | 13.6 | | | 10 | | | 0.04 | |
| Synthetic fuel | 15 | 3.8 | 0.41 | | | | | | |
| | 23 | 4.7 | | | | | | | |
| Combined industrial | 23.6 | 18.1 | 0.23 | 1.09 | 15.5 | 0.28 | 0.5 | 0.01 | 0.14 |
| Vegetable oil | | 25.7 | 0.5 | 0.52 | 20.8 | 0.32 | 0.3 | 0.11 | 0.34 |
| High strength chemical | | 5.4 | 0.35 | 0.88 | 5.1 | | 0.7 | 0.07 | 0.15 |
| Tetraethyl lead | | | | | 1.6 | 0.36 | 0.6 | 0.04 | 0.03 |

TABLE 4.12. (continued).

| Wastewater | T (°C) | BOD Basis K (d⁻¹) | a | a' | COD Basis K (d⁻¹) | a | a' | bX_d (d⁻¹) | b'X_d (d⁻¹) |
|---|---|---|---|---|---|---|---|---|---|
| Organic chemicals | 6.5 | 5.4 | 0.25 | 0.56 | 4.1 | 0.12 | 0.21 | 0.005 | 0.06 |
|  | 19 | 7.2 |  |  |  |  |  |  |  |
|  | 26.5 | 21.3 |  |  |  |  |  |  |  |
| Viscose cellulose |  | 12.3 | 0.6 | 0.4 | 28.9 |  |  |  | 0.09 |
| Synthetic fibers | 11.1 | 6.7 | 0.41 | 0.8 |  |  |  | 0.02 | 0.09 |
|  | 19.2 | 8.2 |  |  |  |  |  |  |  |
| Vegetable tannery | 22 | 1.19 | 1.92 | 2.27 |  | 0.5 | 0.54 | 0.02 | 0.015 |
| Air products and chemicals |  | 10.4 | 0.32 | 0.35 | 12.9 |  |  | 0.01 | 0.15 |
| Hardboard mill | 22 | 4 | 0.5 | 0.87 |  |  |  | 0.03 | 0.1 |
| High strength saline organic chemicals | 6 | 2 | 0.12 | 3.85 | 0.98 | 0.09 | 3.57 | 0.01 | 0.2 |
|  | 24 | 12 |  | 1.36 | 3.4 |  | 1 |  |  |
| Coke |  | 6.31 | 0.08 | 0.59 |  |  |  | 0.008 | 0.003 |
| Chemical | 8.5 | 0.9 | 0.38 | 1.44 |  |  |  |  |  |
|  | 15.2 | 3.4 |  |  |  |  |  |  |  |
|  | 23.9 | 7.7 |  |  |  |  |  |  |  |
| H-coal liquification | 13.4 |  | 0.48 | 0.52 | 3.1 | 0.31 | 0.37 | 0.03 | 0.04 |
|  | 23.8 |  |  |  |  |  |  |  |  |

(continued)

TABLE 4.12. (continued).

| Wastewater | T (°C) | BOD Basis | | | COD Basis | | | | |
|---|---|---|---|---|---|---|---|---|---|
| | | $K$ (d$^{-1}$) | $a$ | $a'$ | $K$ (d$^{-1}$) | $a$ | $a'$ | $bX_d$ (d$^{-1}$) | $b'X_d$ (d$^{-1}$) |
| Textile dye | 10.5 | 2.32 | 0.62 | 1.05 | | 0.32 | 0.51 | 0.03 | 0.042 |
| | 22 | 6 | | | | | | | |
| Kraft and semi-chemical pulp and paper | | 16.5 | | | 7.62 | | | | |

218

$$\mu = \frac{\mu_m S}{K_s + S} \qquad (4.20)$$

and

$$q = \frac{q_m S}{K_s + S} \qquad (4.21)$$

where

$\mu_m$ = maximum specific growth rate, $d^{-1}$
$S$ = reactor substrate concentration, mg/l
$K_s$ = substrate concentration when rate is one-half the maxi-
　　mum rate, mg/l
$q_m$ = maximum specific substrate removal rate, $d^{-1}$
$\mu$ = specific growth rate, $d^{-1}$
$q$ = substrate removal rate, mg/l-d

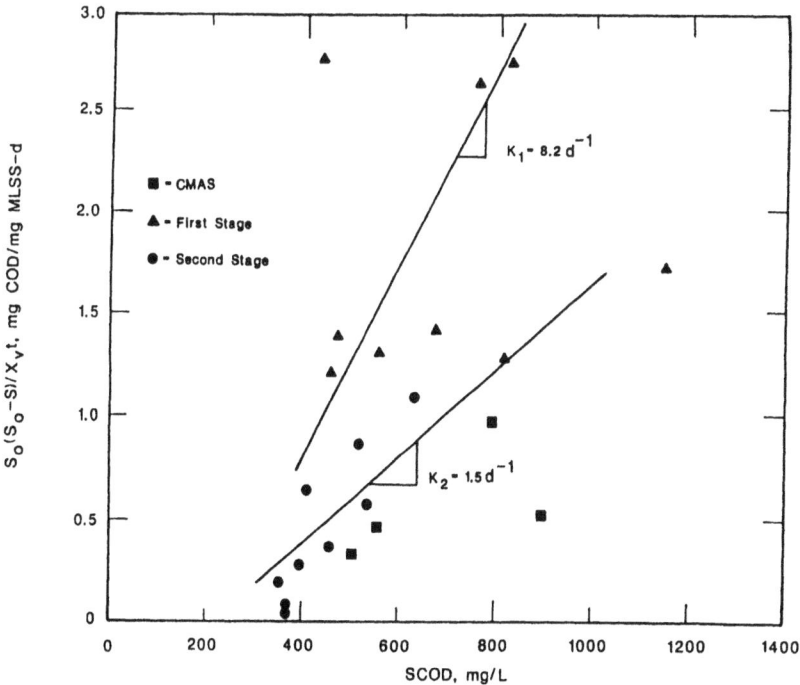

**Figure 4.28** Effect of two-stage operation on reaction rate coefficient for a synthetic fibers wastewater.

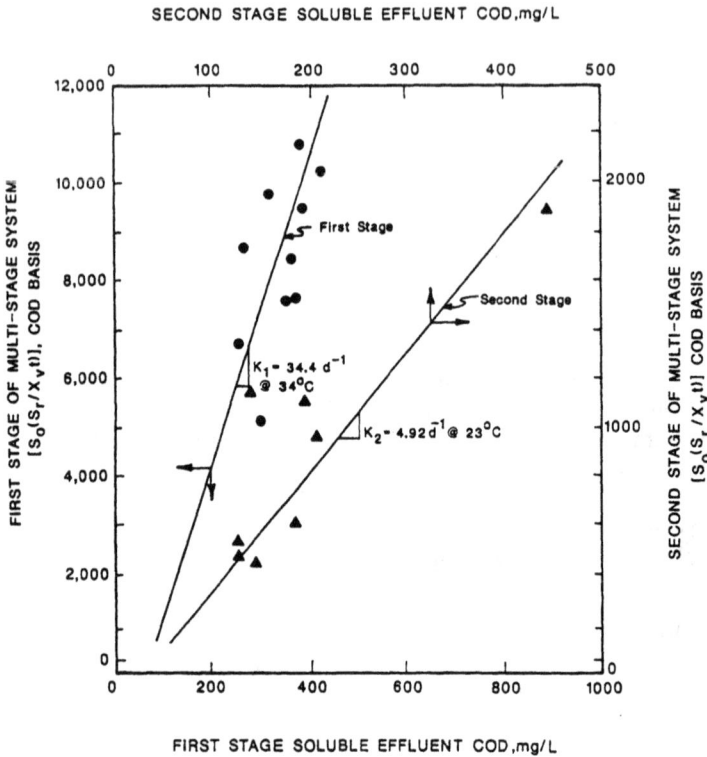

**Figure 4.29** Effect of two-stage operation on reaction rate coefficient for an organic chemicals wastewater.

The specific growth rate, $\mu$, is related to $\theta_c$ by the following relationship:

$$\mu = \frac{1}{\theta_c} + bX_d \qquad (4.22)$$

In a complete mix reactor

$$S = \frac{K_r(1 + bX_d\theta_c)}{[\theta_c(aq_m - bX_d)] - 1} \qquad (4.23)$$

This relationship for dichlorophenol (DCP) is shown in Figure 4.30 [8]. The relative biodegradability of specific organics are shown in Figure 4.31.

For the case of a plug-flow reactor, the performance equation derived from the Monod relationship is:

$$\frac{1}{\theta_c} = \frac{q_m(S_o - S)}{(S_o - S) + CK_s} - bX_d \qquad (4.24)$$

where

$$C = (1 + \alpha) \ln [(\alpha S + S_o)/(1 + \alpha S)]$$

$$\alpha = R/Q$$

Effluent concentrations (S) of specific priority pollutants can be computed from Equations (4.23) and (4.24). This is shown in Example 4.6. Since most effluent permit limits for these compounds are in the mg/l range, it is improbable that bioinhibition will be a factor. The reported maximum biodegradation rates for a variety of organics are shown in Table 4.13 [10,14]. The "$EC_{50}$" values in Table 4.13 [10,14] represent the substrate concentration at which the SOUR is 50 percent of the uninhibited uptake rate.

*Example 4.6*

Compute the SRT required in a complete mix activated sludge plant to reduce phenol from 10 mg/l, to 15 $\mu$g/l, where:

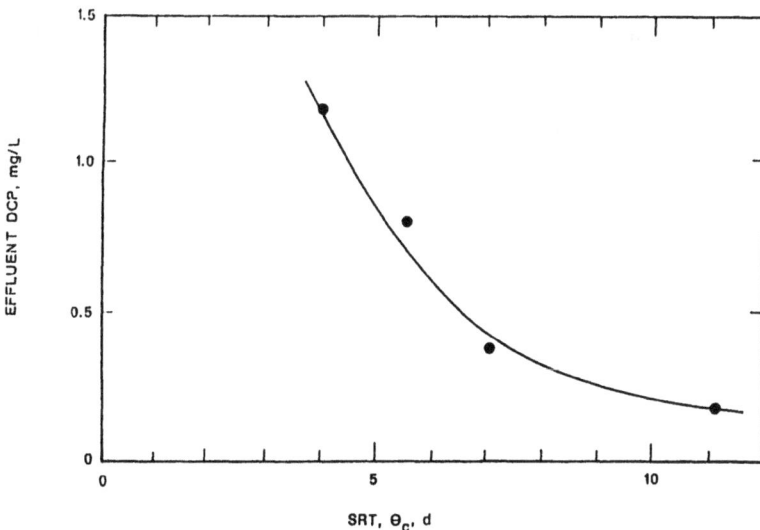

**Figure 4.30** Effect of SRT on DCP removal.

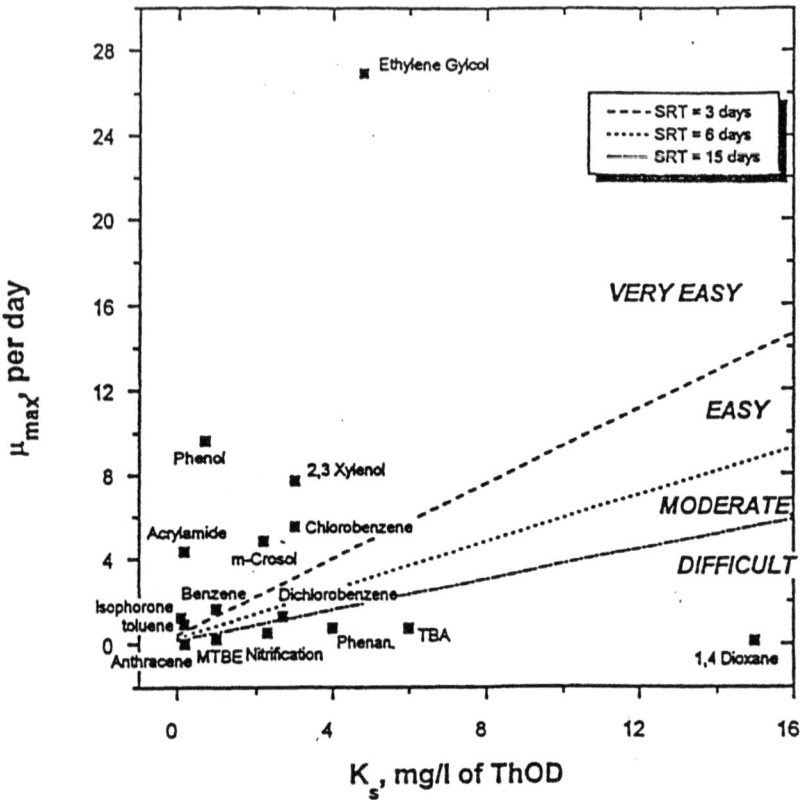

**Figure 4.31** Relative biodegradability of specific organics at different SRT at 25°C (effluent concentration = 0.5 mg/l as COD and $b$ = 0.11 per day) [38].

$q_m = 1.8$ g/g VSS · day at 20°C
$\theta = 1.1$
$K_s = 100$ µg/l
$a = 0.6$
$bX_d = 0.05$ day$^{-1}$ at 20°C
$\quad = 0.033$ at 10°C

Equation (4.23) can be rearranged to yield

$$\theta_c = \frac{K_s + S}{aq_mS - bX_d(K_s + S)}$$

$$= \frac{0.1 + 0.015}{0.6 \cdot 1.8 \cdot 0.015 - 0.05(0.115)}$$

$$= 11.0 \text{ days}$$

TABLE 4.13. Biodegradability and Bio-Toxicity Data [10,14].

| Compound | Biodegradability | Rate of Biodegradation (mg COD/ gVSS-hr) | EC$_{50}$, mg/l | |
|---|---|---|---|---|
| | | | Nitrosomonas | Heterotrophs |
| Cyclohexane | A | — | 97 | 29 |
| Octane | A | — | 45 | — |
| Decane | C | — | — | — |
| Dodecane | D | — | — | — |
| Methylene chloride | D | — | 1.2 | 320 |
| Chloroform | D | — | 0.48 | 640 |
| Carbon tetrachloride | — | — | 51 | 130 |
| 1,1-dichloroethane | — | — | 0.91 | 620 |
| 1,2-dichloroethane | — | — | 29 | 470 |
| 1,1,1-trichloroethane | — | — | 8.5 | 450 |
| 1,1,2-trichloroethane | — | — | 1.9 | 240 |
| 1,1,1,2-tetrachloroethane | — | — | 8.7 | 230 |
| 1,1,2,2-tetrachloroethane | — | — | 1.4 | 130 |
| Pentachloroethane | — | — | 7.9 | 150 |
| Hexachloroethane | — | — | 32 | — |
| 1-chloropropane | D | — | 120 | 700 |
| 2-chloropropane | — | — | 110 | 440 |
| 1,2-dichloropropane | — | — | 43 | — |
| 1,3-dichloropropane | C | — | 4.8 | 210 |
| 1,2,3-trichloropropane | — | — | 30 | 290 |
| 1-chlorobutane | D | — | 120 | 230 |
| 1-chloropentane | D | — | 99 | 68 |
| 1,5-dichloropentane | — | — | 13 | — |
| 1-chlorohexane | D | — | 85 | 83 |
| 1-chlorooctane | — | — | 420 | 52 |
| 1-chlorodecane | D | — | — | 40 |
| 1,2-dichloroethylene | D | — | — | — |
| trans-1,2-Dichloroethylene | — | — | 80 | 1,700 |
| Trichloroethylene | A | — | .81 | 130 |
| Tetrachloroethylene | — | — | 110 | 1,900 |
| 1,3-dichloropropene | — | — | .67 | 120 |
| 5-chloro-1-pentyne | — | — | .59 | 86 |
| Methanol | A | 26 | 880 | 20,000 |
| Ethanol | A | 32 | 3,900 | 24,000 |
| 1-propanol | A | 71 | 980 | 9,600 |
| 1-butanol | A | 84 | — | 3,900 |
| 1-pentanol | A | — | 520 | — |
| 1-hexanol | A | — | — | — |
| 1-octanol | A | — | 67 | 200 |
| 1-decanol | B | — | — | — |
| 1-dodecanol | B | — | 140 | 210 |
| 2,2,2-trichloroethanol | — | — | 2.0 | — |
| 3-chloro-1,2-propanediol | D | — | — | — |

*(continued)*

TABLE 4.13. (continued).

| Compound | Biodegradability | Rate of Biodegradation (mg COD/gVSS-hr) | EC$_{50}$, mg/l | |
|---|---|---|---|---|
| | | | *Nitrosomonas* | Heterotrophs |
| Ethylether | C | — | — | 17,000 |
| Isopropylether | D | — | 610 | — |
| Acetone | B | — | 1,200 | 16,000 |
| 2-butanone | — | — | 790 | 11,000 |
| 4-methyl-2-pentanone | — | — | 1,100 | — |
| Ethyl-acrylate | — | — | 47 | — |
| Butyl-acrylate | — | — | 38 | 470 |
| 2-chloropropionic-acid | A | 24 | 0.04 | 0.18 |
| Trichloroacetic-acid | D | 0 | — | — |
| Diethanolamine | A | 16 | — | — |
| Acetronitrile | A | — | 73 | 7,500 |
| Acrylonitrile | A | — | 6.0 | 52 |
| Benzene | A | — | 13 | 520 |
| Toluene | A | — | 84 | 110 |
| Xylene | A | — | 100 | 1,000 |
| Ethylbenzene | B | — | 96 | 130 |
| Chlorobenzene | D | — | .71 | 310 |
| 1,2-dichlorobenzene | — | — | 47 | 910 |
| 1,3-dichlorobenzene | D | — | 93 | 720 |
| 1,4-dichlorobenzene | D | — | 86 | 330 |
| 1,2,3-trichlorobenzene | — | — | 96 | — |
| 1,2,4-trichlorobenzene | D | — | 210 | 7,700 |
| 1,3,5-trichlorobenzene | — | — | 96 | — |
| 1,2,3,4-tetrachlorobenzene | — | — | 20 | — |
| 1,2,4,5-tetrachlorobenzene | D | — | 9 | — |
| Hexachlorobenzene | D | — | 4 | 350 |
| Benzyl alcohol | A | — | 390 | 2,100 |
| 4-chloroanisole | — | — | — | 902 |
| 2-furaldehyde | B | 37 | — | — |
| Benzonitrile | B | — | 32 | 470 |
| *m*-Tolunitrile | — | — | .88 | 290 |
| Nitrobenzene | A | 14 | .92 | 370 |
| 2,6-dinitrotoluene | — | — | 183 | — |
| 1-nitronaphthalene | — | — | — | 380 |
| Naphthalene | A | — | 29 | 670 |
| Phenanthrene | C | — | — | — |
| Benzidine | D | — | — | — |
| Pyridine | A | — | — | — |
| Quinoline | A | 8.5 | — | — |
| Phenol | A | 80 | 21 | 1,100 |
| *m*-Cresol | A | — | .78 | 440 |
| *p*-Cresol | A | — | 27 | 260 |

TABLE 4.13. (continued).

| Compound | Biodegradability | Rate of Biodegradation (mg COD/ gVSS-hr) | EC$_{50}$, mg/l Nitrosomonas | EC$_{50}$, mg/l Heterotrophs |
|---|---|---|---|---|
| 2,4-dimethylphenol | — | 28.2 | — | — |
| 3-ethylphenol | — | — | — | 144 |
| 4-ethylphenol | — | — | 14 | — |
| 2-chlorophenol | — | — | 2.7 | 360 |
| 3-chlorophenol | — | — | .20 | 160 |
| 4-chlorophenol | A | 39.8 | .73 | 98 |
| 2,3-dichlorophenol | — | — | .42 | 210 |
| 2,4-dichlorophenol | — | 10.5 | .79 | — |
| 2,5-dichlorophenol | — | — | .61 | 180 |
| 2,6-dichlorophenol | — | — | 8.1 | 410 |
| 3,5-dichlorophenol | — | — | 3.0 | — |
| 2,3,4-trichlorophenol | — | — | 52 | 7.8 |
| 2,3,5-trichlorophenol | — | — | 3.9 | — |
| 2,3,6-trichlorophenol | — | — | 42 | 14 |
| 2,4,5-trichlorophenol | — | — | 3.9 | 23 |
| 2,4,6-trichlorophenol | — | — | 7.9 | — |
| 2,3,5,6-tetrachlorophenol | — | — | 1.3 | 1.5 |
| Pentachlorophenol | — | — | 6.0 | — |
| 2-bromophenol | — | — | .35 | — |
| 4-bromophenol | B | — | .83 | 120 |
| 2,4,6-tribromophenol | — | — | 7.7 | — |
| Pentabromophenol | — | — | .27 | — |
| Resorcinol | A | 57.5 | 7.8 | — |
| Hydroquinone | B | 54.2 | — | — |
| 2-aminophenol | — | 21.1 | .27 | .04 |
| 4-aminophenol | — | 16.7 | .07 | — |
| 2-nitrophenol | — | 14.0 | 11 | 11 |
| 3-nitrophenol | — | 17.5 | — | — |
| 4-nitrophenol | A | 16.0 | 2.6 | 160 |
| 2,4-dinitrophenol | — | 6.0 | — | — |

All Nitrosomonas and aerobic heterotroph data corrected for pKa (ionization) and H (gas/liquid partitioning).

$A = \dfrac{BOD}{TOD} > 50\%$, readily biodegradable

$B = \dfrac{BOD}{TOD}$ 25–25%; moderately biodegradable

$C = \dfrac{BOD}{TOD} < 10$–25%; refractory

$D = \dfrac{BOD}{TOD} < 10\%$; non-degradable

What increase in SRT is required to produce the same effluent phenol concentration if the temperature is reduced to 10°C

$$q_m 10° = q_m 20° \cdot 1.1^{-10}$$
$$= 1.8/2.6$$
$$= 0.69$$
$$\theta_c = \frac{0.1 + 0.015}{0.6 \cdot 0.69 \cdot 0.015 - 0.033(0.115)}$$
$$= 47.6 \text{ days}$$

To use Equations (4.23) and (4.24), it is necessary to determine $q_m$, $K_s$, $a$, $b$, and $X_d$ for the wastewater in question. Terms $a$, $b$, and $X_d$ are readily determined by conventional techniques. Evaluation of terms $K_s$ and $q_m$, however, presents a greater challenge. Templeton and Grady [15] have recently shown that the history of the biomass will dictate the value of the coefficients in Equation (4.23). When bacterial cells are grown at a constant specific growth rate in continuous culture, physiological "equilibrium" is established. As a result, the bacterial cells achieve fixed levels of ribonucleic acid (RNA), protein, and other macromolecules unique to their current growth conditions. The values of $q_m$ and $K_s$ will reflect these operating conditions. It is, therefore, important to determine the values of the coefficients with a sludge at the same historical SRT as the wastewater treatment plant and containing the same wastewater constituents. Philbrook and Grady [16] have shown that because of microbial competition each CMAS will have associated with it a unique microbial community. Reactors that have been operated at high specific growth rates will contain communities that are characterized by high $q_m$ and low $K_s$ values. Another explanation for the differences in $q_m$ may be in the change in active mass with SRT. As the SRT is increased, nondegradable mass accumulates, resulting in a progressive decrease in active mass and presumably a decrease in qm as related to the total volatile suspended solids. In order to maintain the integrity of the results, a short-term test such as the fed-batch reactor should be used in order not to change the population dynamics.

The modified FBR test as described by Philbrook and Grady [16] is applicable to the determination of the kinetic coefficients $q_m$ and $K_s$ under field operating conditions. In the test, plant or pilot plant sludge at the desired SRT is placed in a two-liter reactor. Plant wastewater containing the desired priority pollutant is added at a constant rate. In order to determine $q_m$, the addition rate must exceed the degradation rate. Since in many wastewaters the priority pollutant levels are low, the wastewater may have to be spiked to insure a sufficient concentration to meet the conditions of the test. It is important, however, that the concentration levels achieved in the test are below the inhibition threshold. This can be found by the shape of the concentration-time curve. The degradation rate,

$q_m$, is computed as the difference in the slopes of the substrate addition rate and the residual substrate accumulation.

A second FBR test is then conducted with the addition rate of the priority pollutant equal to one-half the maximum rate determined in the first test. The steady state concentration observed in the reactor will be $K_s$. FBR test data for phenol are shown in Figure 4.32. Hoover [17] found a high variability in $q_m$ with sludges operating under the same loading conditions with time. Based on these observations, a routine test program should be established at a treatment plant and values for $q_m$ and $K_s$ interpreted on a statistical basis.

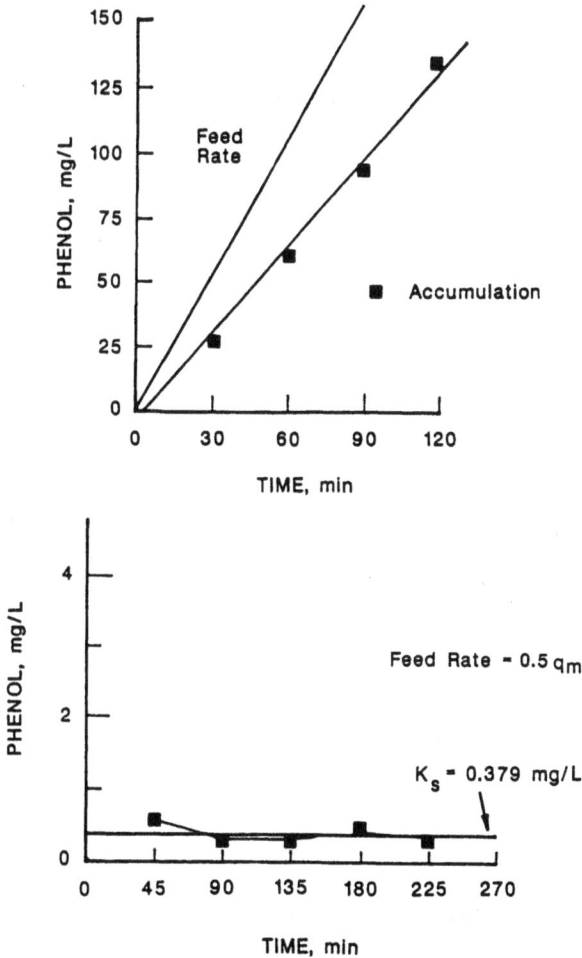

**Figure 4.32** FBR relationships to determine $q_m$ and $K_s$.

In some cases an organic compound that degrades very slowly exhibits effluent toxicity. In this case the SRT must be regulated to reduce the organic concentration to non-toxic levels. The effects of SRT and reduced effluent concentration on the aquatic toxicity of nonylphenols are shown in Figure 4.33.

## SOLUBLE MICROBIAL PRODUCT FORMATION

The kinetics and principals of biodegradation in the activated sludge process have been previously described. These equations indicate that soluble microbial products are generated through the biodegradation of organics and through endogenous degradation of the biomass. The residual SMP are oxidation by-products that are nondegradable in the activated sludge process. Pitter and Chudoba [10] have indicated that depending on cultivation conditions, the nonbiodegradable waste products can amount to 1 to 10 percent of the COD removed. Data for biodegradation of a peptone-glucose mixture and a synthetic fiber wastewater are shown in Figure 4.34. They indicate that approximately 0.20 mg of nondegradable TOC (NDTOC) was produced per mg of influent TOC for the synthetic fiber wastewater. The peptone-glucose containing wastewater produced approximately 0.12 mg NDTOC per mg influent TOC. In both cases the

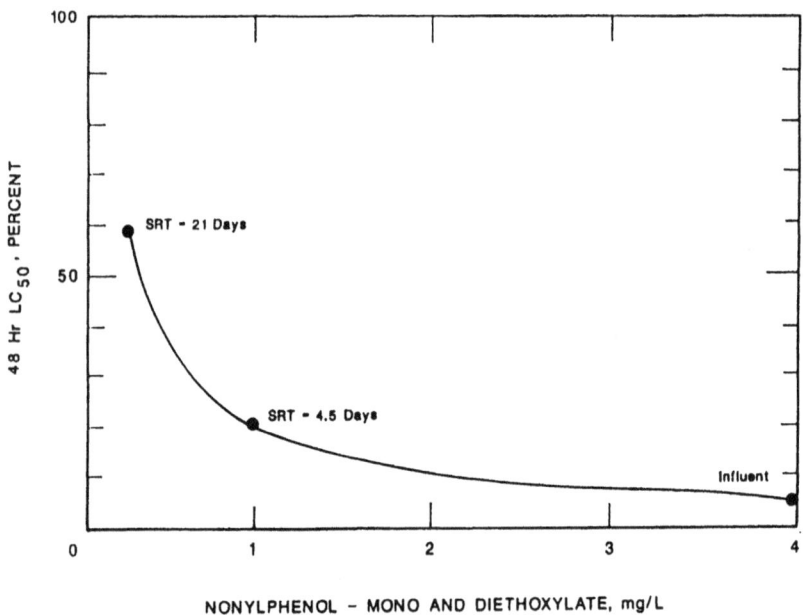

NONYLPHENOL - MONO AND DIETHOXYLATE, mg/L

**Figure 4.33** Effect of SRT on toxicity reduction for nonylphenols.

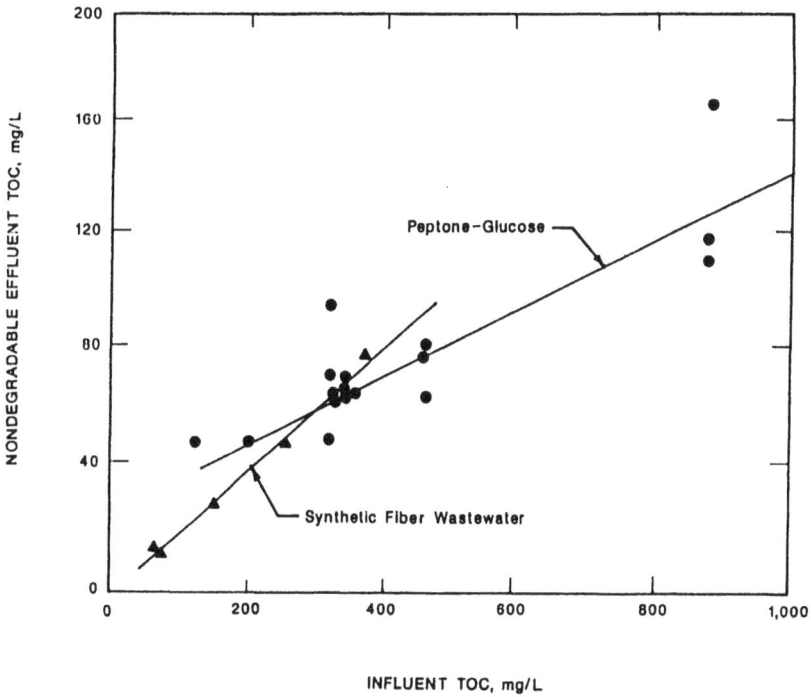

**Figure 4.34** Nondegradable TOC as related to influent TOC.

ratios were constant over the range of influent loading conditions. This indicates that there was a constant metabolic by-product or a portion of the original substrates that were nondegradable.

Many of the metabolic by-products are of high-molecular weight. The molecular weight distribution (expressed as the TOC and COD fractions) of the influent and biologically treated effluent from a plastic additives wastewater and a biologically treated effluent from a glucose synthetic wastewater are presented in Table 4.14. Pitter and Chudoba [10] have

TABLE 4.14. Molecular Weight Distribution of a Biological Effluent.

| Molecular Weight | Plastic Additives Wastewater | | Glucose [18] Wastewater COD (%) |
|---|---|---|---|
| | Influent | Bioeffluent | |
| | TOC (%) | | |
| > 10,000 | — | 11.5 | 45 |
| 500–10,000 | — | 14.5 | 16 |
| < 500 | 100 | 74.0 | 39 |

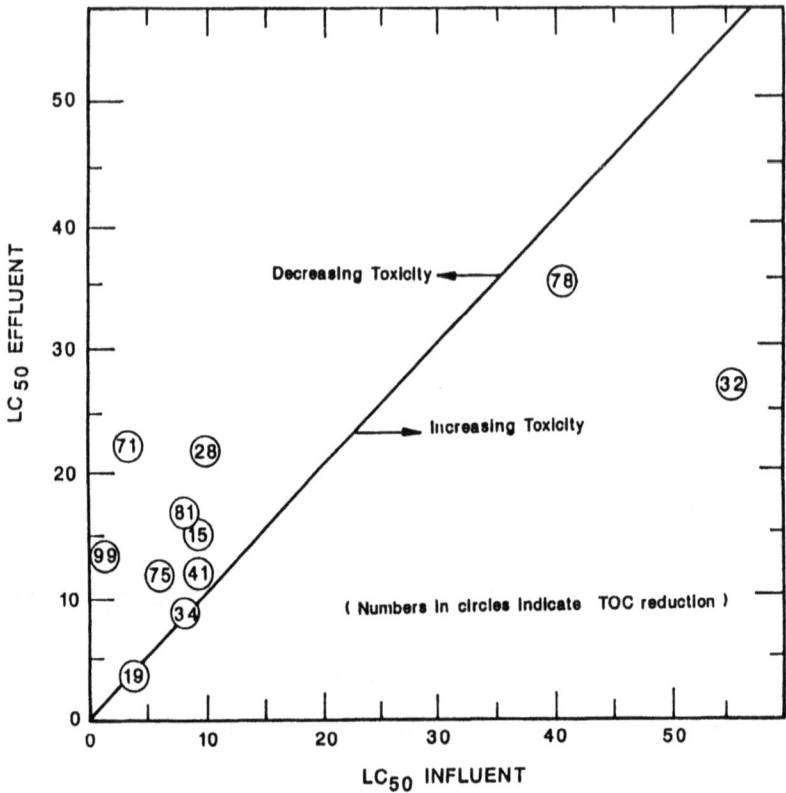

**Figure 4.35** Effect of SMP on effluent toxicity for wastewaters from the plastics and dyestuffs industry.

indicated that approximately 75 percent of the SMP from treatment of a wastewater containing only phenol had a molecular weight above 1,000. It has been further determined that some of these high molecular weight fractions are toxic to some aquatic species. Results showing the toxicity of several wastewaters from the plastics industry before and after biological oxidation are shown in Figure 4.35. While most wastewaters showed reduced toxicity following bio-oxidation, two of the wastewaters exhibited increased toxicity after bio-oxidation, making the oxidation by-products suspect toxicants. There is reason to believe that the high-molecular weight SMP strongly adsorbs on activated carbon. This phenomenon makes granular (GAC) or powdered activated carbon (PAC) an excellent process for toxicity reduction following the activated sludge process when toxicity is caused by SMP.

## ACTIVATED SLUDGE EFFLUENT VARIABILITY

It is apparent that as the organic composition of the wastewater changes, the rate coefficient $K$ will also change. This is not a problem for wastewaters such as those from a dairy or food processing plant since their composition remains substantially unchanged, and hence $K$ will remain nearly constant. Wastewaters generated from multi-product plants with campaign production, however, will experience a constantly changing wastewater composition and hence a highly variable $K$ rate. Daily data from a plant treating a mixture of municipal sewage, a pulp and paper mill wastewater, and two organic chemical wastewaters are shown in Figure 4.36. The high variability in $K$ is due to the variation in wastewater composition.

The rate coefficient combines the effects of all removal mechanisms: biosorption, biodegradation, and volatilization, unless steps are taken to separate the effects of volatilization. This is usually not possible in data reported from industrial wastewater treatment plants. Unusually high "ap-

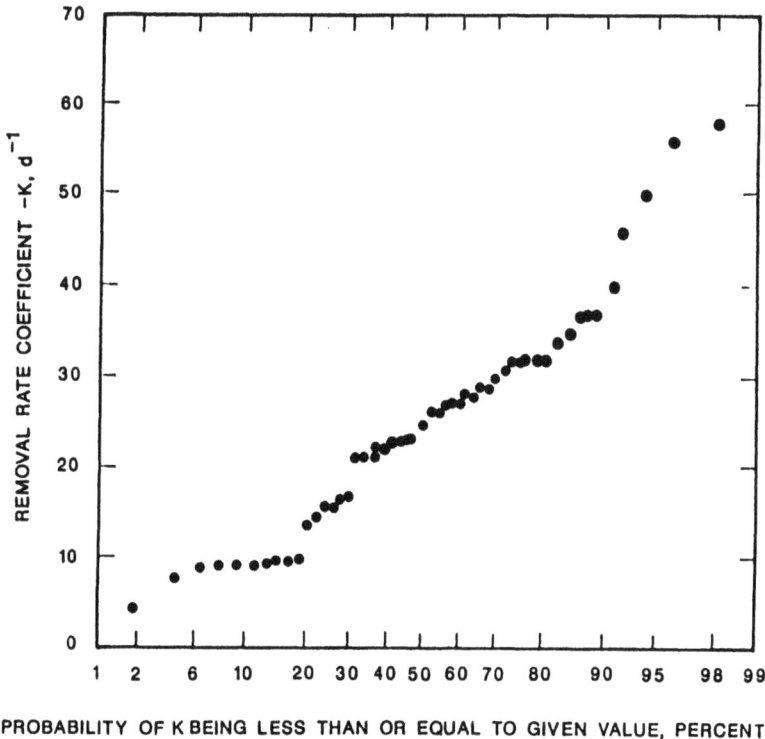

**Figure 4.36** Variability in $K$ as related to wastewater composition.

parent'' reaction rate coefficients may be observed when volatile organics constitute a large portion of the wastewater. Evidence of volatilization is demonstrated in Figure 4.37 for an organic chemical industry wastewater where $K$ increased with an increased influent BOD loading. It was subsequently demonstrated that the high $K$ rates and BOD loadings were due to increased concentration of volatile solvents that exerted an influent BOD but were stripped from the aeration basin before they were metabolized. Volatilization of substrate should be evaluated when calculated $K$ values exceed about 30/d at 20 to 25°C.

Industrial wastewater discharge permits typically contain two limiting conditions: a weekly or monthly average limit, and a daily maximum limit. The treatment process must be designed and operated to reliably satisfy both of these discharge conditions.

A suggested design approach is based on a statistical distribution of the removal rate coefficient and the performance of the equalization basin. For average discharge conditions the mean $K$ value is assumed with the average discharge limit and influent load and substituted into Equation (4.25) as follows:

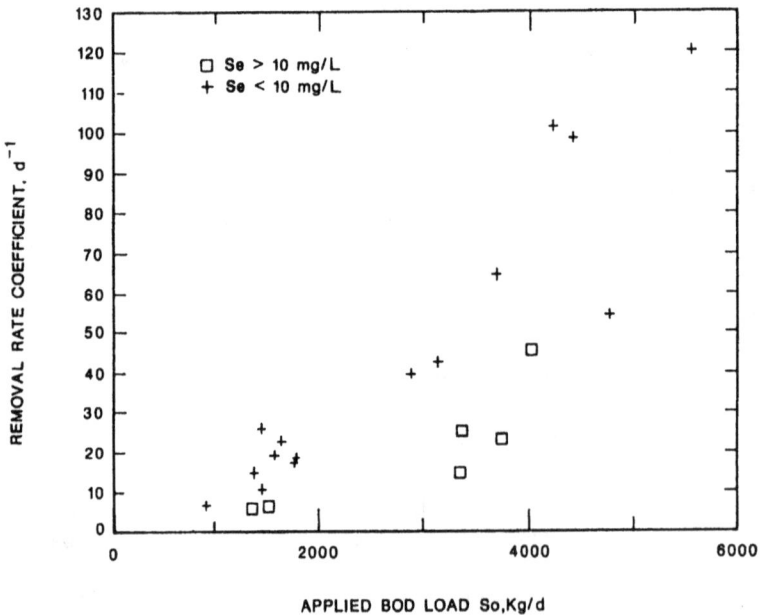

**Figure 4.37** Effect of volatile organics on the reaction rate coefficient $K$ for a pharmaceutical wastewater.

$$\frac{S_{oa} - S_a}{X_v t} = K_{50\%} \frac{S_a}{S_{oa}} \qquad (4.25)$$

where

$S_{oa}$ = average influent BOD, mg/l
$S_a$ = average permit soluble BOD, mg/l
$K_{50\%}$ = the 50 percentile value of $K$, d$^{-1}$

For the maximum permit condition:

$$\frac{S_{om} - S_m}{X_v t} = K_{5\%} \frac{S_m}{S_{om}} \qquad (4.26)$$

where

$S_{om}$ = maximum influent BOD from the equalization basin, mg/l
$S_m$ = maximum permit soluble BOD, mg/l
$K_{5\%}$ = 5 percentile value of $K$, d$^{-1}$

The value of $X_v t$ is calculated for Equations (4.25) and (4.26) and the larger of the two values is used for design and operation. However, if the $X_v t$ value computed for the maximum permit condition exceeds twice the value computed for the average condition, changes in equalization or plant production schedules should be considered in order to reduce the difference.

The variability in effluent BOD, TSS and O&G for a petrochemical wastewater is shown in Figure 4.38. The treatment system includes a gravity oil-water separator, a dissolved air flotation unit, and an activated sludge system. The wastewater characteristics leaving the flotation unit to the activated sludge process are shown in Table 4.15. Daily influent and effluent variability in terms of BOD and TOC for a wastewater from a petroleum refinery complex is shown in Figures 4.39a and 4.39b, respectively.

## BIOINHIBITION OF THE ACTIVATED SLUDGE PROCESS

Many organics will exhibit a threshold concentration at which they inhibit the heterotropic and/or nitrifying organisms in the activated sludge process. This is shown in Figure 4.40 for a wastewater containing increas-

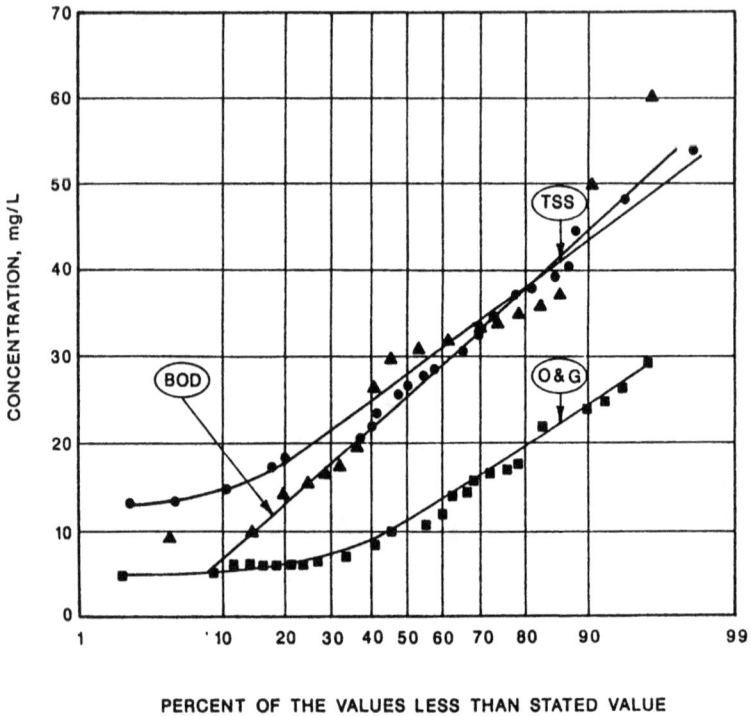

**Figure 4.38** Effluent variability for a petrochemical wastewater.

ing concentrations of inhibitory organics in a batch activated sludge system. Inhibition has been defined by the Haldane equation (or its modifications).

$$\mu = \frac{\mu_m S X_v}{S + K_x S^2/K_I} - b X_d X_v \qquad (4.21)$$

in which $K_I$ is the Haldane inhibition coefficient.

TABLE 4.15. Effluent Characteristics from Dissolved Air Flotation Unit.

| Parameter | Mean (x) | Standard Deviation (s) | $\bar{x} \pm s$ | $\bar{x} \pm 2s$ |
|---|---|---|---|---|
| TSS, mg/l | 30 | 12 | 18–42 | 13[a]–54 |
| BOD$_5$, mg/l | 28 | 13 | 15–41 | 9[a]–54 |
| COD, mg/l | 123 | 42 | 81–165 | 64[a]–207 |
| Oil and grease, mg/l | 12.6 | 7.4 | 5.2–20 | 4.0[a]–27.4 |

[a]Lowest value measured is greater than $\bar{x} - 25$.

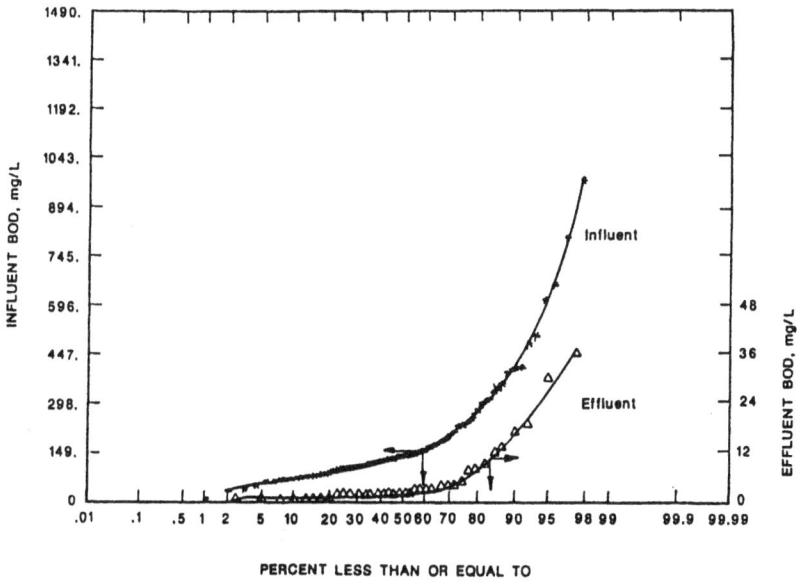

**Figure 4.39a** Variability in influent and effluent BOD for petroleum wastewater.

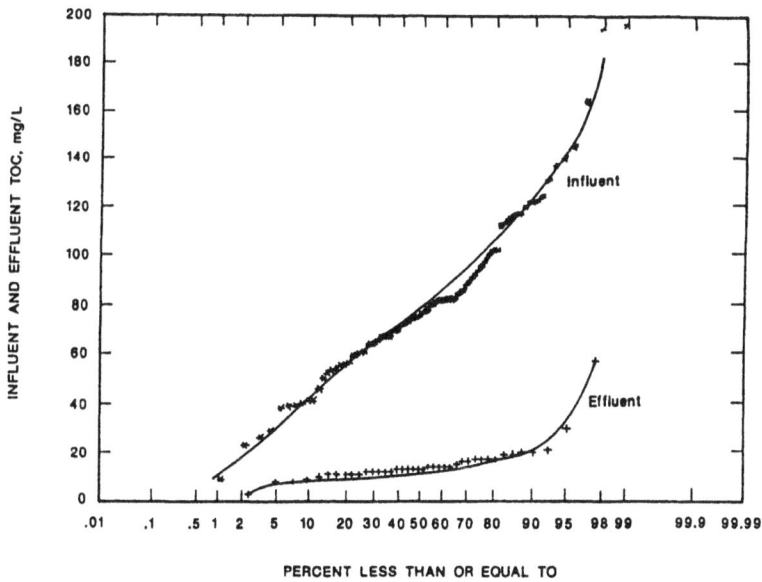

**Figure 4.39b** Variability in influent and effluent TOC for petroleum wastewater.

An example of bioinhibition by a plastics additives wastewater is shown in Figure 4.41. As the concentration of effluent COD (and inhibitory agent) increased, the SOUR decreased, resulting in higher effluent SBOD concentrations. In this case, the inhibition was removed by pretreating the specific wastewater with hydrogen peroxide ($H_2O_2$) thereby effecting detoxification and enhanced biodegradability. These effects are shown in Figure 4.42. An alternative approach to reduce inhibition is to add powdered activated carbon to the mixed liquor to adsorb the toxicant.

Several acidic, aromatic, and lipid soluble organic compounds have been demonstrated to "uncouple" oxidative phosphorylation. The result of this uncoupling effect is uncontrolled respiration and oxidation of primary substrates and intracellular metabolites. At low concentrations, uncoupling is evidenced by highly elevated oxygen utilization rates and no effect on cell growth or substrate removal. At higher concentrations, inhibition and toxicity are demonstrated by dramatic reductions in both oxygen utilization rate and cell growth.

Volskay and Grady [19] and Watkin [20] have shown that inhibition can be competitive (the inhibitor affects the base substrate utilization),

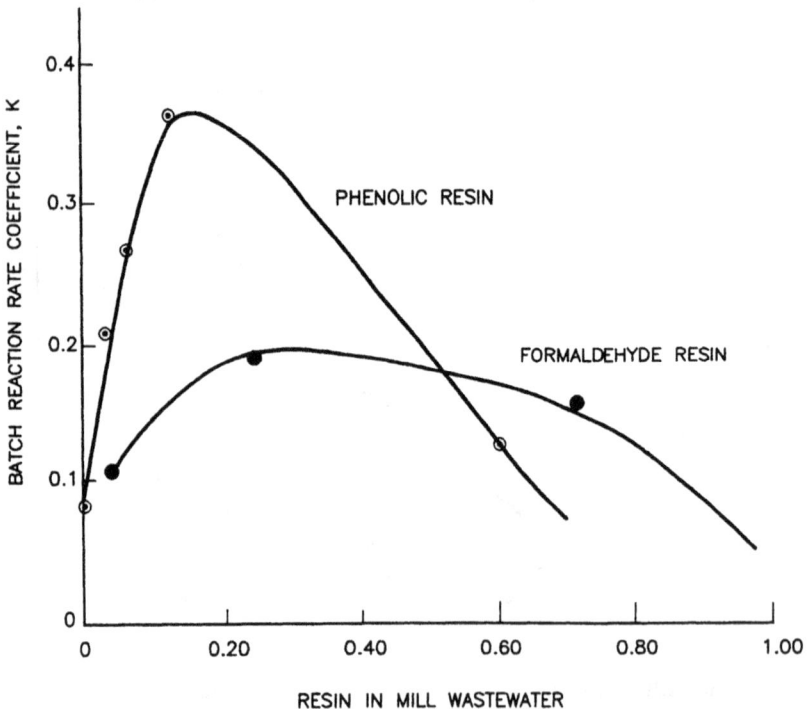

Figure 4.40 Inhibition from resin addition to pulp mill wastewater.

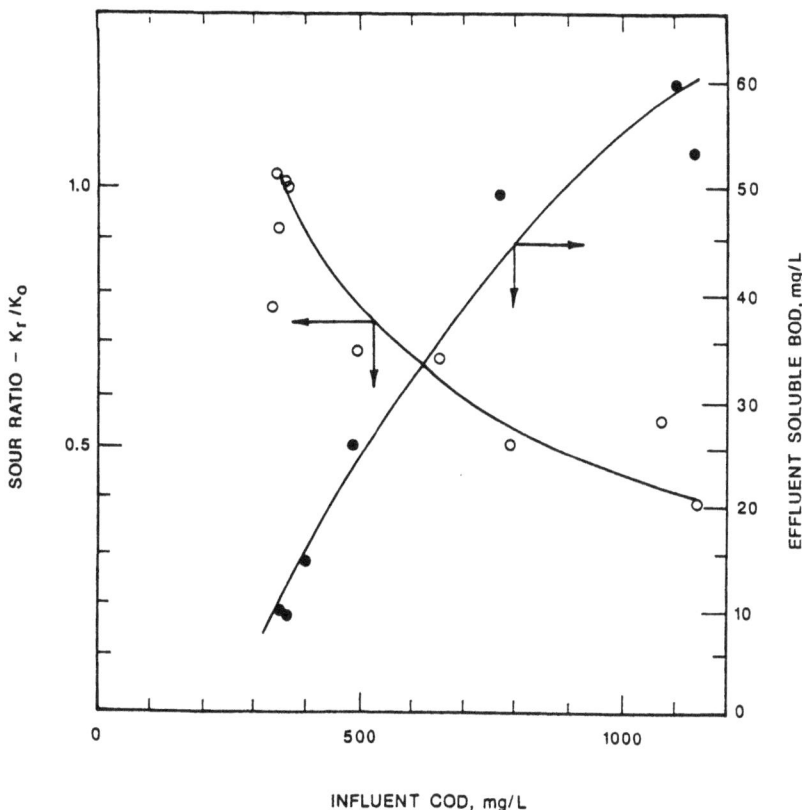

**Figure 4.41** Activated sludge inhibition from a plastics additives wastewater.

non-competitive (the inhibitor rate is influenced), or mixed, in which both rates are influenced. The effects of substrate and inhibitor concentrations on the respiration rate of a microbial culture expressed as a fraction of the rate in the absence of the inhibitor has been shown by Volskay and Grady [19].

While the relationships described above define the mechanism of inhibition, they are of limited use in evaluating most industrial waste-waters. In many cases, the inhibitor itself is not defined, variable sludge and substrate composition will influence inhibition, and interactions will frequently exist between inhibitors. For example, Watkin and Eckenfelder [20] showed a variation in $K_I$ of 6.5 to 40.4 for different sludges and operating conditions treating 2,4 dichlorophenol (DCP) and glucose. Volskay and Grady [19] showed a variation in the concentration of pentachlorophenol, which would cause 50 percent inhibition of

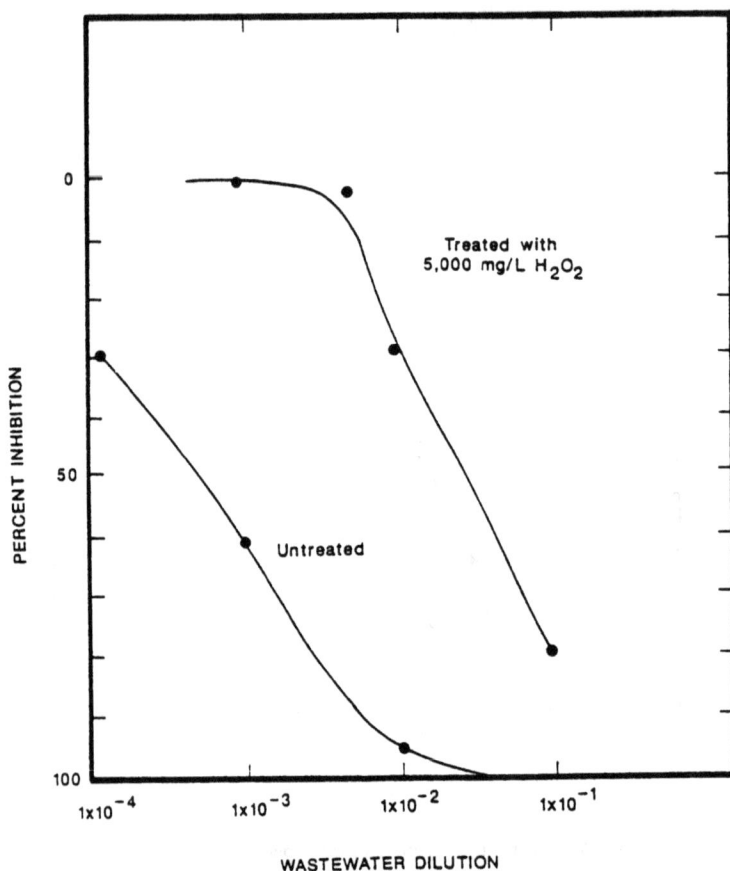

**Figure 4.42** Detoxification of a plastics additives wastewater.

oxygen utilization rates of 2.6 to 25 mg/l. The inhibition constant $K_I$ is likely to be highly dependent on the specific enzyme system involved, which in turn is dependent on the history and population dynamics of the sludge. In some cases, the inhibition constant may be dependent on the particular metabolic pathways that are present in any given microbial population. Therefore, it becomes apparent that each wastewater must be independently evaluated for its bioinhibition effects. Several protocols have been developed for this purpose. These are the fed-batch reactor of Philbrook and Grady [16] and Watkin and Eckenfelder [20], the OECD Method 209 of Volskay and Grady [19], and the glucose inhibition test of Larson and Schaeffer [21]. Depending on the particular wastewater, one or more of these test protocols will be applicable. Each of these methods is discussed in the following subparagraphs.

## OECD METHOD 209

The OECD Method involves measurement of activated sludge oxygen uptake rate from a synthetic substrate to which the test compound has been added at various concentrations. The oxygen uptake rate is measured immediately after addition of the test compound and after 30 minutes of aeration. The $EC_{50}$ value is determined as the concentration of test compound at which the oxygen uptake rate (at 30 min) is 50 percent of the uninhibited oxygen uptake rate. The OECD Method uses 3,5-dichlorophenol as a reference toxicant to insure that the test is working properly and that the biomass has the appropriate sensitivity. The reference $EC_{50}$ value should be between 5 and 30 mg/l for the test to be valid. Volskay and Grady [19] employed this protocol with modifications to determine the toxicity of selected organic compounds. Since many of the compounds were volatile, the test was modified by using more dilute cell and substrate concentrations and by conducting the test in vessels that were sealed by the insertion of a polyfluoroethylene plug. This protocol is recommended for wastewaters containing high concentrations of volatile organics.

## FED-BATCH REACTOR (FBR)

Fed-batch reactors have previously been used to determine nitrification kinetics [22] and removal kinetics of specific pollutants in activated sludge [11]. The essential characteristics of the FBR procedure are that (1) substrate is continuously introduced at a sufficiently high concentration and low flow rate so that the reactor volume is not significantly changed during the test; (2) the feed rate exceeds the maximum substrate utilization rate; (3) the test duration is short and therefore allows simple modeling of biological solids growth; and (4) various acclimated activated sludges are used.

A schematic diagram of the FBR configuration is shown in Figure 4.43. Two liters of test sludge are placed in the FBR. An initial sample is taken for determination of oxygen utilization rate and mixed liquor volatile and total suspended solids prior to the start of the feed flow. The feed flow rate at 100 ml per hour is introduced and aliquots of the reactor contents are withdrawn from the FBR every 20 minutes for the duration of the three-hour test. Suspended solids determinations are made every hour during the test.

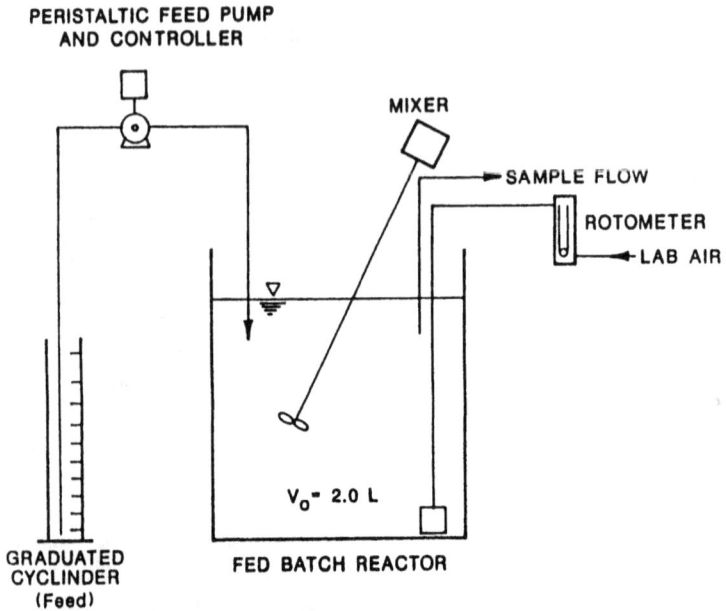

**Figure 4.43** Fed-batch reactor (FBR) configuration.

As discussed under the OECD protocol, the oxygen uptake rate will usually decrease in the presence of inhibition. As long as there is no inhibition, the SOUR should remain constant at a maximum rate. Inhibition, however, will cause a progressive decrease in SOUR. The OUR is determined *in situ* during the course of the FBR test. The same limitations apply to this protocol as to the OECD protocol.

The theoretical responses in a fed-batch reactor to both inhibitory and non-inhibitory substrates are depicted in Figure 4.44. In the specific case that substrate is added at a sufficiently high mass and low volumetric flow rate, the maximum substrate utilization rate will be exceeded and reactor volume change will be insignificant. If the basic premise of the fed-batch reactor holds, i.e., that volume change is negligible, and if the mass feed rate exceeds the maximum substrate utilization, then a substrate concentration buildup will result in the reactor with time. Non-inhibitory substrate response results in a linear residual substrate buildup in the reactor with time. The maximum specific substrate utilization may be calculated as the difference in slopes between the substrate feed rate and the residual substrate buildup rate divided by the biomass concentration. In the case of inhibition, substrate utilization would rapidly decrease, resulting in an upward deflection of the residual substrate concentration curve as shown in Figure 4.44. As inhibition progresses and acute

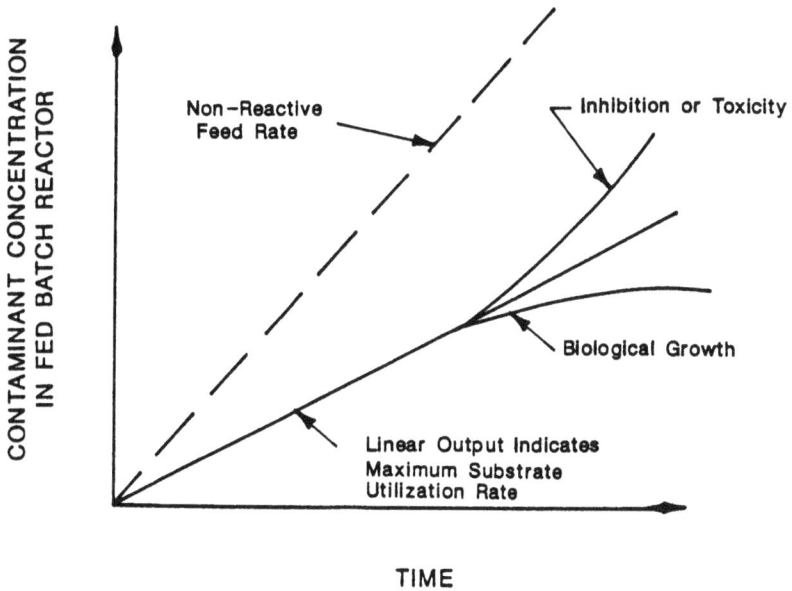

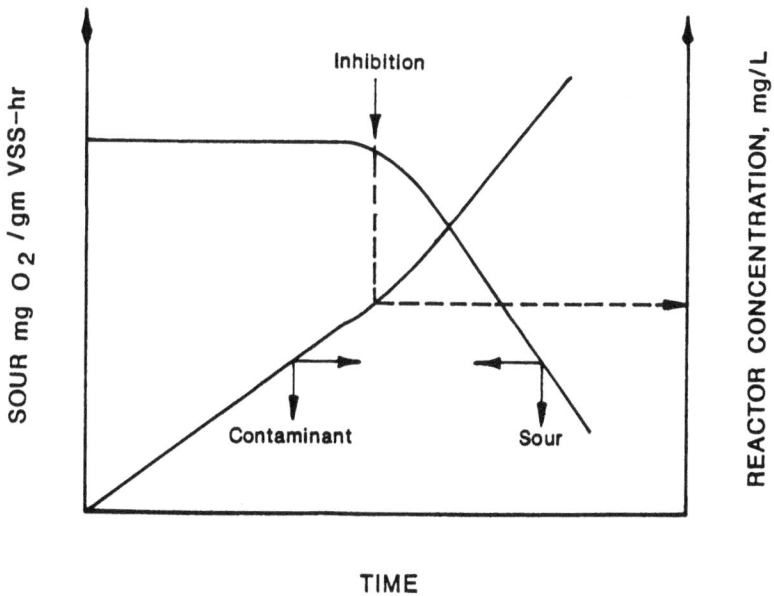

**Figure 4.44** Theoretical fed-batch reactor output with influent substrate mass flow rate greater than $q_{max} \cdot X_v$ and inhibition effects.

toxicity occurs, the trace of the residual substrate concentration should become parallel to the substrate feed rate. The inhibition constant, $K_I$, may be approximated by identifying the inhibitor concentration at the midpoint of the curvilinear portion of the substrate response.

## GLUCOSE INHIBITION TEST

Larson and Schaeffer [21] developed a rapid toxicity test based on the inhibition of glucose uptake by activated sludge in the presence of toxicants. The test was subsequently modified for application to a variety of industrial wastewaters. The procedure is as follows:

(1) Place 10 ml of sample into a centrifuge tube.
(2) Add 1 ml of the stock glucose solution.
(3) Add 10 ml of activated sludge to the centrifuge tube and aerate at low rate.
(4) After 60 minutes, add two drops of HCl and transfer tubes to the centrifuge.
(5) Measure glucose concentration.
(6) Sludge control—Substitute 10 ml of deionized water for the sample in Step (1) and perform Steps (2) through (5) as before.
(7) Glucose control—Place 30 ml of deionized water in a centrifuge tube. Add 1 ml of stock glucose solution. Do not add sludge or aerate. Add two drops of HCl and measure glucose uptake as follows.

The percent inhibition is calculated as follows:

$$\text{Percent Inhibition} = \frac{C - C_B}{C_o - C_B}$$

$C$ = final glucose concentration in sample solution
$C_B$ = final glucose concentration in sludge control sample
$C_o$ = initial glucose concentration (glucose control)

Results of the inhibition effects are presented in Figure 4.42 for $H_2O_2$ treatment of a plastic additives wastewater and in Figure 4.45 for an organic chemicals wastewater.

## INHIBITION OF NITRIFICATION

In many industrial wastewaters, carbonaceous oxidation is not inhibited, but the more sensitive nitrifiers are inhibited. Small doses of powdered activated carbon may adsorb the inhibiting organics, permitting nitrifica-

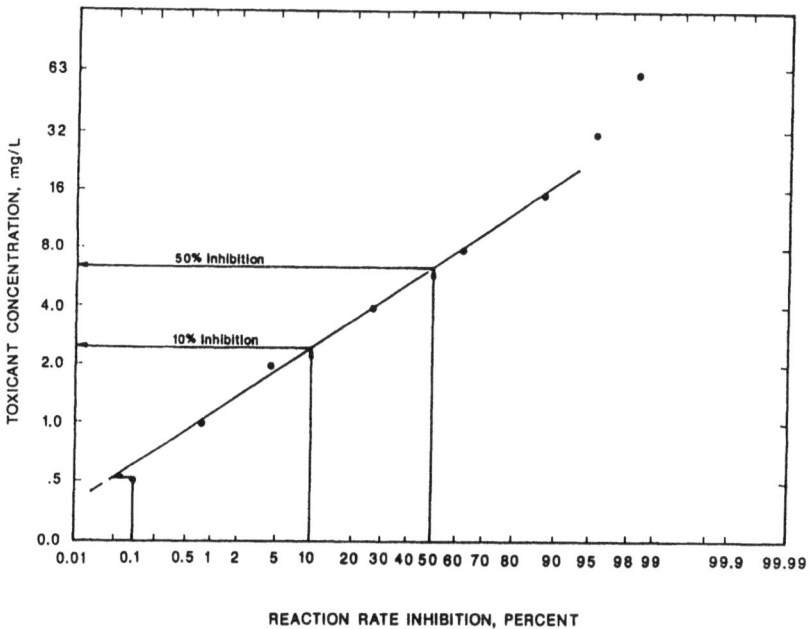

**Figure 4.45** Bioinhibition effects from the glucose inhibition test.

tion to proceed. Nitrification inhibition kinetics can be studied using the FBR procedure and/or the following protocol to evaluate the effect of carbon addition:

(1) Add PAC to the wastewater in a range of dosages (usually 30 to 100 mg/l).

(2) Mix for 2 hours to approach equilibrium.

(3) Add the wastewater-carbon mixture to washed activated sludge containing active nitrifiers.

(4) Aerate for 24 hours with intermediate sampling for analysis of $NH_3$, $NO_2^-$, and $NO_3^-$. Determine the nitrification rate expressed as mg $NO_3^-$ produced/hr-mg MLVSS and compare to a wastewater aliquot that was aerated without PAC. Select the carbon dosage that maximizes the nitrification rate.

## EFFECT OF TEMPERATURE

Temperature affects all biological reactions. The magnitude of the effect is related to the characteristics of the wastewater organics and their physical state (i.e., suspended, colloidal, or soluble). Most activated sludge processes operate in the mesophilic range (4 to 39°C). The effect of temperature on the reaction rate coefficient can be defined by Equation (4.27)

$$K_T = K_{20} \cdot \theta^{(T-20)} \qquad (4.27)$$

where

$K_T$, $K_{20}$ = reaction rate coefficients at temperatures of $T$ and
       20°C, respectively, d$^{-1}$
   $T$ = mixed liquor temperature, °C
   $\theta$ = empirical temperature correction coefficient, dimensionless

Equation (4.27) indicates that the $K$ value increases exponentially with temperature and is applicable over a temperature range of 4 to 31°C. At mixed liquor temperatures between 31 and 39°C the value of $K$ is approximately constant and declines at higher temperatures. The decline in the rate coefficient at temperature above about 40°C is frequently associated with deterioration and dispersion of the biological floc, poor sludge settleability and high effluent suspended solids concentration.

The reduction in the $K$ rate over a temperature range of 96°F to 125°F for a pulp and paper mill wastewater is shown in Figure 4.46. Over this temperature range the ratio of $K_T/K_{96}$ steadily declined.

For industrial wastewaters containing high soluble substrate concentrations, the value of the temperature correction coefficient ($\theta$) has been found to vary from 1.03 to 1.10. Since no consistent correlation has been demonstrated between the value of $\theta$ and wastewater characteristics, it is necessary to determine $\theta$ experimentally. Figure 4.47 shows data obtained for an agricultural chemicals wastewater, analyzed to determine the effect of temperature on the $K$ rate and the value $\theta$.

It should be noted that a significant difference in the effect of temperature on substrate removal has been found for municipal wastewater as compared to industrial wastewaters. This is because a primary substrate removal mechanism for municipal wastewater is the biological entrapment of suspended and colloidal organics—a physical phenomenon that is insensitive to temperature. As a result, the temperature correction coefficient has been found to have a value of 1.015 when considering overall BOD removal in municipal wastewater treatment. The temperature coefficient $\theta$ has a value of 1.04 for the endogenous coefficient $b$.

The temperature effect on removal of specific organic compounds, such as phenol, is frequently much greater than its effect on overall organic removal. Figure 4.48 shows the results from a plant treating a wastewater containing primarily acetone and phenol. The calculated value of the temperature coefficient was 1.1.

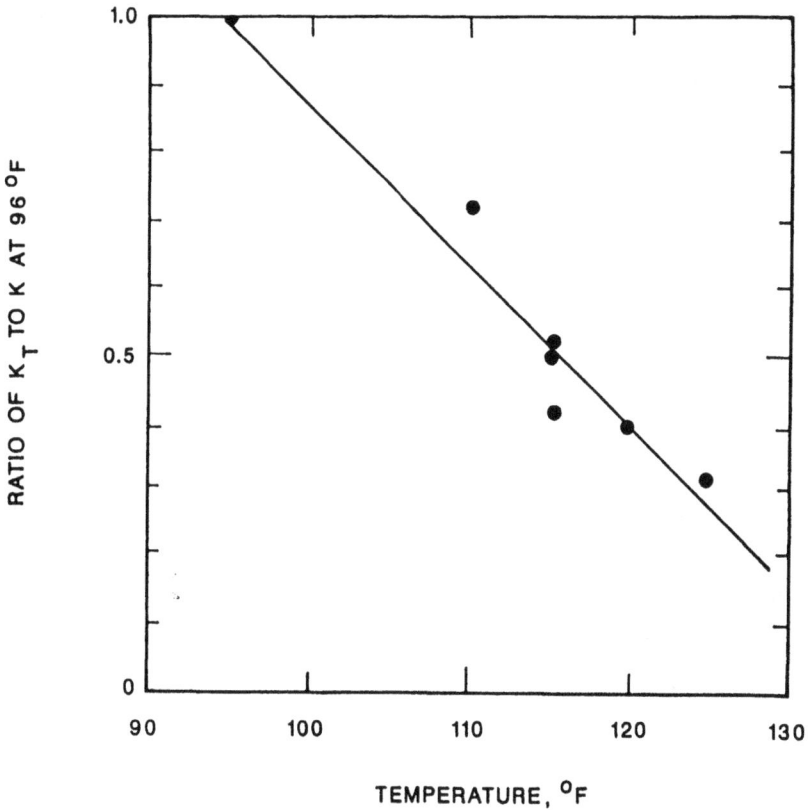

**Figure 4.46** Effect of temperature on the reaction rate coefficient, $K$, for a pulp and paper mill effluent.

The magnitude of the temperature effect on removal of specific compounds creates operating problems when low effluent values are required on a year-round basis. In many cases, permit values are readily achievable during summer operation but difficult or impossible to achieve during winter operation. The application of steam to raise the mixed liquor temperature is frequently required during cold weather conditions. If the substrate removal rate is inhibited by a toxic agent, the extent of inhibition is frequently increased during cold weather operating conditions. Under these circumstances, powdered activated carbon can be seasonally added to reduce the bioinhibitory effect and satisfy the discharge permit limit.

In addition to the temperature effects on the biological reaction rate, two other temperature-related phenomena must be considered. As the mixed liquor temperature decreases, from approximately 25°C towards 5 to 8°C, there may be an increase in effluent suspended solids. These suspended solids

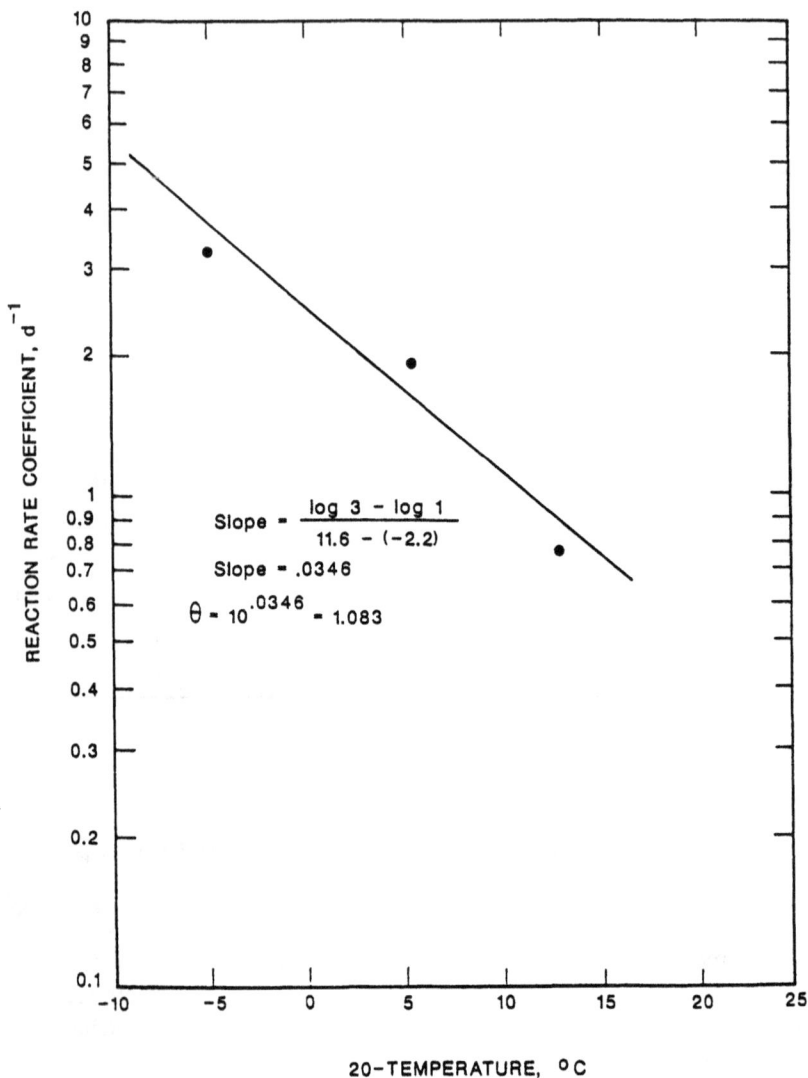

**Figure 4.47** Effect of temperature on the reaction rate coefficient, $K$, for an agricultural chemicals wastewater.

are typically non-settleable, dispersed in nature, and are not removed by conventional clarification processes. For example, a multi-product organic chemicals plant had an average effluent suspended solids value of 42 mg/l during the summer and 104 mg/l during the winter. The second temperature-related effect occurs as the temperature increases above about 35°C. At this temperature range, the biological floc characteristics deteriorate. This phenomenon appears to be related to both the temperature and the characteristics of the wastewater. Floc dispersion was observed at 41°C for a pulp and paper mill wastewater but occurred at 35°C for an agricultural chemicals wastewater. The decrease in mixed liquor zone settling velocity with increased aeration basin temperature for a pulp and paper mill wastewater is shown in Figure 4.49. The increase in effluent suspended solids with mixed liquor temperature for this wastewater is shown in Figure 4.50. Figures 4.51a and 4.51b are photomicrographs of the mixed liquors operated at 96°F and 110°F for this wastewater. At 96°F the flocs were large and protozoa were present.

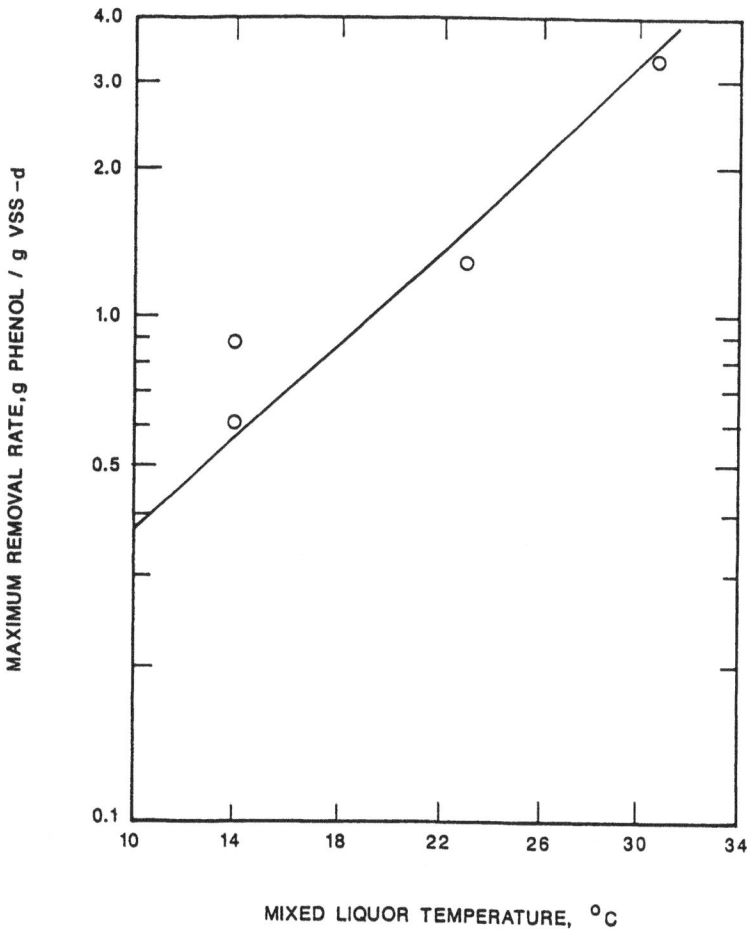

**Figure 4.48** Effect of temperature on the removal rate of phenol for a wastewater containing phenol and acetone.

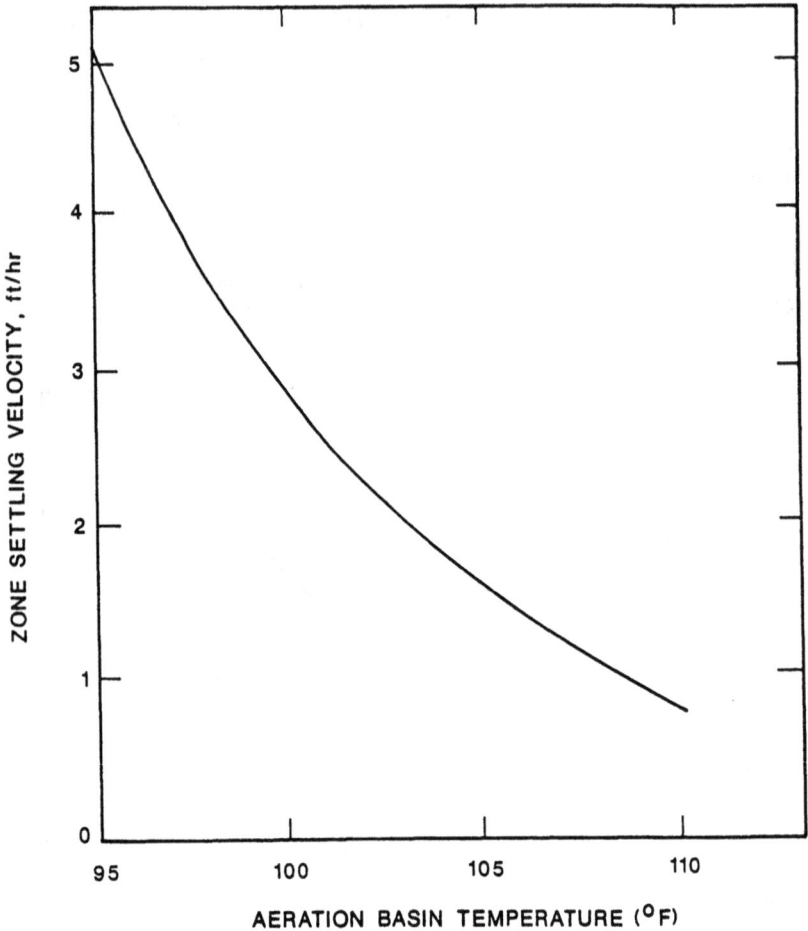

**Figure 4.49** Effect of mixed liquor temperature on the zone settling velocity of activated sludge for a pulp and paper mill wastewater.

At 110°F however, there were no protozoa, the floc was dispersed, and filamentous forms were present.

There have been some efforts to operate the activated sludge process in the thermophilic range (45 to 55°C) when the influent wastewater temperature is already in this range and the BOD is high (2,500 to 3,000 mg/l). Under these operating conditions, the sludge generated was frequently difficult to separate, the mixed liquor was dispersed, and effluent solids concentrations were high.

While the activated sludge process can adapt to a wide range of temperatures it becomes unstable with a sudden temperature change. This causes floc dispersion and an increase in effluent suspended solids. This phenomena is shown in Figure 4.52.

## SLUDGE QUALITY CONTROL

It is of primary importance when treating industrial wastewaters by the activated sludge process that a sludge be produced which will flocculate and settle well. Maintenance of this condition involves engineered and operational controls for filamentous bulking. When treating industrial wastewaters, bulking can be caused by:

- insufficient dissolved oxygen
- insufficient nutrients
- low *F/M* in the case of readily degradable wastewaters

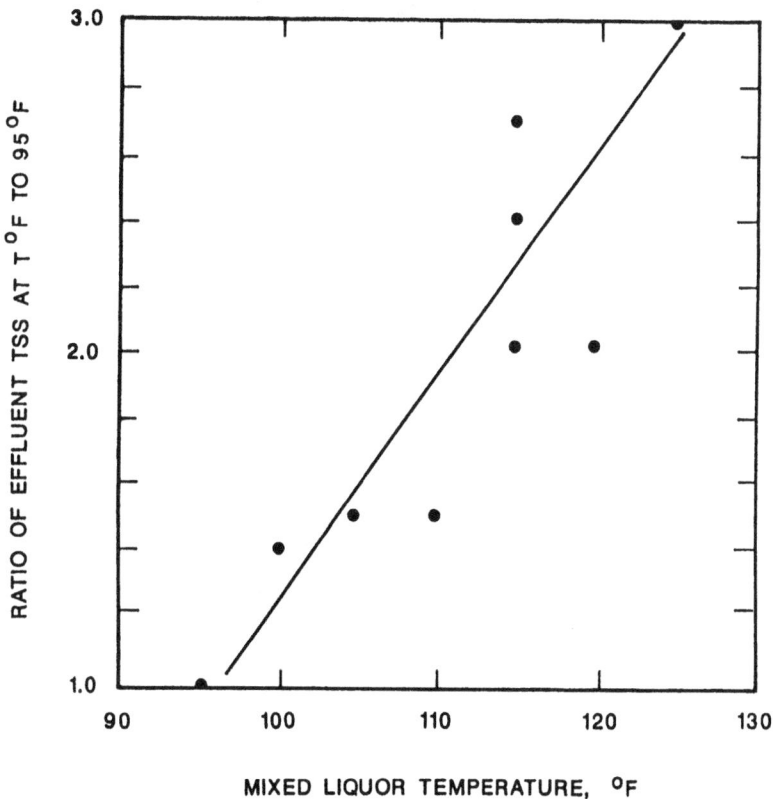

**Figure 4.50** Effect of mixed liquor temperature on effluent TSS for a pulp and paper mill wastewater.

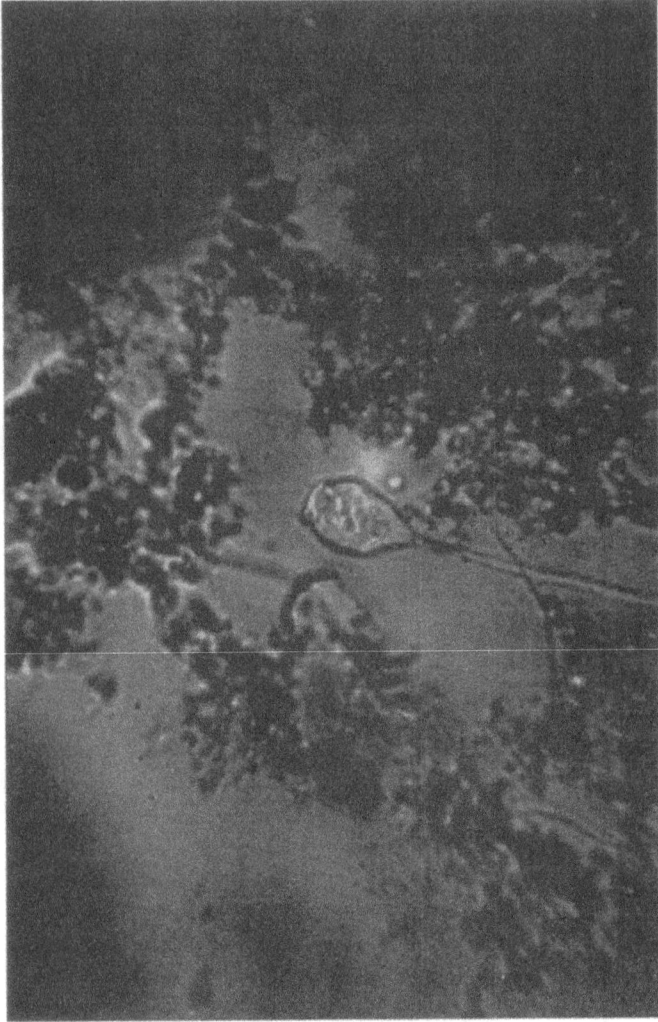

**Figure 4.51a** Photomicrograph of mixed liquor at 98°F for pulp and paper mill wastewater treatment (100×).

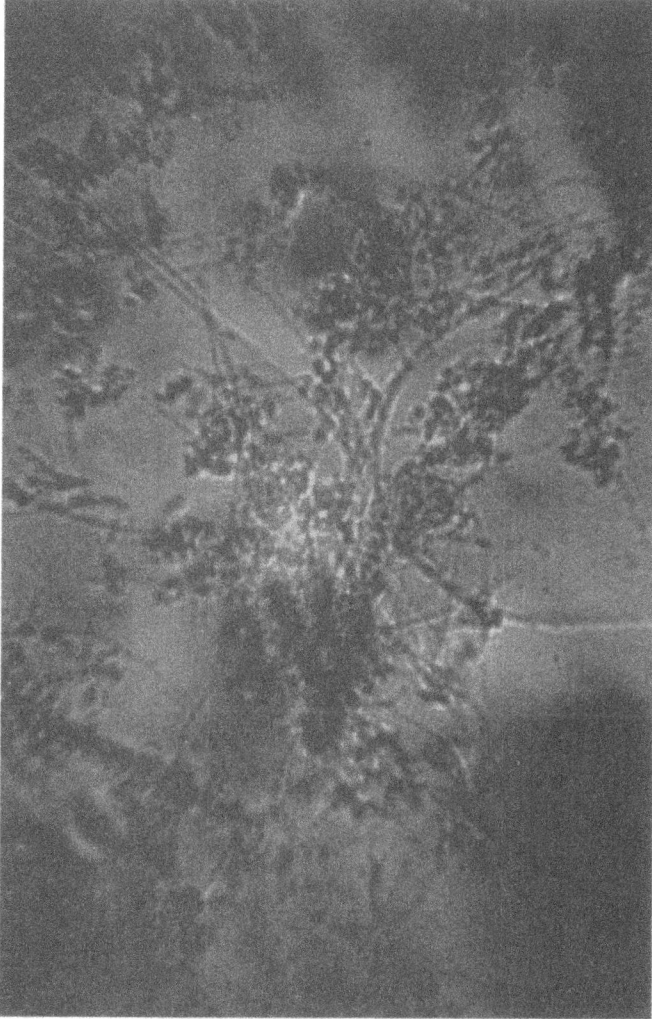

**Figure 4.51b** Photomicrograph of mixed liquor at 110°F for pulp and paper mill wastewater treatment (100×).

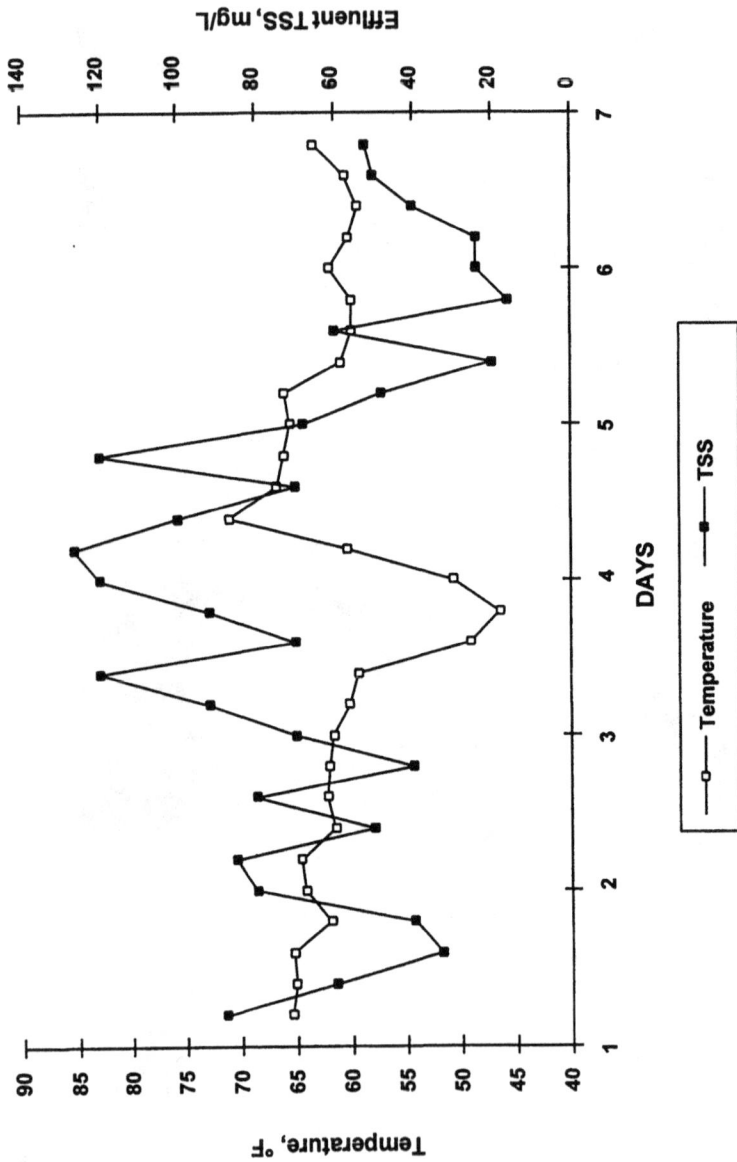

**Figure 4.52** Effect of temperature on effluent TSS for an activated sludge system.

In filamentous sludge bulking there are several factors to consider:

(1) Most filaments can only degrade readily degradable organics. Waste-waters containing complex organics do not experience filamentous bulking since the filaments cannot use these organics as a food source. As a general rule, filamentous bulking will not be experienced with wastewaters having a reaction rate $K$ less than 6 days$^{-1}$.

(2) All things being equal, i.e., adequate oxygen, nutrients and substrate, the floc formers will outgrow the filaments.

(3) Floc formers have the ability to biosorb organics while most filaments do not.

A large number of filaments have been identified in activated sludge systems. The filament types associated with sludge bulking are shown in Table 4.16 [23]. Specific filaments found in various industrial treatment plants as reported by Richard [24] are summarized in Table 4.17.

To illustrate these effects, consider the transfer of dissolved oxygen to the hypothetical biological floc particles illustrated in Figure 4.53. Oxygen must diffuse from the bulk liquid through the floc in order to be available to the organisms within the interior of the floc particle. As it diffuses, it is consumed by the organisms within the floc. If there is an adequate residual of dissolved oxygen (and nutrients and organics) the rate of growth of the floc-forming organisms will exceed that of the filaments and a flocculent well-settling sludge will result. If there is a deficiency in any of these substrates, however, the filaments, having a high surface area to volume ratio, will have a "feeding" advantage over the floc formers and will proliferate due to their higher growth rate under adverse conditions.

TABLE 4.16. Filament Types as Indicators of Conditions Causing Activated Sludge Bulking [23].

| Causative Condition | Filament Types |
|---|---|
| Low dissolved oxygen (for the applied organic loading) | *S. natans*, type 1701 and *H. hydrossis* |
| Low organic loading rate (low *F/M*) | *M. parvicella*, *Nocardia* spp., *H. hydrossis*, types 0041, 0675, 1851, 0803, and 021N |
| Septic wastes/sulfides (high organic acids) | *Thiothrix* supp., *Beggiatoa* supp., *N. limicola II*, and types 021N, 0092, 0914, 0581, 0961 and 0411 |
| Nutrient deficiency—N and/or P (industrial wastes only) | *Thiothrix* supp., *N. limicola III*, type 021N, types 0041 and 0675 |
| Low pH (< pH 6.0) | Fungi |

Note that some filaments occur at several conditions, and particularly at the combinations listed.

TABLE 4.17. Filament Types Found in Industrial Wastewaters [24].

| Wastewater Source | S. natans | Type 1701 | H. hydrossis | Type 021N | Thiothrix I and II | Type 1851 | Type 0581 | Type 0041 | Type 0803 | Type 0675 | Type 0211 | Type 0092 | Type 0914 | M. parvicella | N. limicola | Type 0411 | Nocardia |
|---|---|---|---|---|---|---|---|---|---|---|---|---|---|---|---|---|---|
| Food processing and brewing | ● | ● | ● | ● | ● | ● | | ● | | ● | | ● | | | | | ● |
| Textile | | | | | ● | | | | | | | | | | | | |
| Slaughterhouse and meat processing | ● | ● | ● | ● | ● | | | | | | | | ● | | | | ● |
| Petrochemical | ● | ● | | | ● | | | | | | | | | | | | ● |
| Organic chemicals | | | | ● | ● | | | | | | | ● | ● | | | | ● |
| Pulp and papermill | ● | ● | ● | ● | ● | ● | ● | ● | ● | ● | ● | ● | ● | ● | ● | ● | |

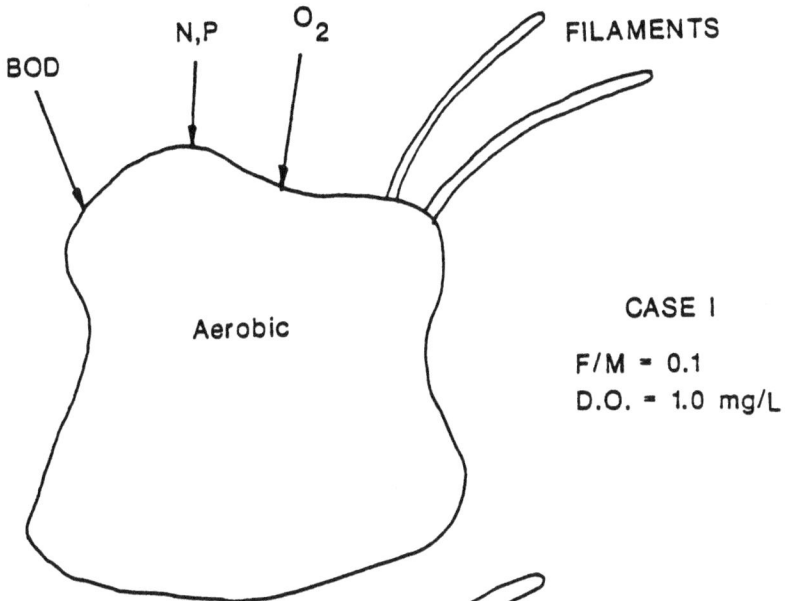

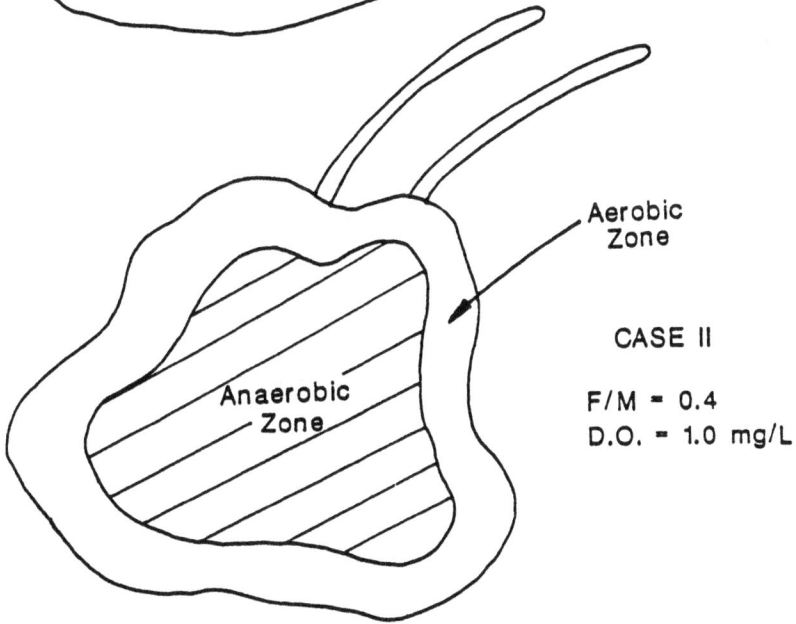

**Figure 4.53** A mechanism of sludge bulking.

Considering Case 1 in Figure 4.53 at an *F/M* of 0.1 d$^{-1}$, the oxygen utilization rate is low and even with a bulk liquid dissolved oxygen concentration of 1.0 mg/l oxygen fully penetrates the floc. Under these conditions, the floc formers will outgrow the filaments. In Case 2 the *F/M* is increased to 0.4 d$^{-1}$ causing a corresponding increase in oxygen uptake rate. If the bulk mixed liquor dissolved oxygen is maintained at 1.0 mg/l, the available oxygen will be rapidly consumed at the periphery of the floc, thus depriving a large interior portion of the floc particle of oxygen. Since the filaments have a competitive growth advantage at low oxygen levels, they will be favored and will outgrow the floc formers. This phenomenon is illustrated in Figure 4.54 using mixed liquor operational data from an activated sludge plant treating a pulp and paper mill wastewater. The solid line in Figure 4.54 represents data generated by Palm et al. [25] as the critical bulk liquid dissolved oxygen concentration relative to the *F/M* applied to the process. The boxed numbers represent the observed mixed liquor SVI under the defined operating condition. As can be seen, the observed performance agrees well with the prediction of Palm et al. [25].

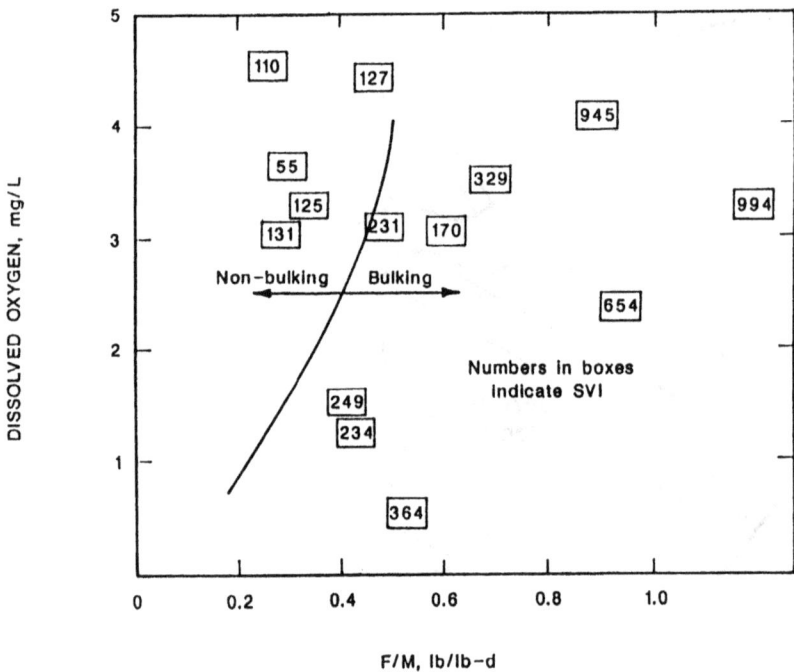

**Figure 4.54** Relationship between *F/M* and dissolved oxygen relative to sludge bulking for a pulp and paper mill wastewater.

In a similar manner, insufficient nitrogen and phosphorus concentrations will result in a nutrient deficiency and filamentous bulking. These effects on the zone settling velocity of the sludge are shown in Figure 4.55 and in Figure 4.56 for a kraft pulp and paper mill wastewater treated in an oxygen-activated sludge plant. In Figure 4.56, the performance prior to the "Christmas Shutdown" showed 90 percent BOD removal, effluent TSS of 60 mg/l and SVI of 40 to 60 ml/g, at an average *F/M* of 0.5 $d^{-1}$. After production startup in early January, the nutrient addition rate was not adequate and resulted in effluent N and P concentrations of 0.4 and 0.2 mg/l, respectively. The SVI increased to over 200 ml/g and caused a decrease in return sludge concentration from 14,000 mg/l to less than 6,000 mg/l. This caused the *F/M* to rise to above 1.0 $d^{-1}$ with a consequent rise in effluent BOD and TSS levels. On February 10, the nutrient addition rates were increased, resulting in improved process performance. Studies in the pulp and paper industry indicated that minimum $NH_3$–N and $PO_4$–P concentrations of 1.5 and 0.5 mg/l, respectively, are required in the final effluent to prevent filamentous bulking due to nutrient deficiency.

In a similar manner the BOD concentration must be sufficient to provide a driving force to penetrate the biological floc. In a complete mix basin, the concentration of soluble BOD in the mixed liquor is essentially equal to the effluent concentration and is therefore low. As a result, substrate penetration of the floc is not achieved, and filaments dominate the interior floc population. In order to shift the population in favor of the floc formers, sufficient substrate driving force must be developed to penetrate the floc and favor their growth. This can be achieved by a batch or plug-flow operating configuration in which a high substrate gradient (driving force) exists and maximum growth occurs at the influent end of the plug-flow basin or in the initial period of each feed cycle of a batch activated sludge process.

Alternatively, a biological selector may be used in which a significant portion of the soluble substrate removal occurs in the selector. Under these conditions, the substrate gradient is high and promotes the growth of floc formers over the filaments. When the wastewater is discharged from the selector, the soluble substrate concentration is relatively low and is available for utilization by the filaments. They do not predominate in the mixed liquor, however, since the principal mass of substrate removed has been initially directed to the growth of floc-forming biomass.

Biosorption can be related to the floc load (*F.L.*) which is defined as the mg of soluble degradable COD ($SCOD_{deg}$) COD applied per g $VSS_b$ in a brief contact period (typically 15 to 20 minutes based on the forward flow rate).

$$F.L. = \frac{S_o}{rX_R + r_R X_v} \qquad (4.28)$$

in which

$S_o$ = soluble degradable COD, mg/l

$X_R$ = recycle VSS, mg/l

$X_v$ = mixed liquor VSS, mg/l

$r$ = sludge recycle ratio $R/Q$

$r_R$ = internal recycle ratio $Q_{R/Q}$

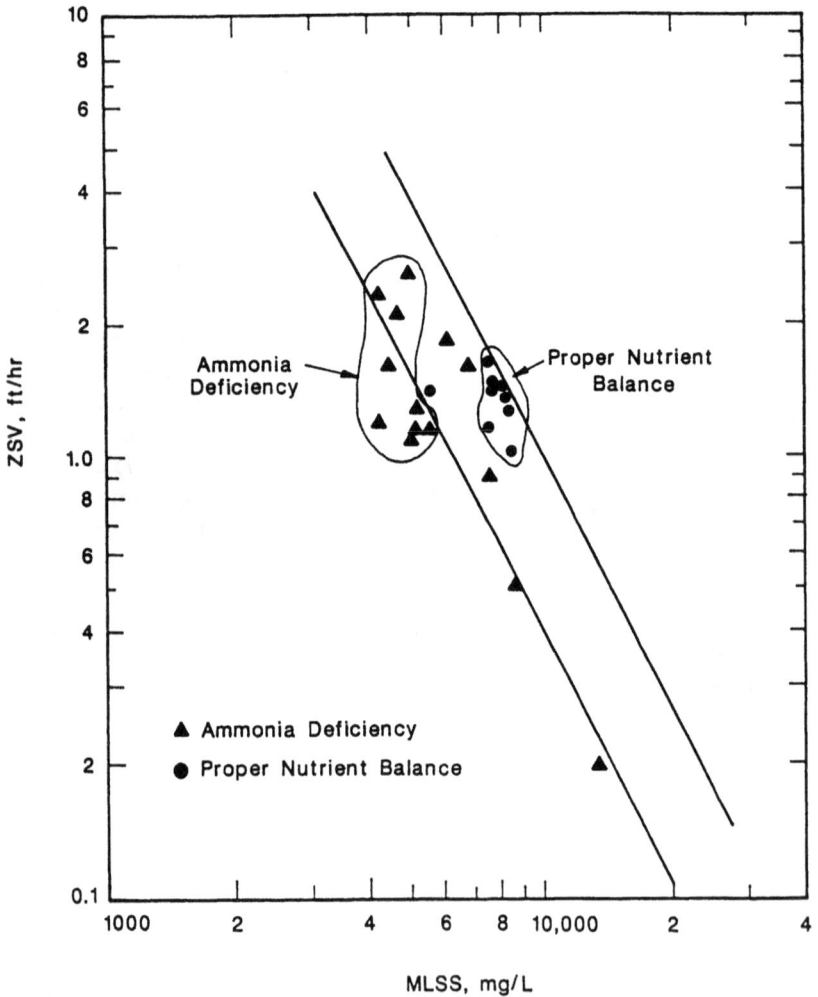

**Figure 4.55** Effect of ammonia deficiency on zone settling velocity.

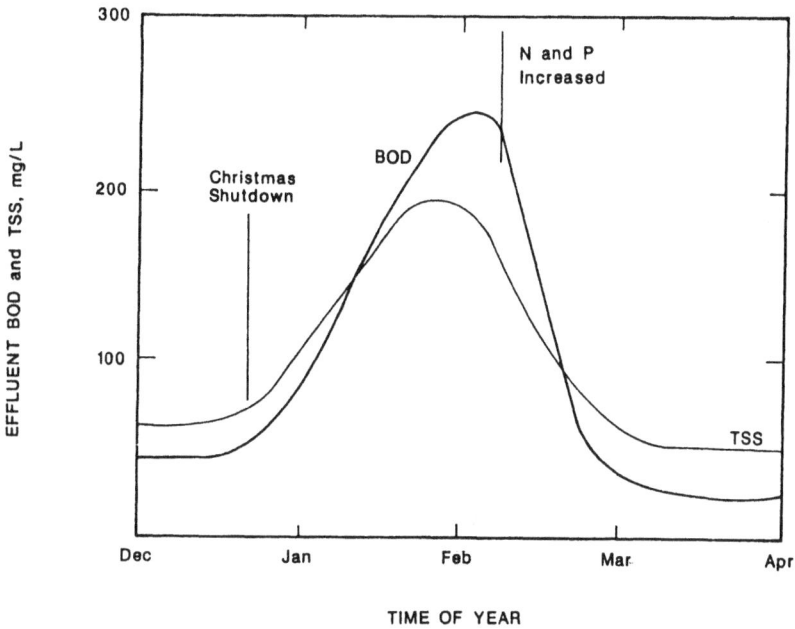

**Figure 4.56** Effect of nutrient deficiency on effluent quality for a pulp and paper mill wastewater.

## Example 4.7: Aerobic Selector Design for Industrial Wastewaters

Biosorption correlation's are shown in Figure 4.57. The design basis is 65 percent removal of sorbable COD in the selector. For the recycle paper case design an aerobic selector for an influent COD of 2000 mg/l and an MLVSS of 4000 mg/l. The floc load is 150 mg COD/g VSS. The recycle sludge concentration is 8000 mg/l VSS.

$$r = \frac{X_v}{X_R - X_v} = \frac{4000}{8000 - 4000} = 1.0$$

rearranging Equation (4.28)

$$\begin{aligned}
r_R &= \frac{S_o - FL_v X_R}{FLX_v} \\
&= \frac{2000 - 0.15 \cdot 1 \cdot 8000}{0.15 \cdot 4000} \\
&= 1.3
\end{aligned}$$

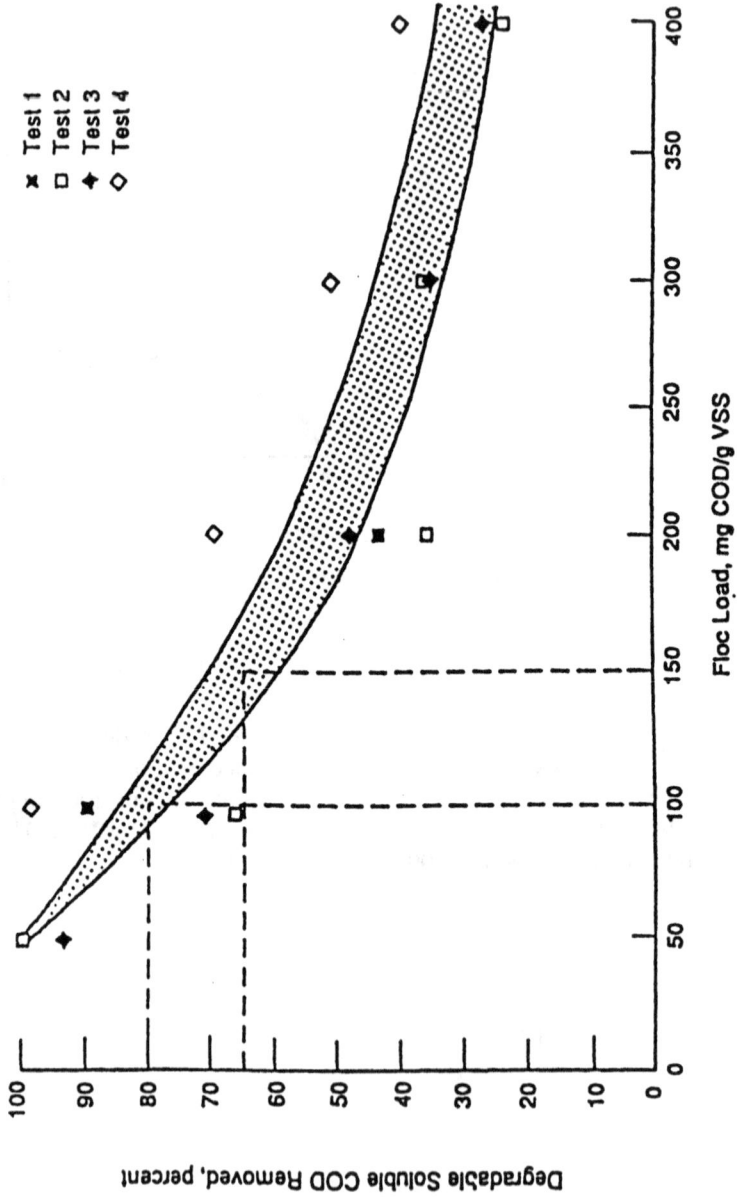

**Figure 4.57** Floc load test results for a pulp and paper mill wastewater.

260

If the return sludge recycle rate is 100% the internal recycle will be 130%. This is shown in Figure 4.58. The selector detention time based on $Q$ will be 1.1 hours.

Results from parallel treatment studies of a grapefruit processing wastewater using an aerobic selector followed by a completely mixed aeration basin versus a plug-flow regime aeration basin are shown in Table 4.18. In both cases the SVI was below 100 ml/g with effluent BOD concentrations ranging from 4 to 18 mg/l. The plug-flow reactor, however, produced a sludge that was more readily dewatered and had superior thickening properties. A parallel complete mix activated sludge process (without selector) operated at the same organic loading rate as these systems produced severe bulking problems and was shut down since it was inoperable.

Filamentous bulking has also been controlled using an anoxic selector prior to a complete mix activated sludge process. Since most of the filamentous organisms cannot function under anoxic conditions, uptake and assimilation of the organics is restricted to the floc-forming organisms. An activated sludge plant treating an organic chemicals wastewater exhibited filamentous sludge bulking due to insufficient dissolved oxygen. An investigation was conducted using an anoxic basin before the aeration basin in one train and increasing the dissolved oxygen to non-diffusion limiting levels in a parallel train. The anoxic basin was operated at a detention time of 2.5 hours, and it was found that the sludge was devoid of filaments and had small flocs. The high DO unit contained some filaments and exhibited large flocs. The settling characteristics under various operating conditions are shown in Table 4.19.

Similar results were developed in a parallel study of anoxic and aerobic selectors for treatment of a kraft pulp and paper mill wastewater. The anoxic selectors had 0.75 hr and 2.5 hr retention times while the aerobic selector had a 30-minute retention time. The effluents from both selectors

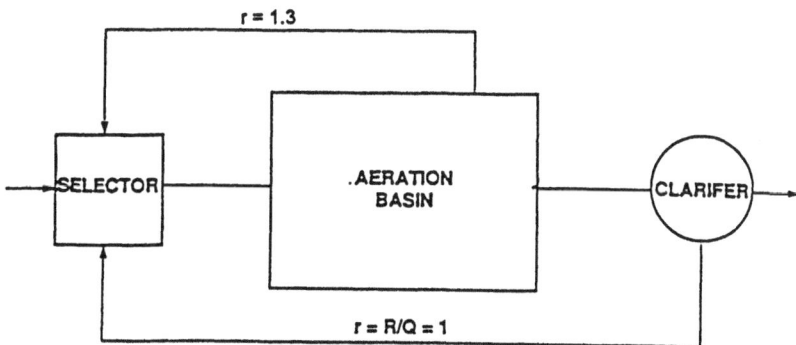

**Figure 4.58** Selector flow sheet.

TABLE 4.18. Treatment of a Grapefruit Processing Wastewater.

| | Aerobic Selector[a] with CMAS | | Plug-Flow Activated Sludge |
| | Reactor No. 1 | Reactor No. 2 | |
|---|---|---|---|
| **Operating Characteristics** | | | |
| Influent BOD, mg/l | 2,543 | 3,309 | 3,309 |
| Effluent BOD, mg/l | 18 | 4 | 6 |
| Influent COD, mg/l | 4,768 | 4,460 | 4,460 |
| Effluent COD, mg/l | 221 | 139 | 135 |
| MLVSS, mg/l | 3,431 | 5,975 | 5,333 |
| SRT, days | 7.2 | 13.2 | 135 |
| Temperature, °C | 22 | 22 | 22 |
| SVI, ml/g | 71 | 67 | 69 |
| F/M, d$^{-1}$ | 0.32 | 0.24 | 0.20 |

**Sludge Characteristics**
Limiting flux to achieve a 1.5% underflow

| | Solids Flux w/o Polymer (lbs/ft²-d) | Solids Flux w/Polymer (lbs/ft²-d) |
|---|---|---|
| Selector w/CMAS | 5.5 | 47 |
| Plug-flow | 24.5 | 62 |

| Specific Resistance | Polymer Dosage (lb/ton) | Specific Resistance (sec²/g × 10⁸) |
|---|---|---|
| Selector w/CMAS | 3.2 | 190 |
| Plug-flow | 2.4 | 104 |

[a]Roc toad = 120 mg CDD/gVSS.


TABLE 4.19. Sludge Characteristics under Various Operating Conditions.

| System | ZSv (ft/hr) | SVI (ml/g) | MLSS (mg/l) | F/M[a] (g/g-d) | Temperature (°C) | Effluent TSS (mg/l) |
|---|---|---|---|---|---|---|
| Low DO | 0.6 | 222 | 3,500 | 0.44 | 33 | 120 |
| Anoxic | 2.0 | 116 | 3,850 | 0.44 | 34 | 50 |
| | 2.0 | 129 | 3.600 | 0.35 | 40 | 140 |
| | 1.4 | 228 | 2,800 | 0.44 | 45 | 130 |
| High DO | 3.6 | 100 | 3,450 | 0.31 | 35 | 40 |
| | 3.5 | 96 | 3,350 | 0.61 | 39 | 160 |
| | 5.5 | 90 | 3,200 | 0.55 | 45 | 200 |
| Low loading | 3.2 | 133 | 2,700 | 0.28 | 24 | 20 |
| | 4.2 | 133 | 2,200 | 0.25 | 35 | 275 |

[a]BOD basis.

were treated by pure oxygen-activated sludge systems. The probability plots of solids flux rates for the three units are presented in Figure 4.59. Photomicrographs (at 100× magnification) of the sludges are shown in Figures 4.60 and 4.61. Note the absence of filaments in the mixed liquor with the anoxic selector (2.5 hr) compared to that with the aerobic selector.

In some cases, increased influent BOD/COD loads result in filamentous bulking that cannot be controlled by process modifications. It has been found that chlorination of the sludge will result in a selective kill of the filaments. At a controlled dosage of chlorine, the filaments with a high surface area to volume ratio will be destroyed (along with those floc-forming organisms on the periphery of the floc). Jenkins et al. [23] have shown that filamentous bulking can be controlled with chlorine dosages of 10 to 12 lbs $Cl_2$/1,000 lbs MLSS-d. It should be noted that if nitrification is desired, the chlorine dosage should be restricted to 4.5 lbs $Cl_2$/1,000 lbs MLSS-d since the nitrifiers are more sensitive and tend to concentrate on the outer floc layers where the dissolved oxygen levels are higher.

When treating low-strength industrial wastewaters that require a short aeration basin detention time (<6 to 8 hrs), chlorine should be applied to the return sludge. For high-strength industrial wastewaters, however, with long hydraulic retention times, the chlorine must be directly applied to the aeration basin. In some cases, hydrogen peroxide has successfully been

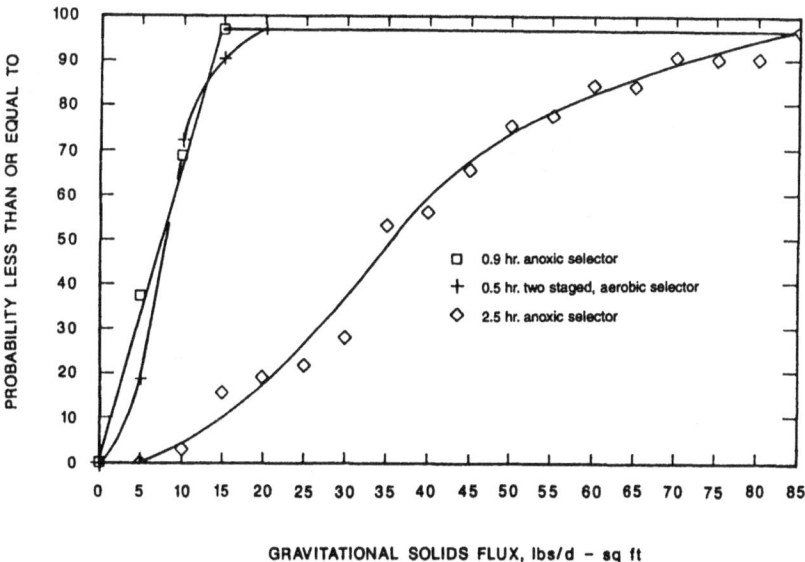

**Figure 4.59** Solids flux rates for anoxic and aerobic selectors treating a pulp and paper mill wastewater.

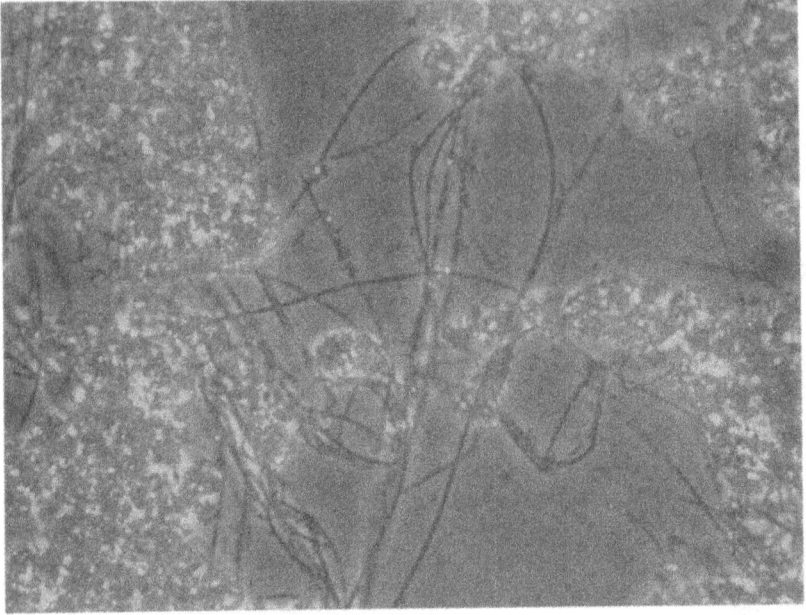

**Figure 4.60** Photomicrograph of mixed liquor from aerobic selector (100×).

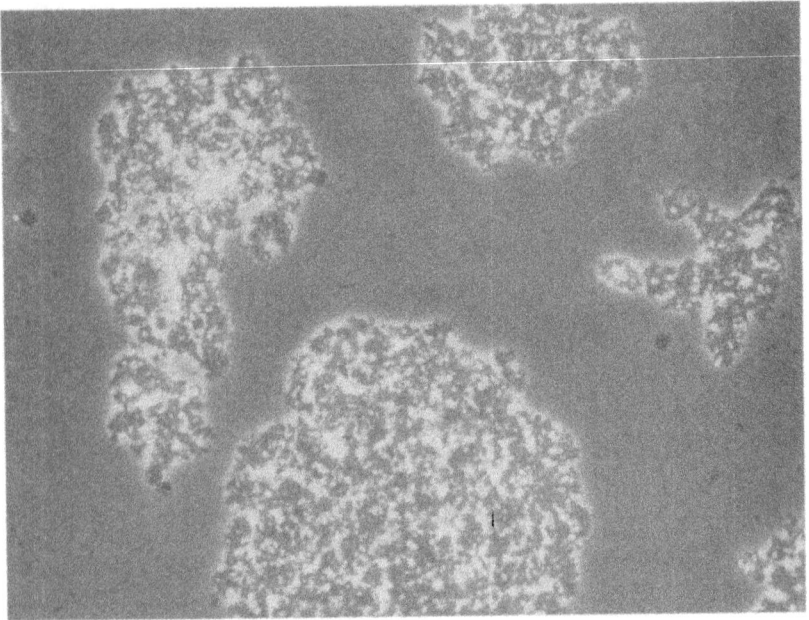

**Figure 4.61** Photomicrograph of mixed liquor from anoxic selector (2.5 hr HRT) (100×).

264

applied for sludge bulking control. It is probable, however, that this relates to low dissolved oxygen filaments, which are suppressed by the oxygen released by the hydrogen peroxide.

## STRIPPING OF VOLATILE ORGANICS

Volatile organic carbon compounds will be released to the atmosphere by oxygenation in the activated sludge process. Depending on the particular VOC, both air-stripping and biodegradation may occur.

The fraction of VOC that is stripped depends upon several factors. Compound specific factors include the Henry's Law Constant, the compound's biodegradation rate, and in some cases the initial concentration of the compound and other substrates. Operational and facility design factors that influence stripping are the method of oxygenation and the power level in the aeration basin. The concentration of non-volatile organics in the wastewater will affect the composition and mass of the sludge and hence the fraction of VOC that is biodegraded and stripped.

The stripping of toluene from municipal activated sludge plants and from an industrial wastewater consisting of creosols and phenolics is shown in Figure 4.62. Higher biodegradation was experienced with a biomass

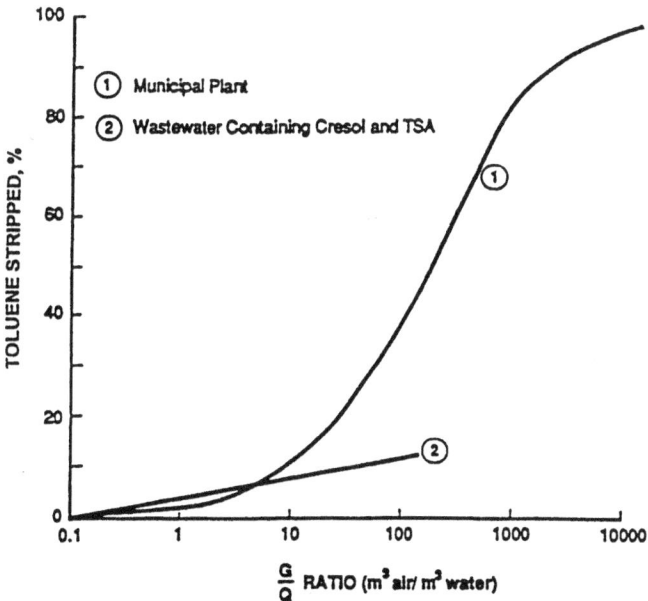

**Figure 4.62** Volatilization of toluene from a municipal and an industrial wastewater treatment plant.

acclimated to creosols. Biodegradation and stripping removal data for a range of VOCs are presented in Table 4.20. As a general rule, as more chlorine atoms are added to the organic compound, the rate of biodegradation decreases and the amount of compound stripped increases. This phenomenon is shown in Figure 4.63 for several chlorinated compounds in the benzene series [20]. It should be noted that all other things being equal, a greater SRT will result in less stripping since a lower power level will usually be employed and the biomass concentration will be higher.

In a diffused air oxygenation system, equilibrium between the gas and liquid phase is quickly reached after the bubble is formed. Under these conditions the amount of VOC that is stripped depends primarily on the gas to liquid ratio. Since under most conditions the volume of gas is relatively small, stripping is minimal. In a mechanical surface aeration system, however, the volume of gas in contact with the liquid is infinite and hence a greater quantity of VOC will be stripped. The effect of power level and the type of aeration device on the stripping of benzene is shown in Figure 4.64. Figure 4.64 was developed assuming an influent COD of 250 mg/l and an influent benzene concentration of 10 mg/l.

## NITRIFICATION AND DENITRIFICATION

Nitrification is a well-defined and established process in the treatment of municipal wastewater. When treating industrial wastewaters, however,

TABLE 4.20. Fate of Selected VOCs in the Activated Sludge Process.

| Compound | SRT (days) | Amount Stripped (%) | Influent Concentration (mg/L) | Reference |
|---|---|---|---|---|
| Toluene | 3 | 12–16 | 100 | [26] |
| | 3 | 17 | 0.1 | [27] |
| Ethylbenzene | 3 | 15 | 40 | [26] |
| | 12 | 5 | 40 | [21][26] |
| | 6 | 22 | 0.1 | [20][27] |
| Nitrobenzene | 6 | < 1 | 0.1 | [20][27] |
| Benzene | 6 | 15 | 153 | [21][26] |
| | 6 | 16 | 0.1 | [20][27] |
| Chlorobenzene | 6 | 20 | 0.1 | [20][27] |
| 1,2-Dichlorobenzene | 6 | 59 | 0.1 | [20][27] |
| 1,2-Dichlorobenzene | 6 | 24 | 83 | [21][26] |
| 1,2,4-Trichlorobenzene | 6 | 90 | 0.1 | [20][27] |
| o-Xylene | 6 | 25 | 0.1 | [20][27] |
| 1,2-Dichloroethane | 3 | 92–96 | 150 | [21][26] |
| 1,2-Dichloropropane | 6 | 5 | 180 | [21][26] |
| Methyl ethyl ketone | 7 | 3 | 55 | [28] |
| | 7 | 10 | 430 | [28] |
| 1,1,1-trichloroethane | 6 | 76 | 141 | [28] |

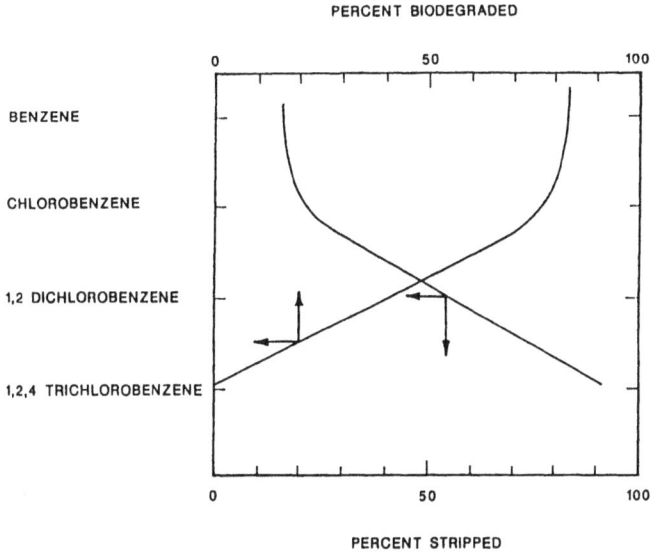

**PERCENT BIODEGRADED**

BENZENE

CHLOROBENZENE

1,2 DICHLOROBENZENE

1,2,4 TRICHLOROBENZENE

**PERCENT STRIPPED**

**Figure 4.63** Relationship between biodegradation and stripping for the chlorinated benzene series.

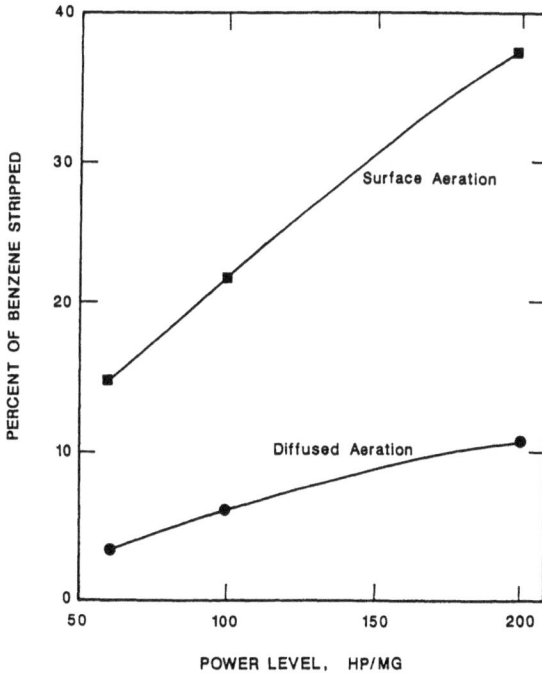

**Figure 4.64** Stripping of benzene as related to power level and type of aerator.

267

nitrification is frequently inhibited or in some instances prevented by the presence of toxic organic compounds or heavy metals. Nitrification results for treatment of an organic chemicals wastewater are presented in Figure 4.65 [29]. These data show that a minimum aerobic SRT of 25 days was required to obtain complete nitrification at 22 to 24°C. The minimum SRT required for complete nitrification of municipal wastewater at these temperatures is approximately 2 to 3 days. An SRT of 55 to 60 days was required for complete nitrification at 10°C versus approximately 7 days for municipal wastewater. It was also shown that at a mixed liquor temperature of 10°C, the nitrifiers were less tolerant to variations in influent composition and temperature than were the heterotrophic organisms responsible for BOD removal and denitrification.

Similar results were obtained for a wastewater from a coke plant in which the nitrification rate was approximately one order of magnitude less than that for municipal wastewater as shown in Figure 4.66. Blum and Speece et al. [14] have identified the toxicity of a variety of organic compounds to nitrification as shown in Table 4.13. In cases where nitrification is significantly reduced or totally inhibited, the application of powdered activated carbon to adsorb the toxic agents may enhance nitrification.

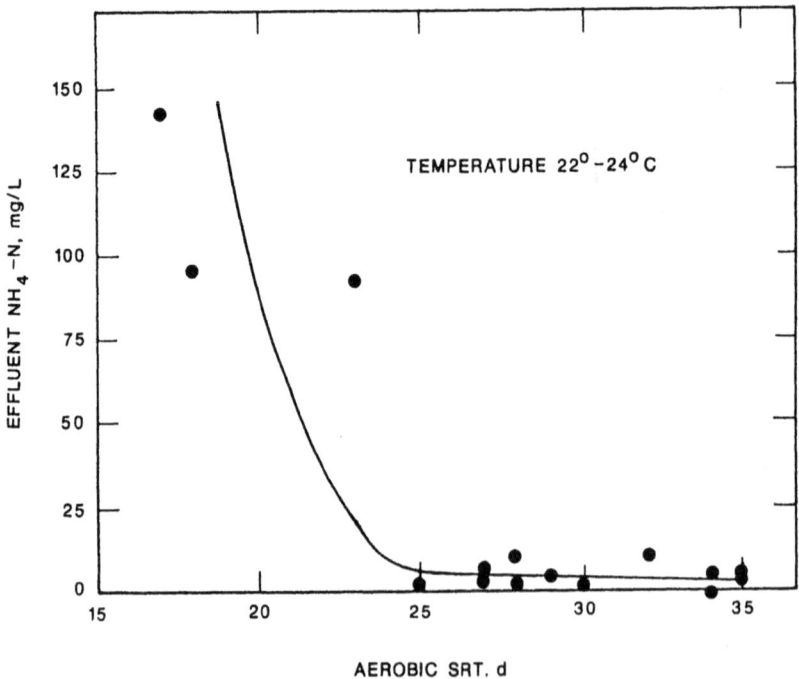

**Figure 4.65** Nitrification relative to the aerobic SRT for an organic chemicals wastewater.

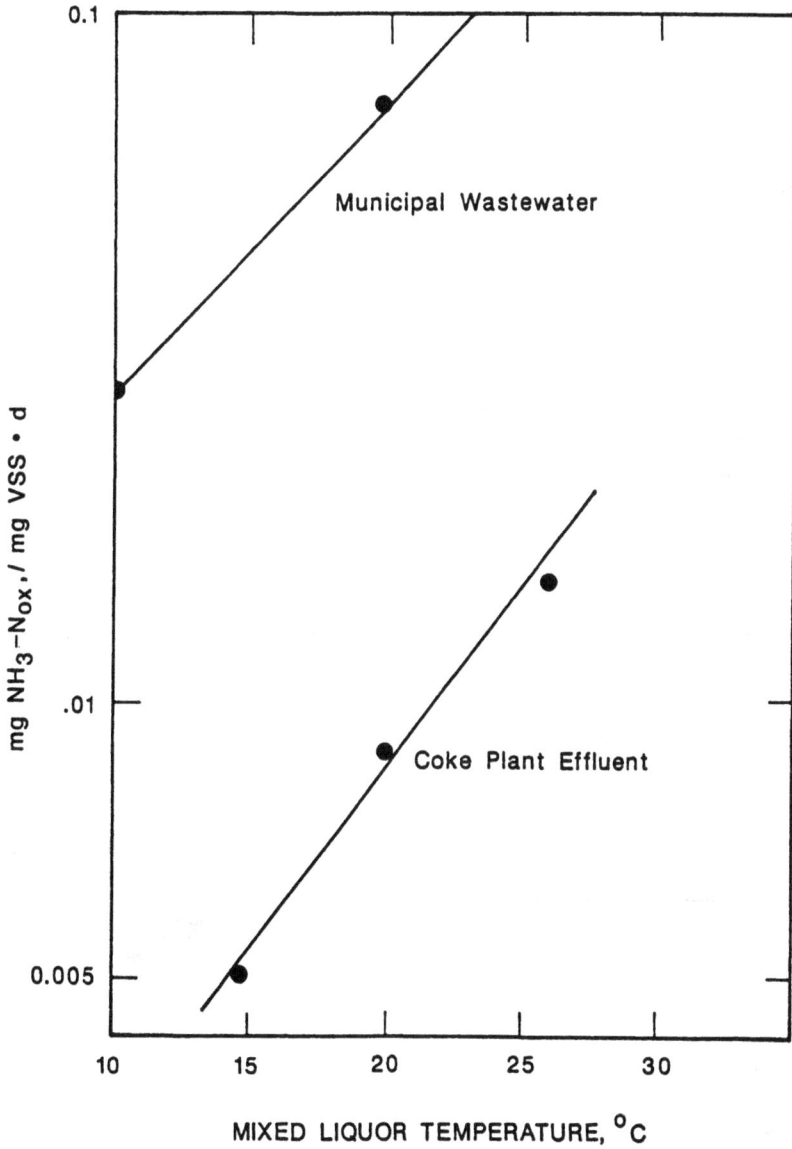

**Figure 4.66** Relationship between nitrification rate and temperature for municipal wastewater and a coke plant effluent.

However, in some cases excessive quantities of PAC are required to achieve single-stage nitrification and a second-stage nitrification step can be successfully employed after a first-stage activated process for removal of carbonaceous material and reduction of nitrifier toxicity. This is shown in Table 4.21 for a plastics additives wastewater in which separate stage nitrification was more effective than single stage PACT® nitrification, which required carbon doses of at least 1,000 mg/l for equivalent nitrification efficiency.

Wastewaters containing high ammonia concentration and negligible BOD can be treated by activated sludge nitrification. As an example, wastewater from a fertilizer manufacturing complex was treated by the activated sludge process. The ammonia nitrogen content of the influent wastewater varied from 339 to 420 mg/l and inorganic suspended solids varied from 313 to 598 mg/l. The TDS was 6,300 mg/l. Because of the high inert suspended solids, the mixed liquor was 20 percent volatile with an SVI of 30 to 40 ml/g. A small fragile floc was generated, which provided an effluent TSS of 55 mg/l. Alkalinity was supplied to the system in the form of sodium bicarbonate. The nitrification rate relationship with temperature is shown in Figure 4.67. The rate is lower than that predicted for nitrification of municipal wastewater. The alkalinity requirements showed considerable variation (Figure 4.68) which was attributed to the presence of $CaCO_3$ in the inert suspended solids.

Ammonia in the non-ionic form ($NH_3$) inhibits both *Nitrosomonas* and *Nitrobacter* as shown in Figure 4.69 [30]. Since the non-ionic form increases with pH, a high pH combined with a high total ammonia concentra-

TABLE 4.21. Single Stage versus Separate Stage Batch Nitrification Testing of Plastic Additives Wastewater.

| Process | Nitrogen Concentrations at Varying Aeration Times | | | | | |
|---|---|---|---|---|---|---|
| | t = 0 | | | t = 26 hrs | | |
| | TKN (mg/l) | NH₃–N (mg/l) | NO₃–N (mg/l) | NH₃–N (mg/l) | NO₃–N (mg/l) | Nitrif. (%) |
| Single Stage @ F/M = 0.10 d⁻¹ | | | | | | |
| • 0 mg/l PAC | 126 | 26 | 1.5 | 95 | 0.2 | < 1 |
| • 500 mg/l PAC | 134 | 31 | 1.2 | 113 | 2 | 1 |
| • 1,000 mg/l PAC | 52 | 33 | 1.3 | 88 | 72 | 23 |
| • 5,000 mg/L PAC | 92 | 34 | 1.1 | 75 | 38 | 41 |
| Separate Stage @ MLVSS = 1,000 mg/l | 105 | 79 | 7 | 45 | 29 | 21 |

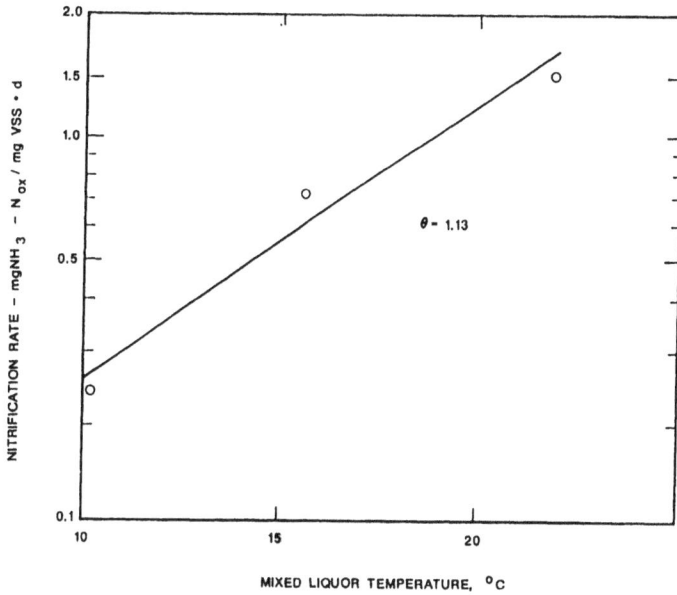

**Figure 4.67** Relationship between nitrification rate and temperature for a fertilizer wastewater.

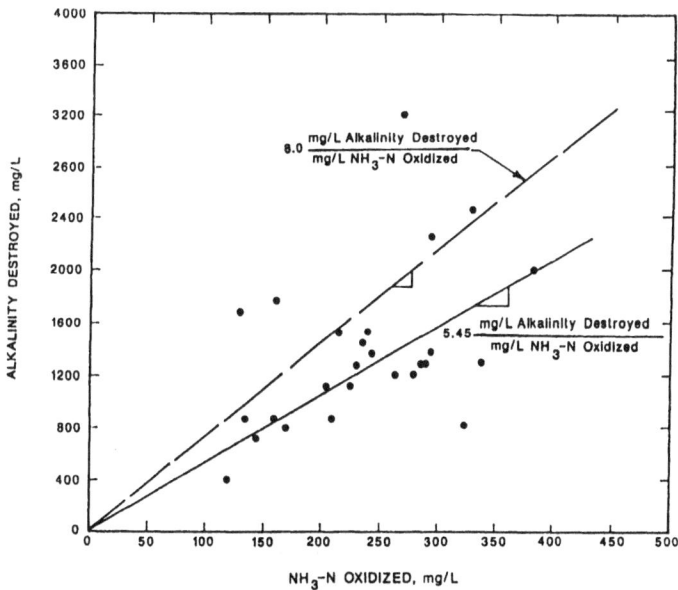

**Figure 4.68** Alkalinity utilization in the treatment of a fertilizer wastewater.

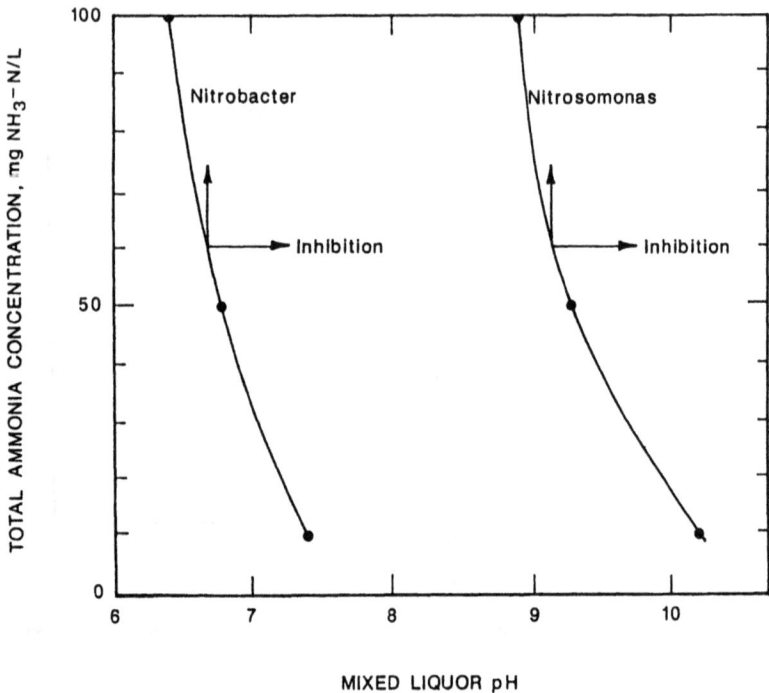

**Figure 4.69** Ammonia inhibition in the activated sludge process.

tion will severely inhibit or prevent complete biological nitrification. Since *Nitrosomonas* is less sensitive to ammonia toxicity than *Nitrobacter*, the nitrification process may only be partially complete and result in accumulation of nitrite ion ($NO_2^-$). This can have severe consequences since $NO_2^-$ is strongly toxic to many aquatic organisms whereas $NO_3^-$ is not. Ammonia toxicity to aquatic organisms is rarely a problem when treating municipal wastewaters since the concentration of ammonia is low and the pH is near neutral. Industrial wastewaters with high ammonia levels and the potential for high pH excursions, however, are a special problem. Under these conditions, it is necessary to control the mixed liquor pH to avoid toxicity due to an ammonia spill or shock load.

Prakasam et al. [31] successfully operated a pilot-scale activated sludge system that achieved greater than 95 percent total nitrogen reduction (nitrification and denitrification) with an influent $NH_4^+$–N concentration in excess of 500 mg/l. The influent wastewater was pond supernatant from anaerobic digestion of poultry wastes. To accomplish this level of nitrogen reduction, the following conditions were provided in the activated sludge process.

- A four-month acclimation period was implemented, during which

time the ammonia loading rate was increased from an initial 0.023 lbs N/lb MLVSS-d to 0.141 lbs N/lb MLVSS-d.
- Methanol was added prior to the denitrification zone at a rate of 4.5 lb/lb $NO_3N$ removed to ensure uninhibited denitrification.
- The process was maintained at a pH level of $8.2 \pm 0.2$.
- A dissolved oxygen residual of 4.0 mg/l was maintained and alkalinity addition using $Na_2CO_3$ was practiced.

Results from this study are summarized in Table 4.22. They demonstrate near-complete total nitrogen removal but poor BOD and TSS removals. The high effluent BOD was due to excess methanol addition and could be lowered by reducing the methanol addition rate or by further biological treatment. The high effluent TSS could be reduced through enhanced clarification using chemical coagulation and flocculation.

Some industrial wastewaters such as those from fertilizer, explosive/ propellant manufacture, and the synthetic fibers industry contain high concentrations of nitrates while others generate nitrates through nitrification. Since biological denitrification generates one hydroxyl ion, while nitrification generates two hydrogen ions, it may be advantageous to couple the nitrification and denitrification processes to provide "internal" buffering capacity. While many organics inhibit biological nitrification, this is not generally true for denitrification. Sutton et al. [29] showed that denitrification rates for an organic chemicals plant wastewater were comparable to those observed using nitrified municipal wastewater. However, biological nitrification of the organic chemicals wastewater was severely inhibited.

Denitrification uses the available BOD as a substrate for carbon and energy and nitrate as an oxygen source. Results of denitrification of an organic chemicals plant wastewater are shown in Figure 4.70 and the temperature effect on denitrification is shown in Figure 4.71.

In similar manner to aerobic oxidation, the denitrification rate will be related to the *F/M* or SRT of the system which in turn determines the fraction of active biomass. Figure 4.72 shows results obtained from the denitrification of municipal wastewater. A similar relationship should be experimentally obtained for industrial wastewaters. As the *F/M* decreases, the rate of denitrification declines. Although denitrification can occur under endogenous conditions (low *F/M*) using internal biomass reserves, it is very slow and requires long retention times. In cases where the wastewater contains constituents of varying degradability, multi-staging of the anoxic basin is desirable.

## ACTIVATED SLUDGE PROCESSES

The objective of the activated sludge process in treating industrial

TABLE 4.22. Removal of Nitrogen in Activated Sludge Nitrification-Denitrification System.

| Operational Mode | Month of Operation | Temp. (°C) | HRT (days) | NH$_4$—N Loading Rate (lb/lb VSS-d) | MLVSS (mg/l) | Influent (mg/l) | | | Effluent (mg/l) | | | |
|---|---|---|---|---|---|---|---|---|---|---|---|---|
| | | | | | | TBOD | TSS | NH$_3$—N | TBOD[a] | TSS$^a$ | NH$_3$—N | NO$_3$—N |
| **Mode I** | | | | | | | | | | | | |
| 2 hrs aeration plus 2 hrs non-aeration | 1 | 16 | 4 | 0.123 | 1,049 | 66 | 61 | 548 | 158 | 223 | 65 | 288 |
| | 2 | 21 | 4 | 0.099 | 1,387 | 66 | 61 | 548 | 158 | 223 | 11 | 185 |
| | 3 | 19 | 4 | 0.044 | 3,245 | 66 | 61 | 548 | 158 | 223 | 9 | 31 |
| | 4 | 22 | 4 | 0.026 | 5,260 | 66 | 61 | 548 | 158 | 223 | 4 | 11 |
| | 5 | 20 | 4 | 0.026 | 5,230 | 66 | 61 | 548 | 158 | 223 | 2 | 4 |
| **Mode II** | | | | | | | | | | | | |
| Front and rear 29% of aeration basin aerated with middle 42% of aeration basin unaerated | 1 | 24 | 4.0 | 0.023 | 4,749 | 86 | 69 | 453 | 138 | 94 | 4 | 136 |
| | 2 | 26 | 2.9 | 0.030 | 3,907 | 86 | 69 | 453 | 138 | 94 | 15 | 1 |
| | 3 | 21 | 3.1 | 0.031 | 4,659 | 86 | 69 | 453 | 138 | 94 | 2 | 23 |
| | 4 | 19 | 2.6 | 0.058 | 4,112 | 86 | 69 | 453 | 138 | 94 | 53 | 18 |
| | 5 | 14 | 3.0 | 0.045 | 3,225 | 86 | 69 | 453 | 138 | 94 | 32 | 7 |
| | 6 | 17 | 3.3 | 0.029 | 4,206 | 86 | 69 | 453 | 138 | 94 | 2 | 10 |
| | 7 | 15 | 3.3 | 0.028 | 4,855 | 86 | 69 | 453 | 138 | 94 | 1 | 1 |

[a]Increase in effluent TBOD and TSS was due to excess methanol addition and poor clarification.

274

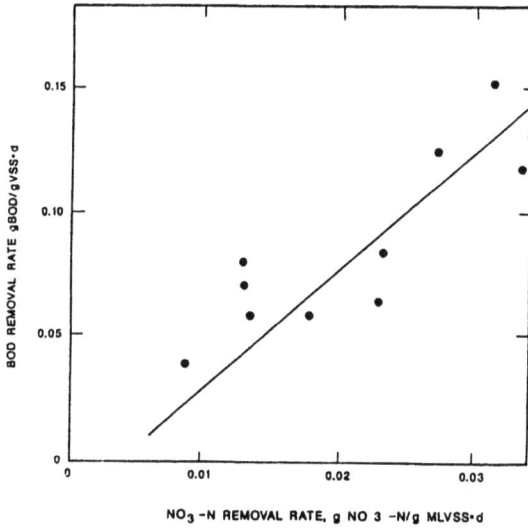

**Figure 4.70** Relationship between nitrate reduction and BOD removal for an organic chemicals wastewater.

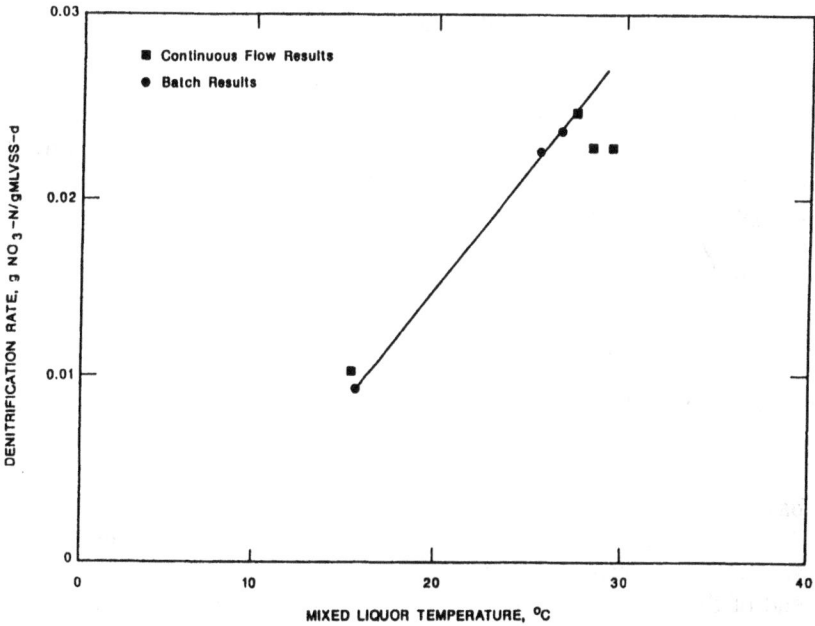

**Figure 4.71** Relationship between denitrification rate and temperature for an organic chemicals wastewater.

275

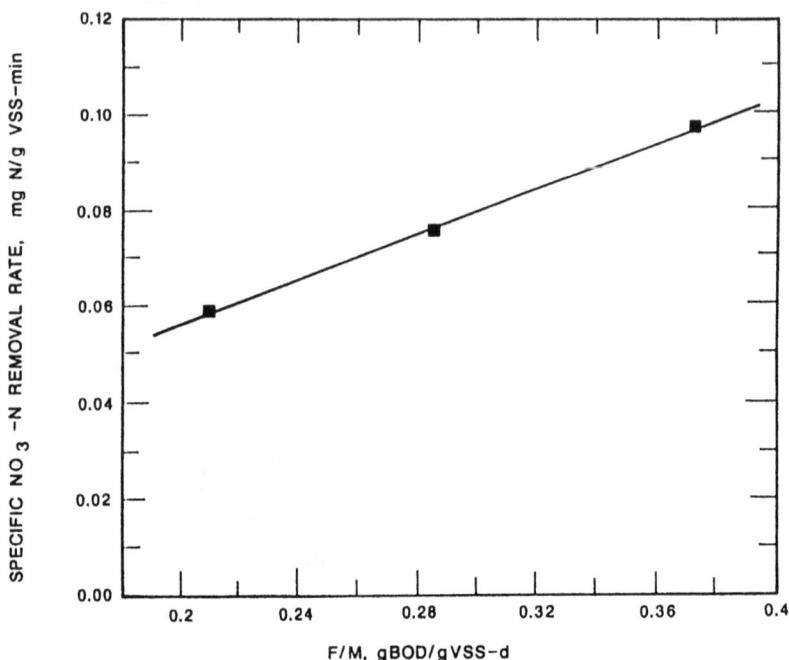

**Figure 4.72** Relationship between F/M and denitrification rate.

wastewaters is to remove soluble and insoluble organics and to convert this material into a flocculent microbial suspension that settles well in conventional gravity clarifiers. A number of modifications of the activated sludge process have been developed to accommodate specific wastewater characteristics and/or operational needs. Generic flowsheets for the treatment of industrial wastewaters are shown in Figure 4.22. As a general rule, the nature of the wastewater will dictate the preferred process modification, primarily for the purpose of maintaining mixed liquor quality.

## PLUG-FLOW ACTIVATED SLUDGE

This activated sludge process modification uses long, narrow aeration basins to provide a mixing regime that approaches plug-flow conditions. A plug-flow regime promotes the growth of a flocculent and well-settling sludge by introducing influent wastewater and return sludge at the head-end of the basin. This provides a high substrate gradient that promotes growth of floc-forming rather than filamentous biomass. If the wastewater contains toxic or bioinhibitory organics, however, they must be removed or equalized prior to entering the head-end of the aeration basin since there is negligible dilution available in the aeration basin itself. The oxygen utilization rate is high at the beginning of the basin and decreases with aeration time. Under normal operating conditions the mixed liquor oxygen utilization rate approaches the endogenous level toward the end of the aeration basin.

Modification of the way in which wastewater and return sludge are brought into contact in a plug-flow system can have a number of benefits. Provision of a separate zone at the inlet, with a volume of about 15 percent of the total aeration volume and a low-energy subsurface mechanical mixer, can achieve controlled anoxic conditions. These conditions generally promote good floc formation and control of filamentous growth. In cases where nitrification occurs, recycle of nitrified mixed liquor (and $NO_3$) from the end of the aeration basin to the anoxic zone at the head-end can achieve significant denitrification and remove some of the influent BOD.

In some cases, a multi-stage process may be employed. This has many of the advantages of plug-flow but does provide intimate mixing and equalization in each stage. Performance data for a multi-stage activated sludge plant treating a kraft pulp and paper mill wastewater is shown in Table 4.23.

## SELECTOR ACTIVATED SLUDGE

A biological selector encourages the growth of floc-forming organisms and therefore maintains sludge quality control. This is particularly important when treating readily degradable wastewaters. Different types and the advantages of biological selectors have been previously discussed.

TABLE 4.23. Multi-stage Activated Sludge Treatment of Bleached Kraft Pulp and Paper Mill Wastewater.

| Parameter | Average | Maximum | Minimum |
|---|---|---|---|
| Flow, MGD | 21.7 | 23.5 | 19.4 |
| Influent TBOD, mg/l | 241 | 260 | 210 |
| Influent SBOD, mg/l | 207 | 225 | 180 |
| Influent TSS, mg/l | 211 | 257 | 164 |
| Detention time, d | 0.132 | 0.147 | 0.121 |
| MLSS, mg/l | 2,077 | 2,219 | 1,810 |
| F/M, lbs BOD/lb MLSS-day | 0.88 | 0.97 | 0.75 |
| Temperature, °F | 72 | 104 | 57 |
| SVI, ml/g | 0.76 | 0.91 | 0.62 |
| SRT, d | 95 | 98 | 93 |
| SLR, lbs/d - sq ft | 21.6 | 23.6 | 19.3 |
| OR-gal/d - sq ft | 941 | 1,021 | 840 |
| Effluent TSS, mg/l | 78 | 99 | 68 |
| Effluent TBOD, mg/l | 41 | 48 | 34 |
| Effluent SBOD, mg/l | 19 | 25 | 13 |

## COMPLETE MIX ACTIVATED SLUDGE

In a complete mix activated sludge process, the wastewater and the return sludge are introduced into the aeration basin at multiple points to facilitate their rapid blending with the basin contents. The objective is to maximize equalization of the influent load within the aeration basin. This process is particularly applicable to wastewaters that contain toxic/bioinhibitory substances or that have highly variable loading patterns. Furthermore, wastewaters with variable pH are neutralized by the basin contents and biological activity. Another advantage of a complete mix process is that the oxygen uptake rate is equalized throughout the basin, thus permitting uniform spacing of the aeration equipment.

Complete mix activated sludge should not be employed when treating readily degradable wastewaters unless a biological selector is installed in front of the aeration basin in order to maintain good sludge settling characteristics. Performance data for CMAS treatment of various industrial wastewaters were presented in Table 4.8 and kinetic parameters in Tables 4.9 through 4.12.

## EXTENDED AERATION

The extended aeration process utilizes a long hydraulic retention time (18 to 24 hrs) and a low *F/M* (high SRT). This results in minimal sludge production but high oxygen requirements per pound of BOD removed. The *F/M* typically will vary from 0.05 to 0.15 $d^{-1}$ (SRT of 20 to 40 days) and MLSS concentrations will range from 3,000 to 5,000 mg/l. The process may be operated in either a complete mix or plug-flow mixing regime. It finds principle application in smaller industries where simplicity of operation and low sludge production are important. It is also applicable for treatment of poorly degradable organics that require high SRT to satisfy discharge limits.

## OXIDATION DITCH SYSTEMS

There are a number of ''loop-reactor'' or oxidation ditch system variants available. In these systems it is necessary to match basin geometry and aerator performance in order to yield an adequate channel velocity for mixed liquor solids transport. The key design factors in these systems relate to the type of aeration that is to be provided. It is normal to design for at least 1 ft/s midchannel velocity in order to prevent solids deposition. The ditch system is particularly amenable to those cases where both BOD and nitrogen removal is desired since both reactions can be achieved in the same basin by alternating aerobic and anoxic zones by control of the oxygenation rate and channel velocity. A typical oxidation ditch aeration basin is shown in Figure 4.73.

## SEQUENCING BATCH REACTORS AND INTERMITTENTLY AERATED AND DECANTED SYSTEMS

In the intermittent treatment approach, a single vessel provides primary biological oxidation and secondary settling processes that are normally associated with multi-tank conventional activated sludge treatment. This process can provide complete nitrification and substantial denitrification with few facility modifications by changing the timing and duration of the aeration and fill-draw cycles.

The SBR operates as a batch fill-and-draw system. Depending on the characteristics of the wastewater, the initial fill period can be rapid as in the case of a readily degradable wastewater to eliminate filamentous sludge bulking or extended in the case of a wastewater that exhibits bio-inhibition. The SBR sequence is illustrated in Figure 4.74.

Plants can be designed on an average *F/M* ratio of 0.05 to 0.20 lb BOD/lb MLVSS-d depending on the quality of effluent that is specified.

**Figure 4.73** Oxidation ditch and final clarifier system.

Influent

| | PURPOSE | OPERATION |
|---|---|---|
| **Fill** | ADD SUBSTRATE | AERATION ON OR OFF |
| **React** | REACTION TIME | AERATION ON |
| **Settle** | CLARIFICATION | AERATION OFF |
| **Decant** **Effluent** | DISCHARGE EFFLUENT | AERATION OFF |
| **Idle** | CYCLE COMPLETE | AERATION ON OR OFF |

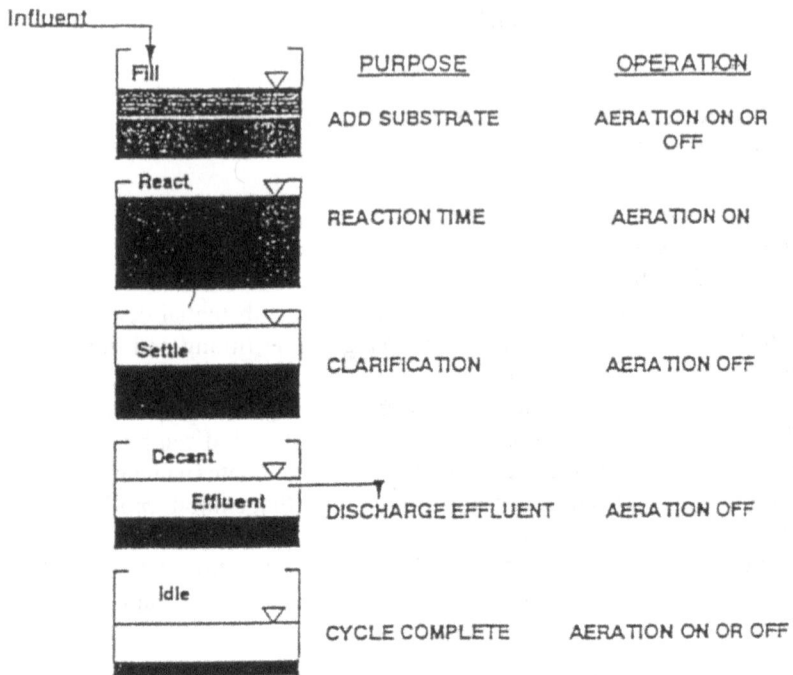

**Figure 4.74** SBR operation sequence.

In calculating the volume occupied by the sludge mass, an upper sludge volume index of 150 ml/g is used. To ensure solids are not withdrawn during decant, a buffer volume of 2 to 3 ft is provided between the final decant water level and the top of the sludge level after settlement.

Operational data for an SBR treating a pulp and paper mill effluent are shown in Figure 4.75. In this case the influent and effluent COD were 1000 mg/l and 63 mg/l respectively, the MLVSS was 4200 mg/l and the SRT was 20 days (32). Design calculations for an SBR are shown in Example 4.8.

*Example 4.8*

Design an SBR to treat a wastewater having a flow rate of 0.50 mgd and a BOD of 500 mg/l. Use an $X_v t$ of 1250 mg-day/L and a feed plus 10 hr aeration period with 1 hr settling and 1 hr effluent decant periods (12 hr total cycle time).

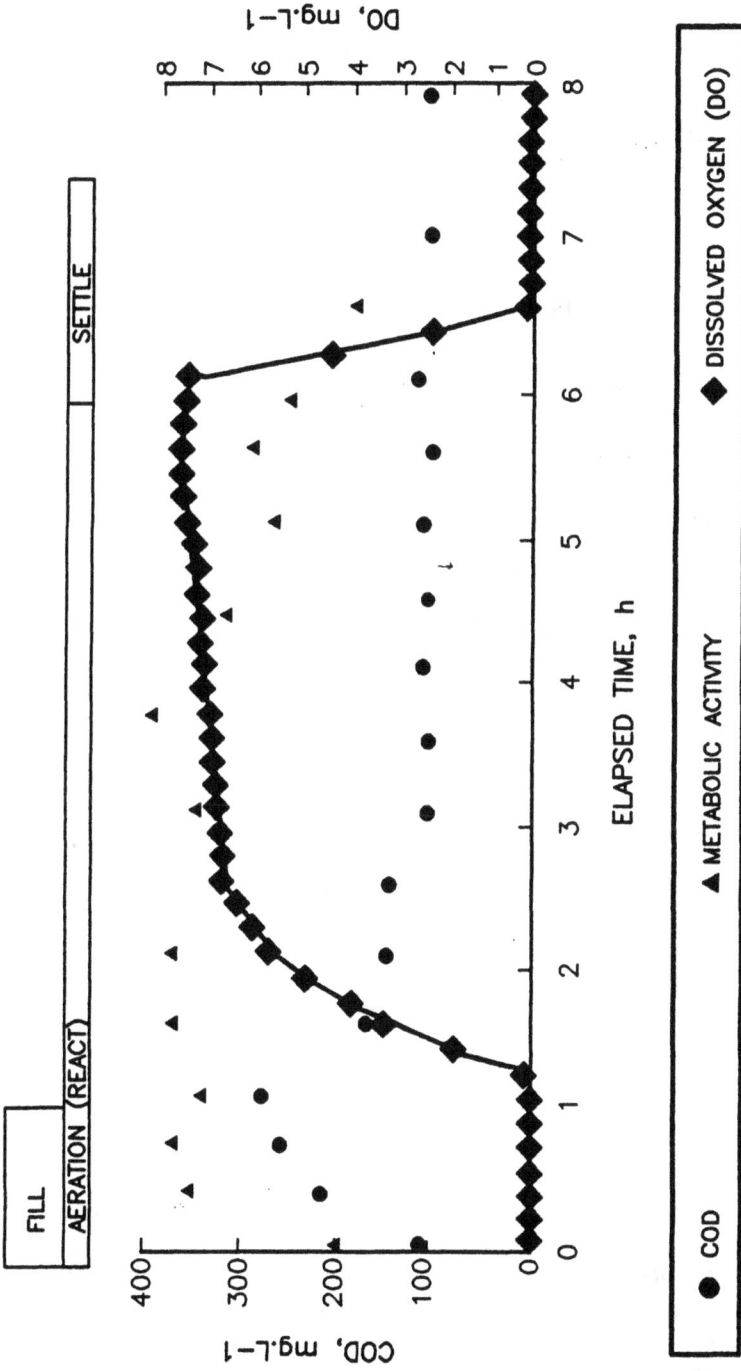

**Figure 4.75** COD and metabolic activity of the activated sludge during an SBR cycle of reactor 20 days sludge age, 8 h cycle time, 6 h aerated react time [32].

$$\text{The } X_v = \frac{1250}{10/24} = 3000 \text{ m/L}$$

The volume treated during one cycle is 0.25 mg and the MLVSS is 0.25 · 3000 · 8.34 · = 6255 lbs

At an SVI of 150 ml/g the volume required for storage of the settled sludge is

$$150 \, \frac{\text{ml}}{\text{g}} \cdot 454 \, \frac{\text{g}}{\text{lb}} \cdot 3.53 \cdot 10^{-5} \, \frac{\text{ft}^3}{\text{ml}} = 24 \text{ ft}^3/\text{lb MLSS}$$

or for MLVSS/MLSS = 0.8, the volume required is

$$= (6255)(2.4)(7.48)(1/0.8)/10^6 = 0.141 \text{ mg}$$

If $\dfrac{\text{MLSS}}{\text{MLVSS}} = 1.25 = 0.141$ mg

Provide 3 ft of freeboard between the settled sludge blanket and the supernatant water level at the end of the decant cycle.
The volume for aeration and settled sludge is

$$0.25 \text{ mg} + 0.141 \text{ mg} = 0.391 \text{ mg}$$

Select an SWD of 16 ft. The area is

$$\frac{391,000}{7.48 \cdot 16} = 3267 \text{ ft}^2$$

and the tank diameter is 65 ft.

Use a total tank depth of 19 ft to include freeboard. The MLVSS under aeration will be

$$\frac{6255}{(3267 \cdot 16 \cdot 7.48/10^6)(8.34)} = 1920 \text{ mg/L}$$

The intermittently aerated and decanted system utilizes a selector zone in which biomass and influent wastewater are mixed. The oxidation reduction potential (ORP) in this zone can be controlled by the aeration intensity, cycle time and system organic loading. In this way a positive to negative ORP is generated as needed to favor bioselective mechanisms for nitrogen and phosphorus removal. The selector zone also provides for exposure of biomass to elevated substrate concentrations that favor the growth of floc-forming microorganisms. An intermittent activated sludge system is shown in Figure 4.76.

Figure 4.77 (by Transenviro) describes the sequences of operation within a single basin. Most facilities incorporate multiple basins providing the opportunity to operate with paired basins. The basic sequences in a cycle of operation include fill-aeration, fill-settle, surface skim, and fill-idle. A moving weir is used to withdraw an upper volume of supernatant from the reactor as the treated effluent and has adjustments to permit changes to the rate and depth of removal. Typical operating cycles for industrial wastewaters are 6 to 8 hours depending on the hydraulic-organic loading relationship with 2 hours used for the settle-surface skim sequence.

During the aeration sequence, influent and aerated mixed solids from the main aeration zone are contacted in the selector zone. Soluble BOD removal is determined by the floc-loading conditions. The fill-aeration sequence is followed by the settle sequence. The settle sequence is completed, at which time the surface skimmer is activated. Residual time in the surface skim sequence is used as a fill-idle sequence with

Figure 4.76 Intermittent activated sludge system.

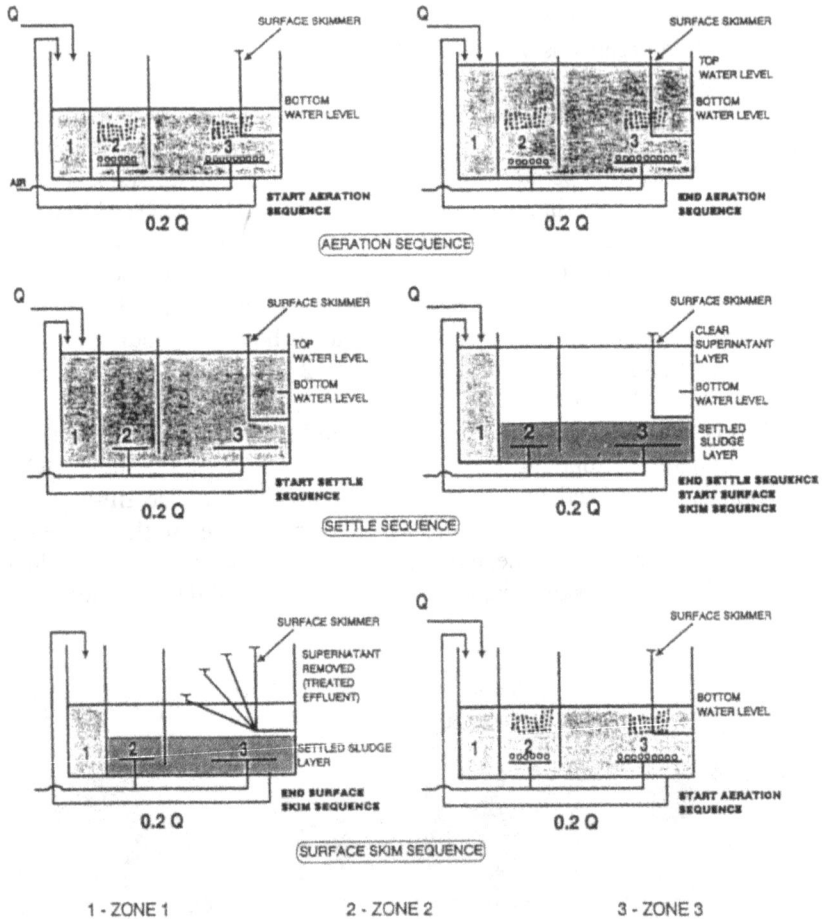

**Figure 4.77** Schematic of cyclic activated sludge sequences.

influent being directed into the basin. Performance data for intermittent systems treating two industrial wastewaters are shown in Table 4.24.

The batch activated sludge process is similar to the intermittent system except that it is usually employed for low-volume, high-strength industrial wastewaters. Wastewater is added over a short time period to maximize biosorption and flocculent sludge growth. Aeration is then continued to provide sufficient $X_v \cdot t$ to achieve the desired effluent BOD/COD concentration. The mixed liquor is then settled and the treated effluent decanted. A schematic of a typical batch activated sludge system is shown in Figure 4.78.

A specialty chemicals wastewater was treated in a batch activated

TABLE 4.24. Performance of Cyclic Activated Sludge Plants.

| Wastewater | Influent, mg/l | | Effluent, mg/l | | | F/M[a] (d⁻¹) | SVI (ml/g) |
|---|---|---|---|---|---|---|---|
| | BOD | COD | BOD | COD | TSS | | |
| Dairy | — | 1,100 | 20 | — | 30 | 0.5 | — |
| Printed circuit board | — | 650 | 6 | 95 | 9 | 0.17 | 50 |

[a]COD basis.

sludge process with the addition of powdered activated carbon and post-coagulation with ferric chloride to produce an effluent of 50 mg/l of BOD and TSS and 1,000 mg/l COD. The daily wastewater flow was 40,000 gal. The sequence of operation was 4 hours addition of wastewater, 23 hours aeration, and 1 hour settling followed by an effluent decant. The performance data for various operating conditions is shown in Table 4.25.

Batch activated sludge plants must be provided with adequate oxygenation capacity such that a minimum residual dissolved oxygen is maintained throughout the aeration cycle. This can be accomplished by increasing the available oxygen transfer capacity or by lengthening the feed cycle. Figure 4.79 demonstrates the impact of a two-hour feed cycle on oxygen demand while treating wastewaters from a hazardous waste disposal facility. Note the difference in oxygen demand depending upon the BOD and COD concentration of the feed wastewater.

## HIGH PURITY OXYGEN SYSTEMS

The high purity oxygen system is a series of well-mixed reactors employing concurrent gas-liquid contact in a covered aeration tank. Feed

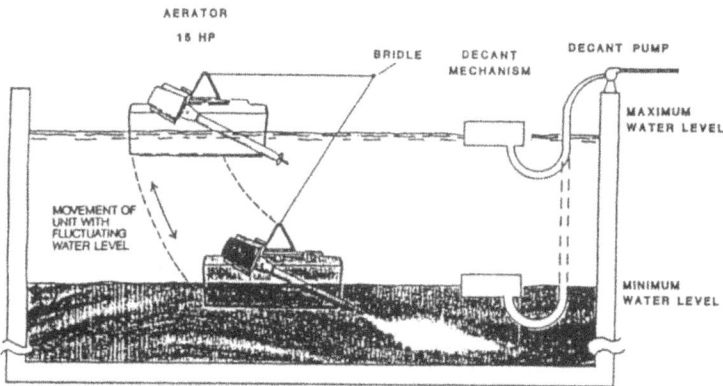

**Figure 4.78** Schematic of batch activated sludge system.

TABLE 4.25. Batch Activated Sludge Treatment Performance for a Specialty Chemicals Wastewater.[a]

| Parameter | Month | | | | | | | |
|---|---|---|---|---|---|---|---|---|
| | 1 | 2 | 3 | 4 | 5 | 6 | 7 | 8 |
| Influent TBOD, mg/l | 5,734 | 5,734 | 5,734 | 5,734 | 7,317 | 7,317 | 7,317 | 7,317 |
| Effluent SBOD, mg/l | 43 | 57 | 49 | 119 | 39 | 156 | 91 | 391 |
| Influent COD, mg/l | 10,207 | 10,207 | 10,207 | 10,207 | 15,242 | 15,242 | 15,242 | 15,242 |
| Effluent COD, mg/l | 920 | 1,992 | 1,456 | 2,067 | 705 | 2,023 | 1,682 | 2,735 |
| Effluent TSS, mg/l | 386 | 828 | 640 | 940 | 250 | 657 | 640 | 700 |
| MLSS, mg/l | 9,246 | 2,430 | 5,520 | 3,108 | 10,300 | 2,025 | 9,761 | 4,572 |
| HRT, d | 16.8 | 16.7 | 15.6 | 7.7 | 14.9 | 14.6 | 6.7 | 6.5 |
| SRT, d | 50 | 50 | 50 | 50 | 30 | 30 | 30 | 30 |
| PAC, mg/l | 1,500 | — | 500 | — | 2,000 | — | 2,000 | — |
| Feed time, hrs | 4 | 4 | 4 | 4 | 4 | 4 | 4 | 4 |
| Aeration time, hrs | 23 | 23 | 23 | 23 | 23 | 23 | 23 | 23 |
| Settling time, hrs | 1 | 1 | 1 | 1 | 1 | 1 | 1 | 1 |
| SVI (ml/g) | 19 | 157 | 32 | 74 | 65 | 75 | 19 | 74 |
| F/M (COD basis) | 0.11 | 0.3 | 0.19 | 0.19 | 0.21 | 0.60 | 0.51 | 0.68 |

[a]Monthly average value.

286

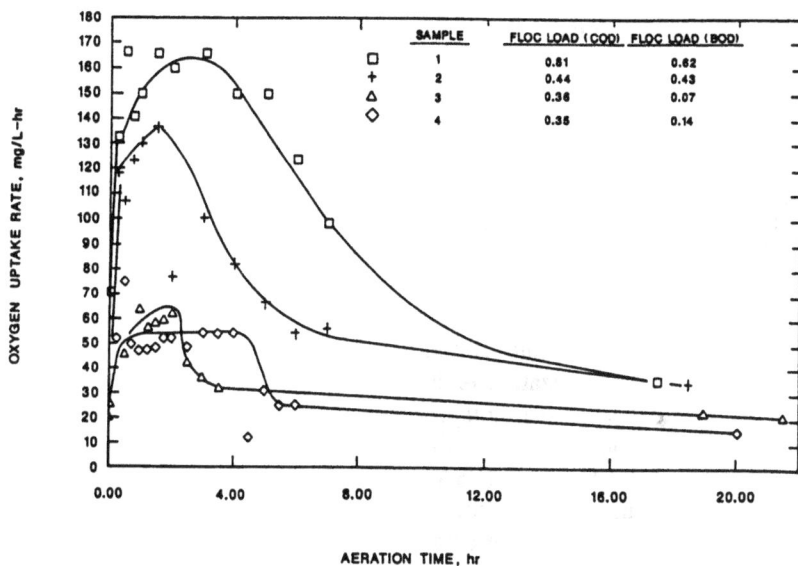

**Figure 4.79** Oxygen demand characteristics in a batch activated sludge process.

wastewater, recycle sludge, and oxygen gas are introduced into the first stage. The oxygen gas is fed at low pressure (approximately 1.5 in $H_2O$). Two gas-liquid contacting systems can be employed: submerged turbine aeration, and surface aeration. With turbine aeration, recirculating gas blowers pump the gas through a shaft to a rotating sparger. The pumping action of the impeller on the same shaft as the sparger promotes adequate liquid mixing and yields relatively long residence times for the dispersed oxygen bubbles. Gas is recirculated within a stage at a rate that is usually higher than the rate of gas flow between stages. A slight pressure drop occurs from stage to stage to prevent gas back-mixing. Since the relative liquid mixing and oxygen transfer requirements vary from stage to stage, each stage is equipped with an independent mixer-compressor combination designed to provide the required level of mixing and oxygenation.

The gas-liquid contact mechanism provided by surface aerators eliminates the need for gas recirculating compressors and associated piping. The mixing intensity required to maintain the sludge in suspension is provided by a low-speed, low-shear impeller. Oxygen gas is automatically fed to either system on a demand basis (based on headspace gas pressure) with the entire unit operating, in effect, as a respirometer. As the organic loading rate increases, the oxygen gas pressure decreases, resulting in an automatic increase in feed-oxygen flow.

Due to the high mixed liquor solids maintained in the oxygen system, the major portion of soluble BOD removal, and thus the highest oxygen demand, occurs in the first stage, which requires the highest mixer and compressor horsepower. The subsequent stages are then utilized to stabilize a sludge that has progressively decreasing oxygen demand. Effluent mixed liquor from the system is settled and the clarifier underflow is returned to the first stage for blending with the feed. The exhaust gas from the final stage is vented to the atmosphere. The system normally operates with a vent-gas composition of 30 to 50 percent oxygen. Due to the net transfer of gas to the liquid, the vent-gas flow rate will be only 10 to 20 percent of the gas feed rate. Based upon economic considerations, about 90 percent oxygen utilization is desired.

Two basic oxygen generation processes are employed: a traditional cryogenic air separation process for large installations (greater than 20 tpd) and a pressure swing adsorption (PSA) system for smaller installations. With larger installations, deep tank construction with submerged turbine aeration is preferable, while a surface aerator-PSA combination is the most cost-effective for smaller plants. The power requirements for the surface and turbine aeration equipment vary from 0.08 to 0.14 horsepower per thousand gallons depending on the waste strength, mixing requirement, feed oxygen purity and the capacity of the aeration equipment. The oxygenation systems are typically designed to maintain 6 mg/l dissolved oxygen in the mixed liquor at peak load conditions. During unusually severe peak loads, additional oxygen can be transferred to the liquid if the dissolved oxygen level decreases to 1 mg/l. Liquid oxygen storage is provided for backup purposes with the same supply capacity as the installed plant. It is therefore possible to double the feed-oxygen flow to the aeration tank if needed to satisfy unusual organic loading events. This results in an increased gas phase oxygen partial pressure and increased oxygen transfer, but reduced oxygen utilization. This is not an economical mode of operation over extended time periods. A schematic diagram of a high purity oxygen plant is shown in Figure 4.80.

The maintenance of a fully aerobic floc will maximize the endogenous rate coefficient $b$ and thereby minimize the excess sludge under moderate to high $F/M$ loading conditions. The sludge settling rate will also be at a maximum under fully aerobic conditions. These conditions are promoted under the high DO operating levels that are available with the pure oxygen-activated sludge process especially for high-strength wastewaters.

## TREATMENT OF INDUSTRIAL WASTEWATERS IN MUNICIPAL ACTIVATED SLUDGE PLANTS

Municipal wastewater is unique in that a major portion of the organics are present in suspended or colloidal form. Typically, the BOD in municipal

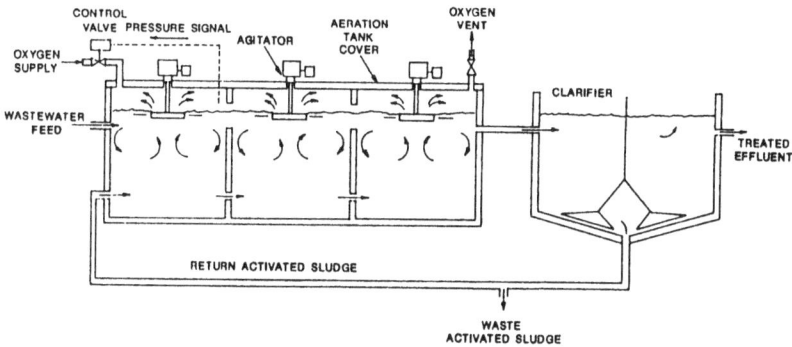

**Figure 4.80** High purity oxygen flowsheet.

sewage will be 33 percent suspended, 33 percent colloidal, and 33 percent soluble. The BOD of industrial wastewaters, however, is frequently 100 percent soluble. In an activated sludge plant treating municipal wastewater, the suspended organics are rapidly enmeshed in the flocs, the colloids are adsorbed on the flocs, and a portion of the soluble organics are absorbed. These reactions occur in the first few minutes of wastewater-biomass contact in the aeration basin. By contrast, for readily degradable industrial wastewaters, such as food processing, a portion of the BOD is rapidly sorbed and the remainder removed as a function of time and biological solids concentration. Very little sorption occurs in refractory wastewaters. The kinetics of the activated sludge process of the municipal plant will therefore vary depending on the percentage and type of industrial wastewater in the combined wastewater flow to the plant.

The percentage of biological solids in the aeration basin will also vary with the amount and nature of the industrial wastewater contribution. For example, domestic wastewater without primary clarification will yield a sludge that is approximately 47 percent biomass at a 3-d SRT. Primary clarification will increase the biomass percentage to about 53 percent. Increasing the sludge age will also increase the biomass percentage as influent volatile suspended solids undergo degradation and synthesis. Similarly, the addition of soluble industrial wastewater will increase the biomass percentage in the activated sludge.

As a result of these considerations, there are a number of factors that must be considered in the process design for activated sludge treatment of combined municipal and industrial wastewaters.

## EFFECT ON EFFLUENT QUALITY

Soluble industrial wastewaters will affect the overall substrate removal rate coefficient, $K$ of the blended wastestreams. Refractory wastewaters such as those from tannery and chemical facilities will reduce the overall rate while readily degradable wastewaters such as those from food processing and brewery facilities will increase the rate. It has been shown that for process design purposes, the mean value of the rate coefficient can be computed as a substrate mass-weighted average of the individual components constituting the wastewater. This is illustrated in Example 4.9.

*Example 4.9*

A wastewater is to be pretreated in an activated sludge process and then blended with a domestic wastewater for final treatment in a second activated sludge basin as shown below:

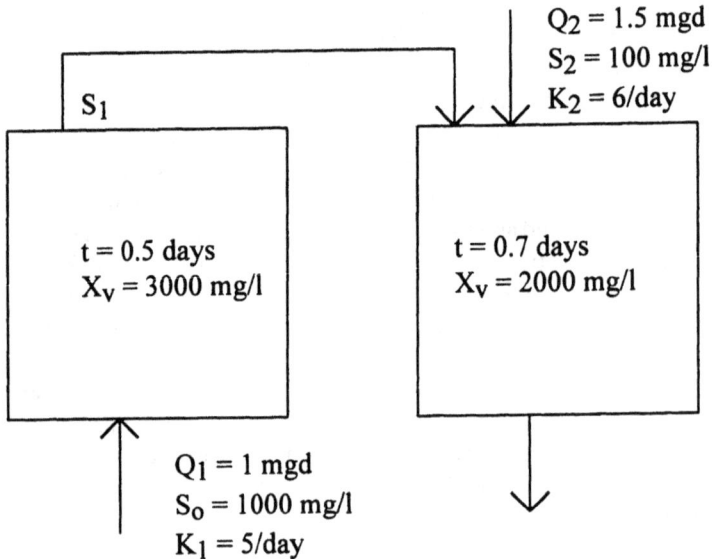

$$Q_2 = 1.5 \text{ mgd}$$
$$S_2 = 100 \text{ mg/l}$$
$$K_2 = 6/\text{day}$$

$$S_1$$

$$t = 0.5 \text{ days}$$
$$X_v = 3000 \text{ mg/l}$$

$$t = 0.7 \text{ days}$$
$$X_v = 2000 \text{ mg/l}$$

$$Q_1 = 1 \text{ mgd}$$
$$S_o = 1000 \text{ mg/l}$$
$$K_1 = 5/\text{day}$$

The effluent SBOD from the first aeration basin is

$$(S_e)_1 = \frac{1000^2}{1000 + 5 \cdot 3000 \cdot 0.5}$$
$$= 118 \text{ mg/l}$$

The reaction rate, $K_2$, of the pretreated industrial wastewater in the second basin is

$$K_2 = 5\left(\frac{118}{1000}\right) = 0.59/\text{day}$$

The influent concentration to the second basin is

$$(S_o)_2 = \frac{118.1 + 100 \cdot 1.5}{(1.0 + 1.5)}$$
$$= 107 \text{ mg/l}$$

The average rate coefficient, $\bar{K}$, after blending the two wastewaters is

$$\frac{1}{\bar{K}} = \frac{\dfrac{1}{0.59}\,[118.1] + \dfrac{1}{6}\,[100 \cdot 1.5]}{[118.1] + [100 \cdot 1.5]}$$
$$\bar{K} = 1.2/\text{day}$$

The SBOD from the second basin will be

$$(S_e)_2 = \frac{107^2}{107 + 1.2 \cdot 2000 \cdot 0.7}$$
$$= 6.5 \text{ mg/l}$$

## EFFECT ON SLUDGE QUALITY

Readily degradable wastewaters will stimulate filamentous bulking while refractory wastewaters will frequently suppress filamentous bulking. Municipal wastewater itself is subject to filamentous bulking under certain conditions. Addition of a readily degradable wastewater will enhance this potential, implying the use of a selector or a plug-flow configuration may be warranted. Depending on the wastewater mixture, bioinhibitory wastewaters may be effectively treated when mixed with municipal wastewaters in a complete mix configuration.

## EFFECT OF TEMPERATURE

Addition of an industrial wastewater that has a high soluble substrate load will increase the temperature coefficient, $\theta$, of the combined process. This will decrease process efficiency at reduced mixed liquor operating temperatures.

## SLUDGE HANDLING

An increase in soluble organics will increase the percentage of biological sludge in the waste sludge mixture. This generally will decrease thickening and dewaterability, decrease cake solids, and increase chemical conditioning requirements. An exception is pulp and papermill wastewaters, in which pulp and fiber serve as an effective sludge conditioner that enhances dewatering rates but increases the mass of solids for ultimate disposal. An evaluation should therefore be made of changes in sludge handling requirements and production rates that may result from the addition of industrial wastewater.

## BIOINHIBITION AND AQUATIC TOXICITY

As previously discussed, many industrial wastewaters exhibit inhibition of the activated sludge process, particularly with respect to nitrification. Similarly they may be a new source of potential aquatic toxicity if the toxicants "pass through" the municipal treatment works. It is therefore essential that the industrial wastewater be fully evaluated for compatibility with the existing activated sludge process to insure future permit compliance.

## NUTRIENT (N AND P) REQUIREMENTS

Since many industrial wastewaters are nutrient deficient, the BOD : N : P ratio of the combined wastewater should be determined. If the industrial load contribution is not large, the background excess nutrient concentration of the municipal wastewater may provide the required nutrient balance.

## APPLICATION OF POWDERED ACTIVATED CARBON (PACT®)

The PACT® process is the addition of powdered activated carbon to the activated sludge process. A flow schematic of a PACT® system is shown in Figure 4.81. Powdered activated carbon is mixed with the influent wastewater or fed directly to the aeration basin. The carbon bio-sludge mixture is settled and the sludge recycled in the same manner as the conventional activated sludge process. The waste activated sludge contains carbon, biological sludge, and residual influent solids.

The PACT® process offers an advantage for upgrading the performance of existing activated sludge treatment facilities since it can usually be integrated into the facility at lower capital cost than a fixed bed granular carbon adsorption system. Since the addition of PAC enhances sludge

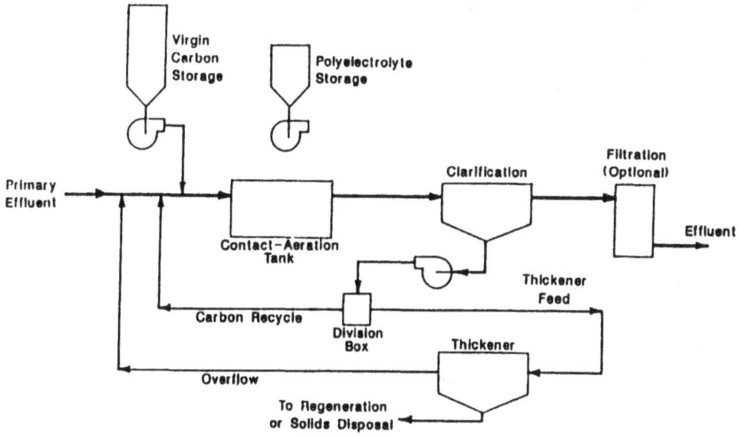

**Figure 4.81** Flow diagram for PACT® wastewater treatment system.

settleability, secondary clarifier area requirements based on conventional hydraulic and solids flux loading rates will usually be adequate, even at high carbon dosages. One problem associated with the application of PAC is that the resulting carbon-biomass sludge is abrasive. Therefore appropriate materials of construction must be provided for pumps, tankage, rake mechanisms, and sludge handling equipment. The high density of the resulting sludge will also require increased torque limits on the final clarifier rake mechanism and subsequent sludge processing equipment.

The PAC mixed liquor sludge will be a mixture of biological floc, carbon particles, and biomass growing on the carbon. Both degradable and nondegradable organics are adsorbed on the carbon, but it is assumed that at high SRT most of the degradable organics will be biologically oxidized. In general, the adsorption capacity of the carbon in the PAC sludge will be higher than that predicted by a batch adsorption isotherm. For this reason, continuous flow treatment studies should be conducted to develop process design and operating criteria for PACT® applications. The comparative results of carbon adsorption capacity in a continuous flow PACT® reactor versus their respective isotherm data for an organic chemicals wastewater are shown in Figure 4.82. Guidelines for design of PACT® systems have been developed by Zimpro. Typical process design criteria are SRT = 5 to 20 d, MLSS = 15 to 20 g/l and PAC feed rates of at least 100 mg/l. Actual operating conditions however, depend on effluent requirements and the purpose of PAC addition. Secondary clarifier area requirements are usually based on solids loading rate, which is controlled by the PAC dosage and SRT required for discharge permit compliance.

PAC is applied to the activated sludge process when conventional

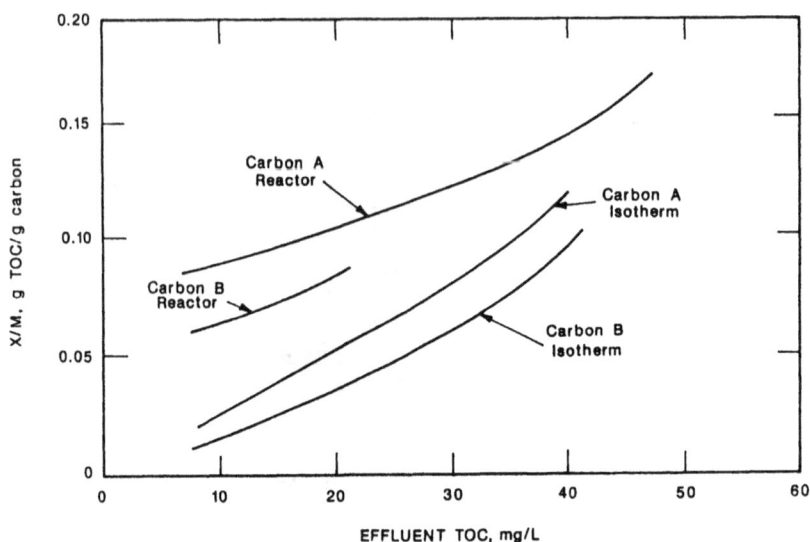

**Figure 4.82** Performance comparison between continuous PAC reactors and their adsorption isotherms.

biological oxidation cannot meet effluent requirements. This usually involves the removal of color, nondegradable COD or TOC, effluent aquatic toxicity, or bioinhibition of heterotrophic or nitrifying organisms. PACT® treatment of an agricultural chemicals wastewater demonstrated that varying PAC dosages were required to achieve effluent requirements for STOC and color. The PAC dosage however, had no effect on acute toxicity or SBOD of the effluent. These results are summarized in Table 4.26.

TABLE 4.26. Effects of Varying PAC Dosages on Activated Sludge Treatment of an Organic Chemicals Wastewater.

| Parameter | Influent | Effluent Quality at PAC Dosage (mg/l) | | | | Tertiary PAC Contactor at 1,000 mg/l PAC[a] |
| | | 0 | 250 | 500 | 1,000 | |
|---|---|---|---|---|---|---|
| SBOD, mg/l | 460 | 21 | 26 | 23 | 20 | NA[b] |
| STOC, mg/l | 380 | 146 | 128 | 121 | 91 | 70 |
| Color, APHA | 1,130 | 1,140 | 750 | 540 | 300 | 240 |
| TSS, mg/l | NA | 20 | 54 | 27 | 25 | NA |
| SOUR, mg/g · hr | NA | 4.9 | 6.3 | 5.4 | 4.2 | NA |
| 48 hr LC$_{50}$[c], % | NA | 19 | 32 | 32 | 33 | 32 |

[a]Treating effluent from activated sludge reactor without PAC addition.
[b]Not analyzed.
[c]With *Daphnia pulex*.

Furthermore, tertiary treatment with carbon (following activated sludge) provided slightly better effluent STOC and color quality than PACT® treatment at the same carbon utilization rate. TOC removal with PAC for several case studies is shown in Table 4.27.

Nondegradable TOC or COD is removed by PAC, but the removal efficiency depends on the chemical characteristics of the TOC and COD material. Much of the color present in textile, pulp and paper, and dyestuff wastewaters is nonbiodegradable but adsorbable on activated carbon. This is shown in Figure 4.83. Results of the application of PAC to a dyestuffs wastewater for removal of color are shown in Figure 4.84.

The concentrations of certain heavy metals are effectively reduced in the PACT® process. Metals removal may occur by adsorption of an organic that has complexed the metal, by surface precipitation with sulfide occurring on the carbon due to its high sulfur content, or by co-precipitation with the biological floc. Results of metals removal by PACT® from an organic chemicals wastewater are shown in Figure 4.85.

Another major application of the PACT® technology is reduction of the aquatic toxicity of a biologically treated effluent. Although no consistent relationship has been established between reduction of aquatic toxicity and reduction of other parameters, such as TOC, $NH_3$-N, or metals, in the majority of cases PAC will effect a reduction of aquatic toxicity. Toxicity reduction results from PACT® treatment of an organic chemicals effluent are shown in Figure 4.86.

As previously discussed, biological nitrification is sensitive to the presence of heavy metals and certain organic compounds. The application of PAC can reduce or eliminate this inhibition by reducing the "free" concentration of the toxic agent through adsorption and subsequent biodegradation. Results reported by Briddle et al. [34] on a coke plant effluent are summarized in Table 4.28. The application of PAC to a chemical plant wastewater to enhance nitrification is shown in Figure 4.87.

TABLE 4.27. Operating Characteristics for PACT® Process Case Studies.

| | Case Study No. | | | | |
|---|---|---|---|---|---|
| Operating Parameter | 1[a] | 2 [33] | 3 [33] | 4 [33] | 5 [33] |
| Aeration System | | | | | |
| HRT, d | 0.32 | .75 | 2.3 | 3.8 | 4.2 |
| SRT, d | 54 | 40 | 5.8 | 20 | 19.3 |
| MLSS, g/l | 34.8 | 10–12 | — | — | — |
| PAC dosage, mg/l | 114 | 170 | 2,270 | 850 | 1,140 |
| TOC, influent, mg/l | 174 | — | 2,470 | 2,330 | 2,490 |
| TOC, removal, % | 81 | — | 85.3 | 98.9 | 98.4 |

[a]Zimpro, 1984.

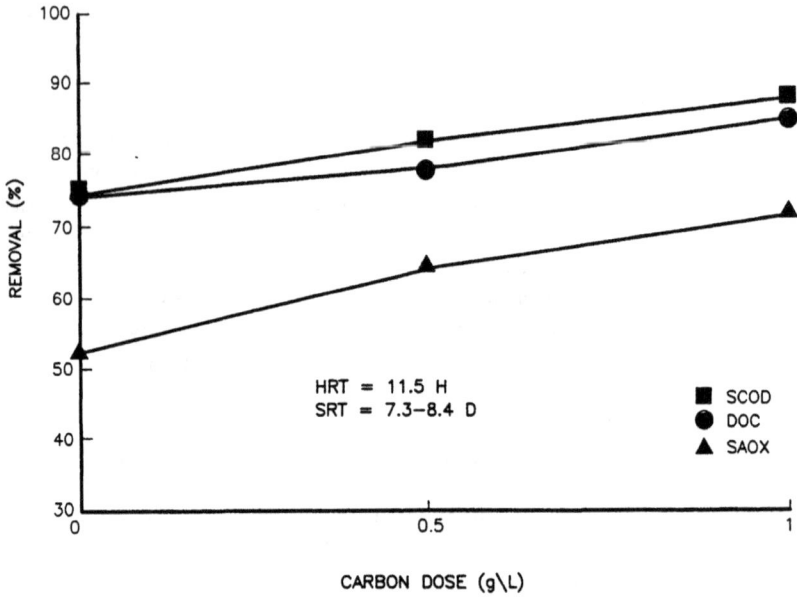

**Figure 4.83** Treatment performance variation with carbon dose [35].

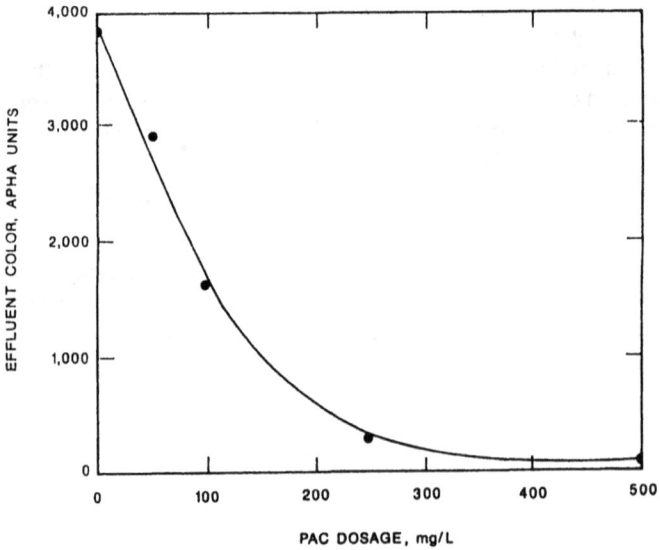

**Figure 4.84** Color removal from dyestuffs wastewater with PAC.

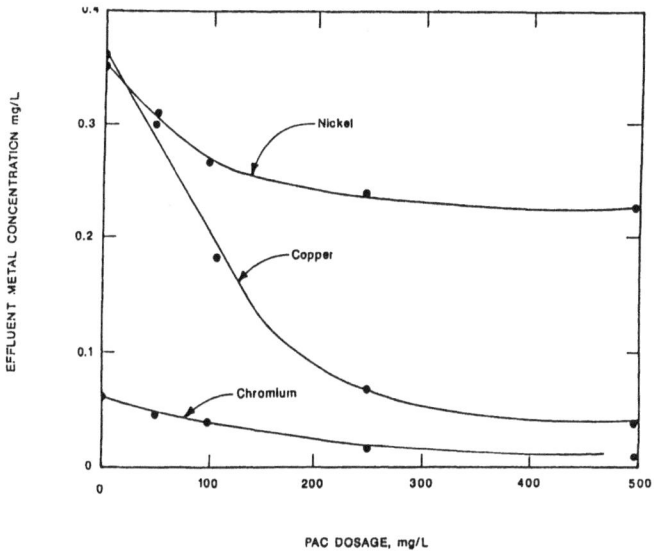

**Figure 4.85** Metals removal from organic chemicals wastewater using PAC.

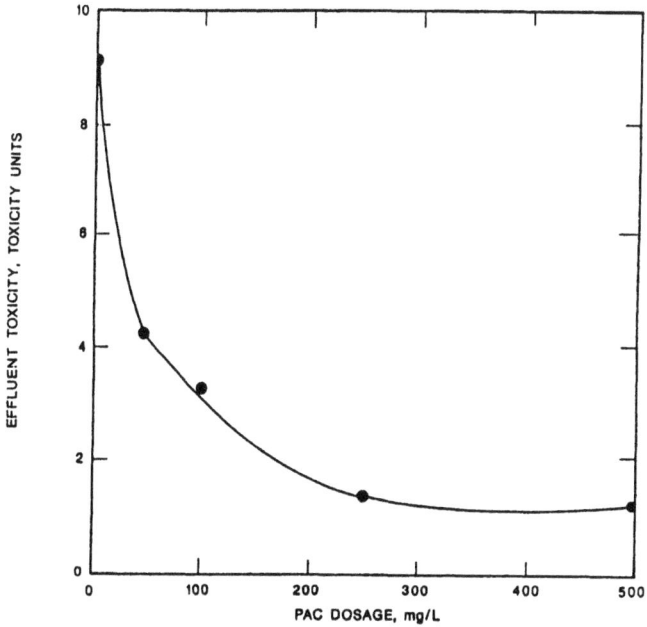

**Figure 4.86** Toxicity reduction in an organic chemicals wastewater using PAC.

**297**

TABLE 4.28.  Effect of PAC on Nitrification of Coke Plant Wastewaters [34].

| PAC Dosage (mg/l) | SRT (days) | Effluent Characteristics | | | | |
|---|---|---|---|---|---|---|
| | | TOC (mg/l) | TKN (mg/l) | NH₃—N (mg/l) | NO₂—N (mg/l) | NO₃—N (mg/l) |
| 0 | 40 | 31 | 72.0 | 68 | 4.0 | 0 |
| 33 | 30 | 20 | 6.3 | 1 | 1.0 | 9.0 |
| 50 | 40 | 26 | 6.4 | 1 | 1.0 | 13.0 |

The addition of PAC to the activated sludge process not only results in improved removal of BOD, $NH_3$, and TOC through improved biodegradation and metals and color removal through adsorption, but it also reduces the loss of some VOC to the gas phase. The mechanism for VOC removal is adsorption and enhanced biodegradation (on the carbon). This removal strategy can be significant when considering control of VOC from the activated sludge process.

**FINAL CLARIFICATION**

Final clarifier performance is related to the solids flux on the clarifier,

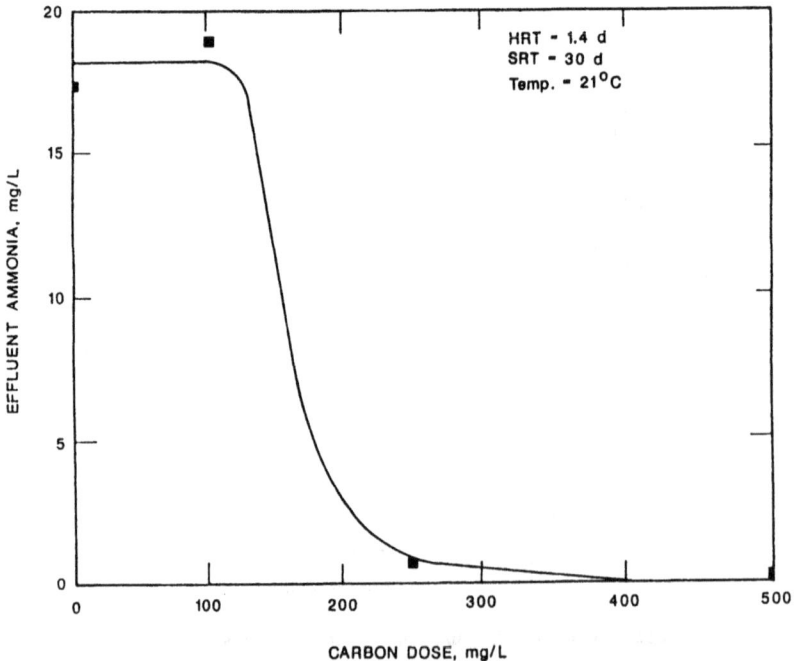

Figure 4.87 Effect of carbon dose on effluent ammonia concentration.

the return sludge concentration and the SVI of the sludge. Multi-variable performance relationships have been developed for municipal wastewater treatment by Daigger and Roper [35]. It has been found, however, that industrial sludges frequently exhibit different settling properties than municipal sludges. A comparison of the limiting solids loading rate, correlated as a Daigger-Roper diagram, of a municipal sludge and a pulp and paper mill sludge at the same SVI, is shown in Figure 4.88. The zone settling velocities for a variety of industrial sludges are summarized in Figure 4.89.

Based on these data it is apparent that mass flux relationships must be developed for specific industrial wastewaters in order to define secondary clarifier area requirements. Sludge zone settling velocity (ZSV) tests at constant MLSS of several mixed liquors indicated a variation in ZSV depending on test cylinder diameter even when all test cylinders were equipped with rotating sludge rakes. These results are shown in Figure 4.90. They indicate that the traditional practice of designing secondary clarifiers based on settling data developed in 1.0 L stirred test cylinders (diameter = 2.5 in) incorporates a 30 percent to 100 percent margin of safety depending upon the particular sludge characteristics.

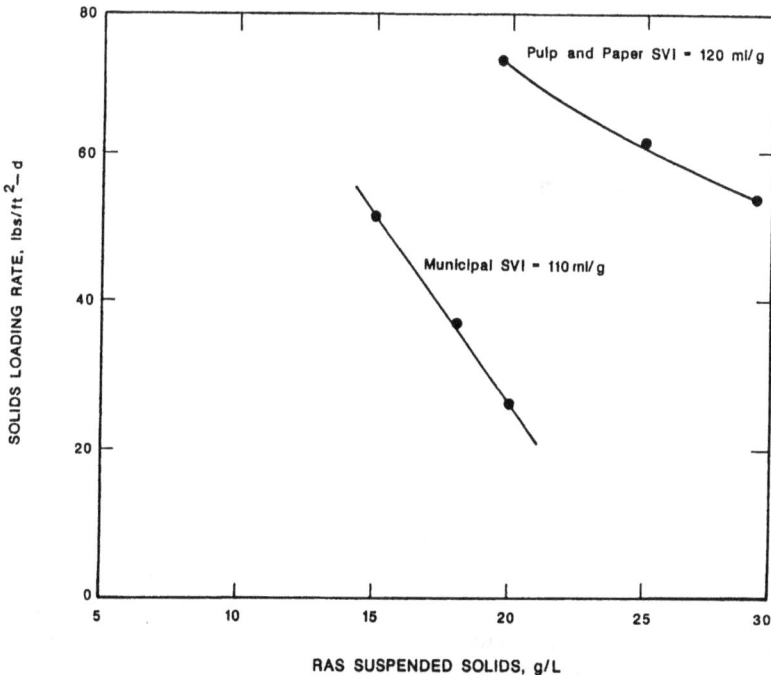

**Figure 4.88** Flux comparison between a municipal and a pulp and paper wastewater activated sludge.

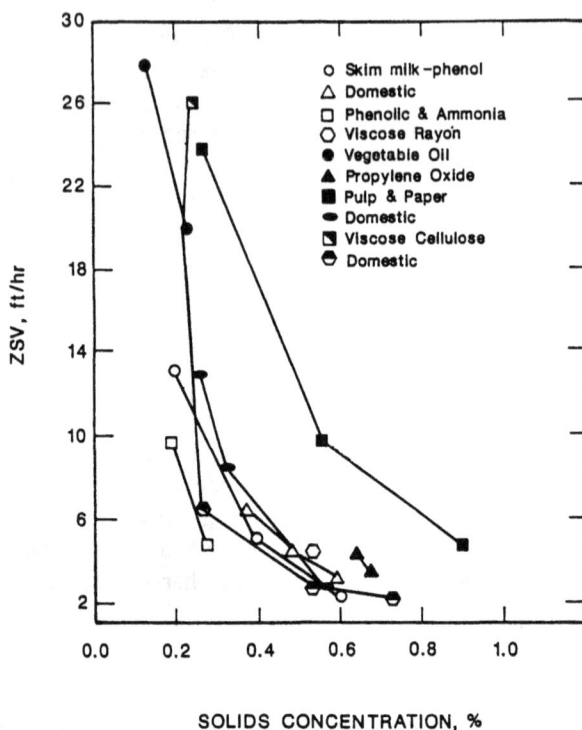

**Figure 4.89** Zone settling velocity characteristics of industrial sludges.

## EFFLUENT SUSPENDED SOLIDS CONTROL

Carryover of suspended solids in secondary clarifier effluent can be due to several causes. These are:

- floc shear due to high aeration power levels
- poor clarifier hydraulics
- high wastewater TDS concentration
- low or high mixed liquor temperature
- rapid change in mixed liquor temperature
- low mixed liquor surface tension

High mixed liquor turbulence levels created by turbine type or mechanical surface aerators can cause floc breakup that results in high effluent suspended solids. Argaman and Kaufman [36] quantified this effect through batch treatability tests as shown in Figure 4.91. This problem can frequently be solved by reducing the aeration basin power level and/or by installing a flocculation zone between the aeration basin and the final clarifier. Results of flocculation

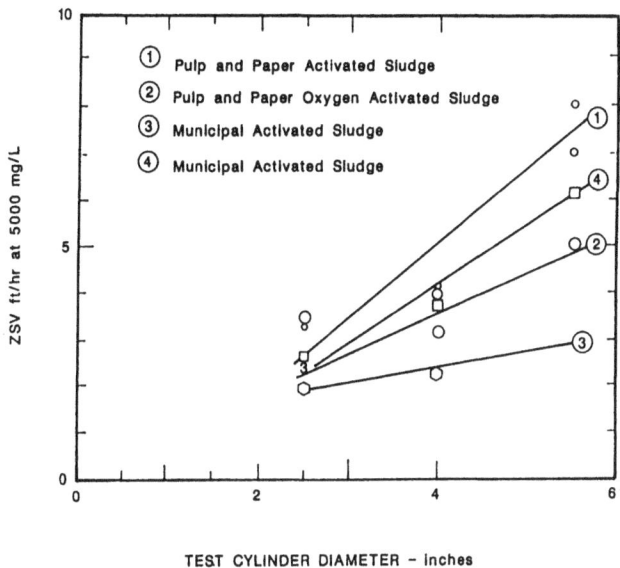

**Figure 4.90** Effect of test cylinder diameter on ZSV.

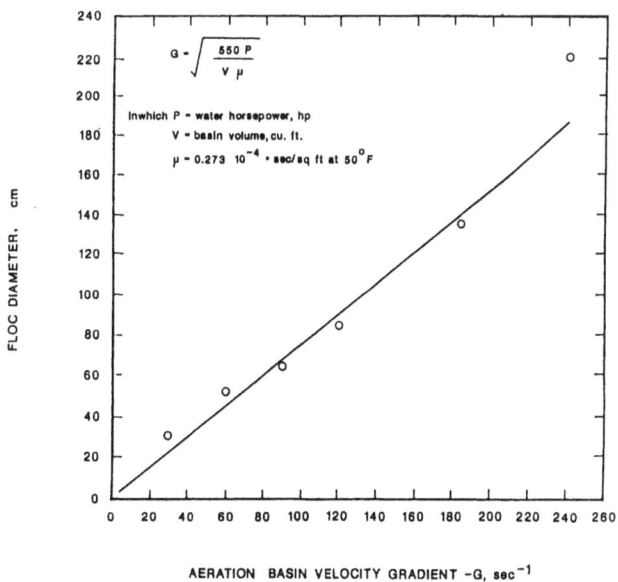

**Figure 4.91** Relationship between floc size and aeration basin power level.

**301**

of mixed liquor from an activated sludge plant treating a pulp and paper mill wastewater at two aeration basin power levels are shown in Table 4.29. These data indicate that flocculation times of 1 to 3 minutes were effective at reducing the settled effluent TSS concentration.

High effluent suspended solids levels will frequently result from a poor clarifier hydraulic design that causes density currents and/or short-circuiting. These conditions result in an upwelling of floc solids at the clarifier peripheral weir. Flow patterns and isoconcentration profiles for TSS are shown in Figures 4.92a and 4.92b for two equally sized circular clarifiers having 12 ft and 9 ft side-water-depths (SWD) and operating at their design flow rates. The shallower clarifier had more solids carryover and consistently exceeded its discharge limit. This problem can be minimized by installing a Stamford baffle, which redirects the upflow of solids away from the effluent weir. A Stamford baffle and its performance are shown in Figure 4.93.

High wastewater TDS levels due to inorganic salts will frequently result in floc dispersion and an increase in effluent suspended solids. Treatability studies of an agricultural chemicals wastewater are summarized in Table 4.30. They demonstrate that wastewaters at 1 to 2 percent TDS had negligible impact on activated sludge settleability or organics removal performance. However, at TDS concentrations greater than 1.6 percent, a significant deterioration in effluent TSS quality was observed. In one wastewater with a TDS of 4.5% the biomass consisted almost entirely of dispersed bacteria which was only retained on a 0.45 micron filter. Table 4.31 summarizes results from this plant. The process was flow through with a hydraulic detention time of 5 days. The effect of influent TDS on activated

TABLE 4.29. Floc Shear Test Results for Pulp and Paper Mill Wastewaters.

| Flocculation Time (min) | Settled TSS[a] (mg/l) |
|---|---|
| 686 Bhp/MG Basin Volume | |
| 0 | 81 |
| 1 | 30 |
| 3 | 28 |
| 5 | 26 |
| 7 | 27 |
| 360 Bhp/MG Basin Volume | |
| 0 | 64 |
| 1 | 28 |
| 3 | 22 |
| 5 | 19 |
| 7 | 21 |

[a]Supernatant TSS following flocculation and 15 min settling period.

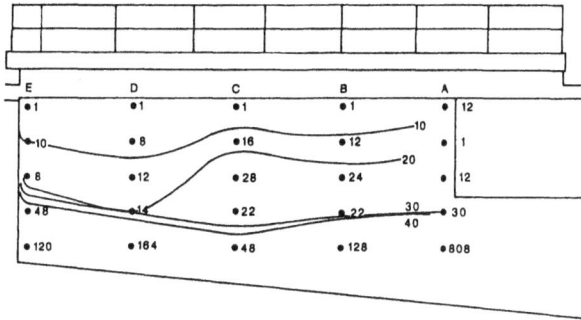

**Figure 4.92a** Isoconcentration profiles for TSS distribution in 12 ft SWD secondary clarifier.

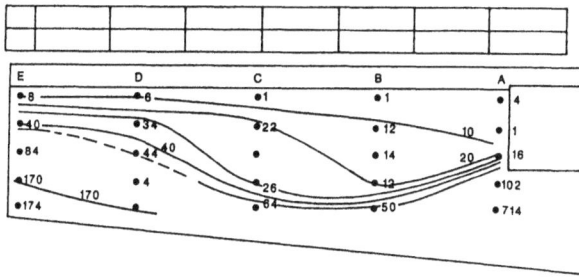

**Figure 4.92b** Isoconcentration profiles for TSS distribution in 9 ft SWD secondary clarifier.

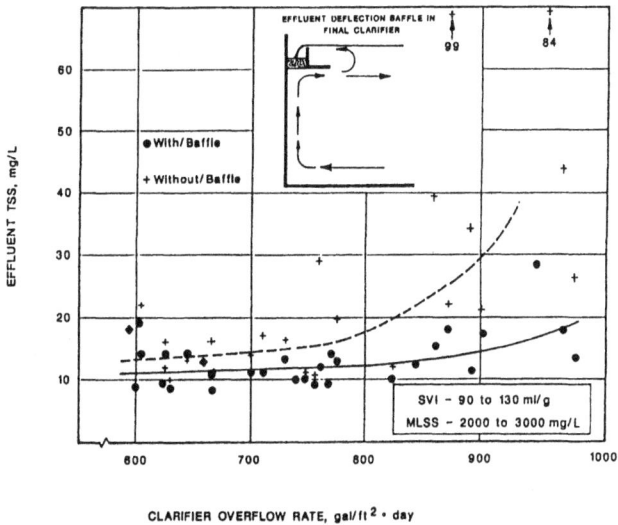

**Figure 4.93** Effect of Stamford baffle on clarifier effluent TSS concentration.

303

TABLE 4.30.  Effects of Total Dissolved Solids on Activated Sludge Treatment of Agricultural Chemicals Wastewater.[a]

| Unit No. | Effluent TDS (mg/l) | Sludge Settleability | | Organics Removal | | Settled TSS[b] (mg/l) |
|---|---|---|---|---|---|---|
| | | Flux Rate (lb/d · sq ft) | SVI (ml/g) | BOD (%) | TOC (%) | |
| 1 | 10,600 | 48 | 61 | 94 | 55 | 32 |
| 2 | 13,200 | 51 | 49 | 96 | 53 | 34 |
| 3 | 15,600 | 51 | 47 | 96 | 57 | 38 |
| 4 | 20,200 | 55 | 46 | 93 | 53 | 101 |

[a]Units operated at 25°C and $F/M$ = 0.2 d$^{-1}$.
[b]Following 30 min settling period.

sludge effluent suspended solids for several petroleum refinery wastewaters is shown in Figure 4.94. Since these solids do not settle, it is necessary to add chemical coagulants to effect their separation. The coagulants can be added between the aeration basin and the final clarifier or directly to the final clarifier if adequate mixing, flocculation, and contact time are provided in the clarifier flocculation zone. It is important that turbulence be controlled in the flocculation chamber to avoid excessive floc shearing before entering the clarification zone.

Coagulants for bio-flocculation include alum, iron salts, and cationic polymer. The application of ferric chloride ($FeCl_3$) for clarification of a specialty chemicals wastewater is shown in Table 4.32 When treating large volumes of wastewater, cationic polymer is usually more cost-effective when alum or iron dosages exceed 50 mg/l due to lower additional sludge volumes. In treatment of two surfactant manufacturing wastewaters, the effluent from extended aeration activated sludge treatment contained high levels of freon extractable organics (O&G). This resulted in an emulsified effluent that required large doses of iron salts (>1,000 mg/l) or other coagulants to reduce effluent TSS and O&G below pretreatment limits. The treatability results for the two wastewaters are summarized in Table 4.33. For "Wastewater II" the O&G concentration actually increased from 75 mg/l to 622 mg/l across the activated sludge treatment process. This resulted in a significant increase in coagulant demand for O&G control.

The effect of seasonal variation in mixed liquor temperature on the

TABLE 4.31.

| Influent | Effluent | | | | |
|---|---|---|---|---|---|
| COD mg/l | COD$_t$ mg/l | COD$_s$ mg/l | VSS[a] mg/l | VSS[b] mg/l | TDS mg/l |
| 6437 | 1182 | 181 | 50 | 597 | 44000 |

[a]1.5 micron filter.
[b]0.45 micron filter.

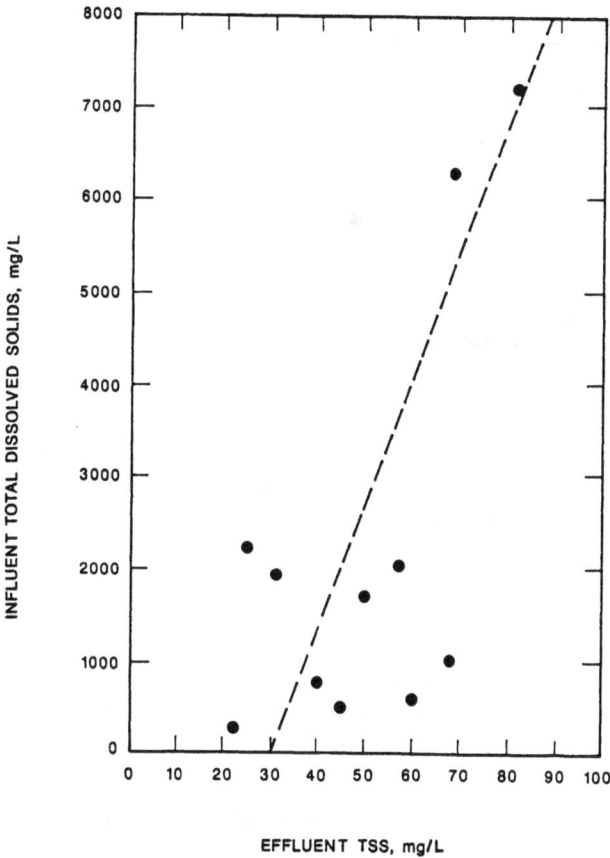

**Figure 4.94** Effect of TDS on effluent suspended solids for petroleum refinery wastewaters.

polymer dose required to produce an effluent suspended solids concentration of 40 mg/l is shown in Figure 4.95. for a multi-product organic chemicals wastewater. As can be seen, decreasing aeration basin temperature resulted in increased polymer dosage to satisfy the required TSS discharge limit. A rapid change (<24 hrs) in mixed liquor temperature will also result in floc dispersion and an increase in effluent suspended solids. This is a temporary effect, however, and mixed liquor settling characteristics will recover when the temperature is stabilized.

Activated sludge also exhibits poor flocculating characteristics at both very low and very high *F/M* conditions. This effect is demonstrated by increased effluent TSS concentrations as shown in Figure 4.96 for an organic chemicals wastewater. At *F/M* values between 0.08 d$^{-1}$ and 0.36 d$^{-1}$ the biomass was well flocculated and the effluent TSS was low, as indicated by the probability plot. At low *F/M* (0.04 d$^{-1}$) however,

TABLE 4.32. Ferric Chloride Coagulation Activated Sludge Effluent.

| Dose[a] (mg/l) | Settled TSS[b] (mg/l) | pH (units) |
|---|---|---|
| 0 | 175 | 7.7 |
| 100 | 114 | 7.6 |
| 200 | 54 | 7.4 |
| 400 | 11 | 6.9 |
| 600 | 5 | 6.7 |

[a] Dose expressed as $FeCl_3 \cdot 6H_2O$.
[b] Following 30 minute settling period.

bio-flocculation deteriorated due to excessive endogenous respiration and the effluent TSS increased. Similarly at high *F/M* (0.70 $d^{-1}$) bio-flocculation and effluent TSS quality were poor.

Dispersed suspended solids also increase with a decrease in surface tension and the presence of chemical dispersants. At one de-inking mill the effluent suspended solids value was directly related to the surfactant usage in the mill. Similarly, activated sludge floc formation was poor when chemical dispersants were used in a Latex rubber manufacturing process to prevent premature flocculation of the rubber.

TABLE 4.33. Coagulation Requirements Following Activated Sludge Treatment of Surfactant Wastewater.

| Coagulant Dosage | Effluent Characteristics | | |
|---|---|---|---|
| | TSS (mg/l) | TBOD (mg/l) | O&G (mg/l) |
| Wastewater I—Influent O&G = 1,060 mg/l | | | |
| pH 6.2, no coagulants | 295 | 194 | 159 |
| pH 6.2, 650 mg/l $FeCl_3$ + 50 mg/l polymer | 230 | 65 | 85 |
| pH 6.2, 1,300 mg/l $FeCl_3$ + 50 mg/l polymer | 190 | 61 | 57 |
| Wastewater II—Influent O&G = 75 mg/l | | | |
| pH 6.1, no coagulants | 175 | 191 | 622 |
| pH 6.1, 650 mg/l $FeCl_3$ + 10 mg/l polymer | NA | 54 | 204 |
| pH 6.1, 1,300 mg/l $FeCl_3$ + 10 mg/l polymer | 108 | 39 | < 25 |

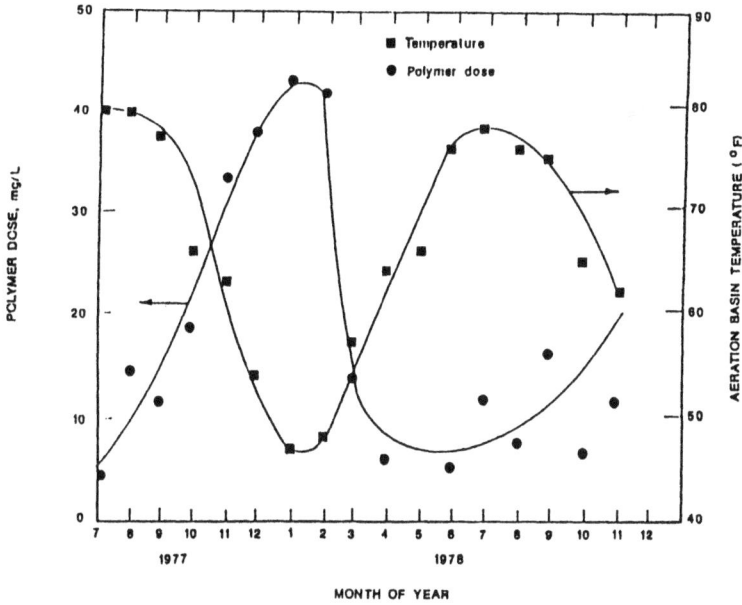

**Figure 4.95** Effect of aeration basin temperature and polymer dose required to achieve 40 mg/l suspended solids [30].

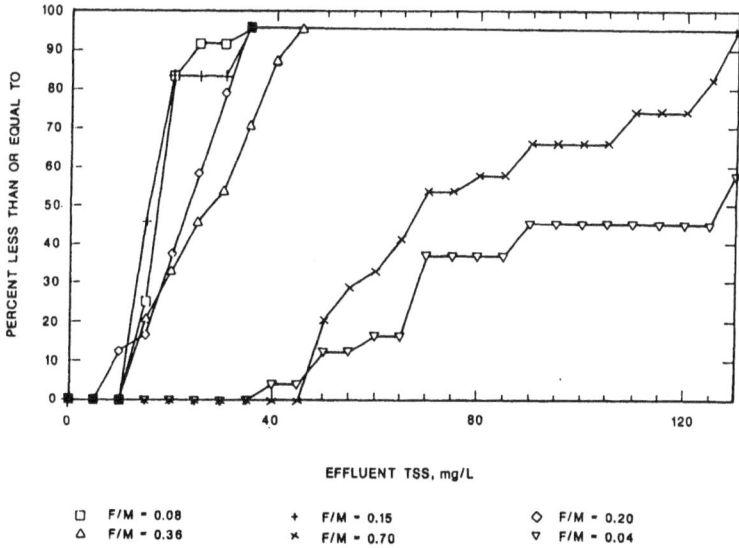

**Figure 4.96** Effect of *F/M* on effluent suspended solids for an organic chemicals wastewater.

**307**

## AEROBIC DIGESTION OF BIOLOGICAL SLUDGES

Aerobic digestion of waste activated sludge involves the oxidation of cellular organic matter through endogenous metabolism. The oxidation of cellular organics expressed as the degradable volaltile suspended solids has been found to follow first-order kinetics. Under batch or plug-flow conditions,

$$\frac{(X_d)_e}{(X_d)_o} = e^{-k_d t} \tag{4.29}$$

where

$(X_d)_e$ = degradable volatile solids after time $t$
$(X_d)_o$ = initial degradable volatile solids
$k_d$ = reaction rate coefficient, d$^{-1}$
$t$ = time of aeration, d

The kinetic parameters can be determined from a batch oxidation test in the laboratory as shown in Figure 4.97.

If the total volatile suspended solids are considered, Equation (4.29) becomes

$$\frac{X_e - X_n}{X_o - X_n} = e^{-k_d t} \tag{4.30}$$

where

$X_o$ = initial VSS,
$X_e$ = VSS after time $t$,
$X_n$ = nondegradable VSS,

In a continuous flow a completely mixed reactor Equation (4.30) is modified to

$$\frac{X_e - X_n}{X_o - X_n} = \frac{1}{1 + k_d t} \tag{4.31a}$$

and the required retention time is

$$t = \frac{X_o - X_n}{k_d(X_e - X_n)} \tag{4.31b}$$

For $n$ completely mixed reactors in series,

$$\frac{X_t - X_n}{X_o - X_n} = \frac{1}{(1 + k_d t_n)^n} \tag{4.32}$$

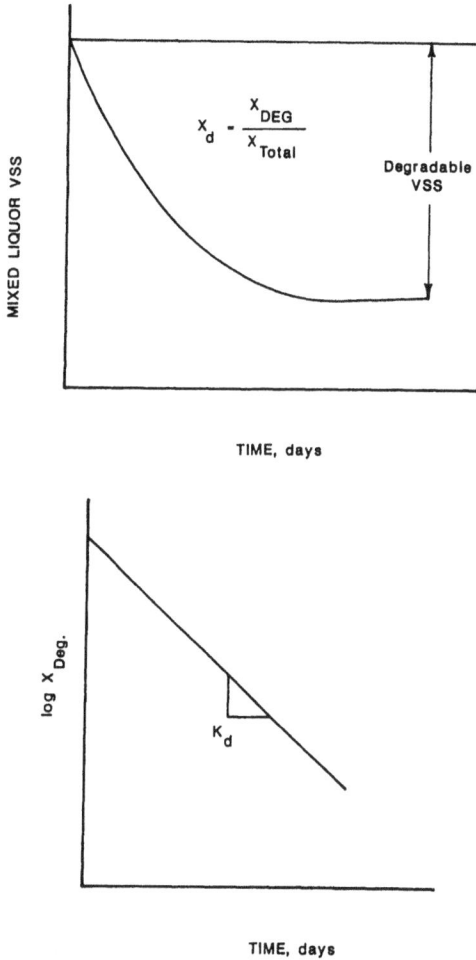

**Figure 4.97** Determination of the degradable fraction and the endogenous coefficient.

In accordance with the kinetic relationship, multiple reactors in series are more efficient than one mixed reactor. For example to achieve 90 percent removal of degradable solids at 20°C would require 9.7 days in a single stage digester and 7.2 days in a three stage digester. Temperature will affect the rate coefficient $k_d$, as shown in Figure 4.98.

The oxygen requirements for aerobic digestion can be estimated as 1.4 lb of oxygen consumed for each pound of VSS destroyed. Nitrogen and phosphorus will be released by the oxidation process. Under mesophilic digestion conditions, nitrification will usually occur because of the long sludge ages in the reactor. Oxygen and alkalinity must be available for this oxidation. Under thermophilic operation, nitrification will be inhibited.

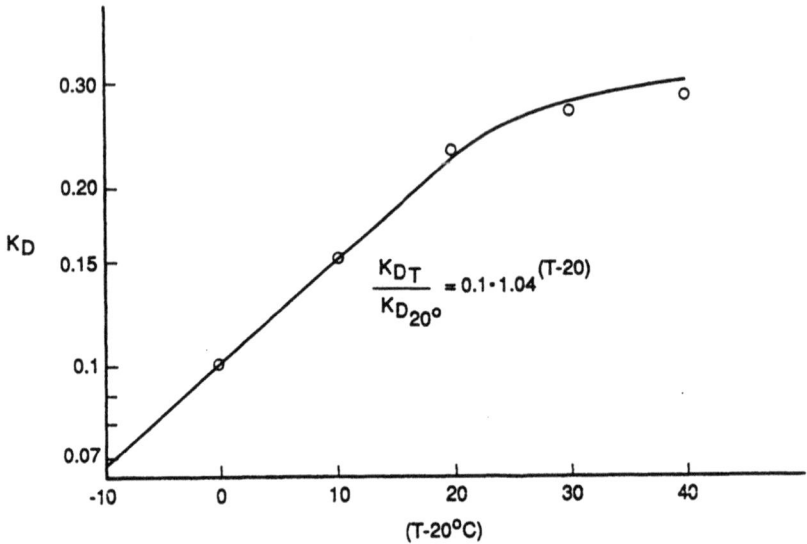

**Figure 4.98** Temperature effects on aerobic digestion.

Conventional aerobic digestion design employs secondary clarifier underflow (0.5–1.5 percent solids) in one or more completely mixed aeration basins. Power levels of 15 to 20 scfm/1000 cf using diffused air or 100 hp/million gal using surface mechanical aerators are usually adequate for providing both mixing and oxygen requirements. Prethickening the sludge offers a number of advantages, namely reducing the basin volume requirements and increasing the temperature due to the exothermic heat of reaction. Andrews and Kambhu have estimated the heat of combustion as 9000 Btu/lb VSS destroyed. Aerobic digestion requirements will depend on the operating sludge age in the aeration process. As the sludge age is increased, more of the degradable biomass is oxidized in the aeration basin and hence less will be oxidized in the aerobic digester. Calculations for aerobic digestion are illustrated in Example 4.10.

*Example 4.10.*

The following data applies:

$a = 0.6$
$S_r = 690$ mg/L
$b = 0.1$/day
$Q = 5$ mgd

Equation (4.6) can be rearranged

$$1 + bX_d\theta_c = aS_r/\Delta X_v$$

For a 10-day SRT $\Delta X_v = 10,340$ lbs/d and for a 30 day SRT $\Delta X_v = 6920$ lbs/d. Design an aerobic digester to yield a final degradable fraction of 0.37.

| | |
|---|---|
| $\theta_c = 10$ days | $\theta_c = 30$ days |
| $X_o = 10,341$ lbs/d | $X_o = 6922$ lbs/d |
| $X_d = 0.67$ | $X_d = 0.5$ |
| $X_d = 6928$ lbs/d | $X_d = 3461$ lbs/d |
| $X_N = 3413$ lbs/d | $X_N = 3461$ lbs/d |
| $X_{DR} = 2000$ lbs/d | $X_{DR} = 2000$ lbs/d |
| $X_t = 5413$ lbs/d | $X_t = 5461$ lbs/d |

$$k_d = 0.155$$

Required Detention Time

$$t = \frac{10341 - 3413}{0.155(5413 - 3413)} \qquad t = \frac{6922 - 3461}{0.155 \cdot 2000}$$
$$= 22.3 \text{ days} \qquad\qquad = 11.1 \text{ days}$$

## OXYGEN REQUIREMENTS

$$(X_D - X_{DR})$$

$$(6928 - 2000)\,1.4 \qquad (3461 - 2000)\,1.4$$
$$= 6900 \text{ lbs/d} \qquad\quad = 2045 \text{ lbs/d}$$

## NITRIFICATION (SEE FIGURE 4.15)

$$N_{ox} = (NiX_o - NeX_t)$$
$$N_{ox} = 0.091 \cdot 10341 - 0.076 \cdot 5413$$
$$N_{ox} = 0.076 \cdot 6922 - 0.07 \cdot 5461$$

530 lbs/d          144 lbs/d

## OXYGEN REQUIREMENTS

$$O_2 = 2295 \text{ lbs/d} \qquad O_2 = 623 \text{ lbs/d}$$
$$\text{Total } O_2 = 9195 \text{ lbs/d} \qquad 2668 \text{ lbs/d}$$

Alkalinity Requirement

$$3790 \text{ lbs/d} \qquad 1030 \text{ lbs/d}$$

## DENITRIFICATION

Assume sludge at 10,000 mg/L
At 10,000 lbs/d, Q = 0.119 mgd

$$V = 0.119 \cdot 22.3 \text{ days} = 2.58 \text{ mg}$$

$O_2$ uptake rate for VSS oxidation

$$\frac{6580 \text{ lbs/d}}{8.34 \cdot 2.58 \cdot 24} = 12.8 \text{ mg/l/hr}$$

Denitrification rate is 0.75 oxygen uptake rate

$$= 12.8 \cdot 0.75 = 9.6 \text{ mg/l/hr}$$
$$NO_3 - N = 507 \text{ lbs/day or } 23.6 \text{ mg/l}$$

in the aeration volume

$$t_{DN} = 23/9.6 = 2.46 \text{ hrs.}$$

Recently a focus has been on auto thermal aerobic digestion (ATAD). In this process the exotheral heat of combustion of the volatile solids increases the temperature in the reactor to the thermphilic range, i.e., 55°C. This is illustrated in Example 4.11.

*Example 4.11*

Feed VSS = 2.5%

$$k_d = 0.30$$

$$X_d = 0.67$$

VSS reduction = 40%

$$t = \frac{25000 - 8250}{0.32 \,(15000 - 8250)} = 8 \text{ days}$$

Assume 50,000 gal/day sludge

$$V_D = 400,000 \text{ gal}$$

lbs/day VSS reduced is

$$25000 \cdot 0.40 \cdot 0.05 \cdot 8.34 = 4170 \text{ lbs/d}$$

The heat generated is BTU = 9300 BTU/lb $\cdot$ 4170 lbs/d = $38.7 \times 10^6$ BTU/d.
lbs $H_2O$/day is

$$50,000 \cdot 8.34 \text{ lbs/gal} = 0.417 \times 10^6 \text{ lbs/d}$$

and the temperature increase will be 92°F or 33°C.
Oxygen requirements

$$4170 \text{ lbs/d} \cdot 1.4 = 5838 \text{ lbs/d}$$

$O_2$ uptake rate

$$\frac{5838}{0.4 \cdot 8.34 \cdot 24} = 73 \text{ mgL/hr}$$

## LABORATORY AND PILOT PLANT PROCEDURES FOR THE DEVELOPMENT OF PROCESS DESIGN CRITERIA

Wastewater characterization should be based on the equalized wastewater. Depending on the nature of the wastewater and the permit requirements, the following parameters should be evaluated:

- BOD and/or COD or TOC
- total and volatile suspended solids
- oil and grease

- priority pollutants
- toxicity (bioassay)
- nitrogen
- phosphorus

For wastewaters that do not contain aquatic toxicity, the following stepwise procedure is applicable to developing the necessary process design data.

(1) Adjust the BOD : N : P ratio to 100 : 5 : 1 neglecting the organic nitrogen at this time. Although organic nitrogen may be hydrolyzed to ammonia in the activated sludge process, it is initially neglected in order to insure adequate nutrients in the experimental phase and then reevaluated in the final process design.

(2) Evaluate the wastewater for bulking potential. Readily degradable wastewaters promote filamentous bulking. This can usually be evaluated by operation of a complete mix reactor to establish the proliferation of filaments during a 5 to 8 day period of operation.

(3) Develop an acclimated mixed liquor. For a wastewater with a low or non-bulking potential, use a CMAS reactor. For a wastewater with a high bulking potential, acclimate the mixed liquor in a batch reactor or a sequencing batch reactor (SBR). Determine the bioinhibition potential using the FBR procedure. Adjust the initial feed rate of the wastewater to be less than 50 percent of the inhibition threshold concentration. Operate the reactor at an *F/M* of 0.2 to 0.4/d. As acclimation proceeds, gradually increase the feed rate until the full waste strength is being treated.

(4) For wastewaters with high bulking potential, either a plug-flow, SBR or a biological selector should be used. Guidelines for choosing the type of selector (aerobic, anoxic, or anaerobic) and the loading condition were presented in Chapter 1.

## REACTOR OPERATION

At least three reactors should be operated in parallel over an SRT range of 3 to 12 days for a readily degradable wastewater and 10 to 40 days for a more bio-refractory wastewater. The SRT should be maintained by daily wasting of an appropriate mixed liquor volume (i.e., for a 10-day SRT, one-tenth of the reactor volume is wasted daily). The waste sludge mass is computed as the VSS in the wasted reactor volume plus the VSS in the reactor effluent. The sampling and analytical schedule for the reactors are summarized in Table 4.34.

At the end of the treatability study, the degradable fraction $X_d$ and the

TABLE 4.34.  Sampling and Analysis Schedule for Biological Treatment
Process Design Study.

| Analysis | Frequency |
|----------|-----------|
| BOD | 3/week |
| COD | Daily |
| $O_2$ uptake rate | Daily |
| MLVSS | Daily |
| Dissolved oxygen | Daily |
| pH | Daily |
| Temperature | Daily |
| Nitrogen[a] | 2/week |
| Phosphorus | 2/week |
| Bioassay[b] | Weekly |
| Specific pollutants | Weekly |

[a]For a nitrogen-deficient wastewater. If nitrification is desired, TKN, $NH_3$—N, $NO_2$—N and $NO_3$—N
should be run 2 times per week.
[b]The test species will depend on permit conditions.

endogenous decay coefficient $b$ are determined. Sludge from each reactor
is washed and aerated and the concentration of VSS is measured every 2
to 3 days until there is no further reduction in VSS. The degradable
fraction and the endogenous decay coefficient can be calculated as shown
in Figure 4.97.

Figure 4.99 presents the methods for graphical analysis of the treatability
reactor data. The rate coefficient, $K$, is determined by plotting $S_o(S_o - S)/$
$X_v t$ versus $S$. If the influent wastewater has a variable composition, $K$ will
not be constant. The oxygen coefficient, $a'$, is determined as the slope of
the plot of $O_2/X_d X_v t$ versus $S_r/X_d X_v t$ and the endogenous respiration coefficient $b'$ is the intercept. The sludge yield coefficient, $a$, is determined from
a plot of $\Delta X_v/X_d X_v t$ versus $S_r/X_d X_v t$. It should be noted that $\Delta X_v$ consists
of the sludge wasted per day plus that in the process effluent. The design
criteria for the final clarifier are determined from zone settling velocity
measurements on the sludge. The size of the settling cylinder has a significant effect on the zone settling velocity as previously discussed.

If the wastewater exhibits aquatic toxicity, the study procedure should
be modified to eliminate toxic wastewater streams or to consider modifications to the activated sludge process. A work flow diagram to achieve this
is shown in Figure 4.100.

This procedure is applied to all major wastewater streams. If heavy
metals are present, they are removed by precipitation as a first step. The
sample is then subjected to a priority pollutant scan and a bioassay. An
FBR test is run to determine biodegradability and inhibition to the activated
sludge process. Acclimated sludge should be used for this test. If the
wastewater is nonbiodegradable and toxic, it should be diverted for source

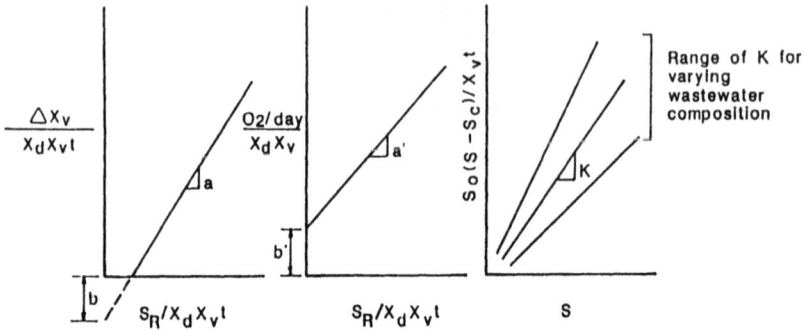

**Figure 4.99** Graphical methods for determining values of process design parameters.

control (or physical-chemical treatment). If it is biodegradable, a long-term bio-oxidation is performed using a 48 hr aeration period in order to remove all the biodegradable constituents in the wastewater. A bioassay is then run on this effluent. If the effluent is still toxic, additional treatment using granular or powdered carbon should be evaluated. Alternatively the wastewater stream can be diverted for source control or physical chemical treatment. If the effluent is non-toxic, degradation and stripping of volatile organic substances is evaluated (if appropriate). If priority pollutants are present, the FBR procedure should be used to develop design criteria.

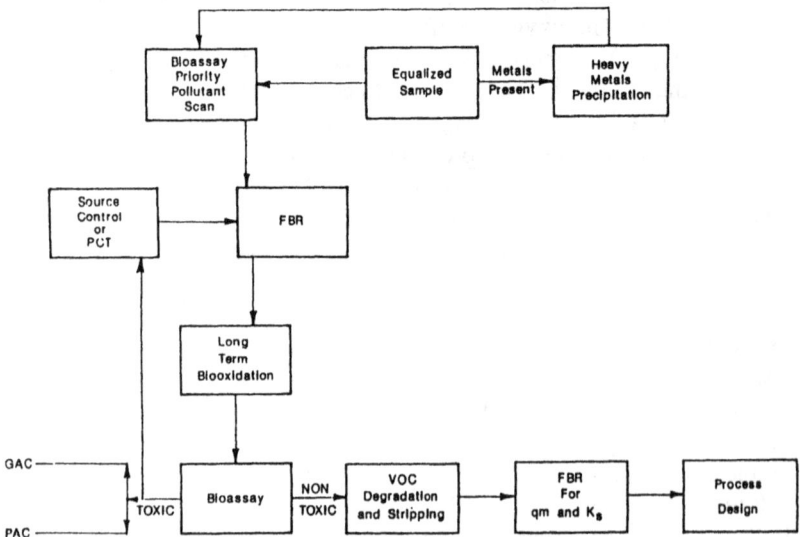

**Figure 4.100** Protocol for the treatability screening of toxic and hazardous wastewaters.

## ACTIVATED SLUDGE DESIGN PROCEDURE FOR SOLUBLE WASTEWATER USING COMPLETE MIX ACTIVATED SLUDGE

### DATA REQUIRED

Average influent flow, mgd
Maximum influent flow, mgd
Average influent BOD, mg/l
Maximum influent BOD, mg/l[1]
Average influent COD, mg/l
Maximum influent COD, mg/l[2]
Winter operating temperature, °C
Summer operating temperature, °C
Average $K$ @ 20°C (50%), d$^{-1}$
Minimum $K$ @ 20°C (5%), d$^{-1}$
MLVSS ($X_v$) during winter operation, mg/l
MLVSS ($X_v$) during summer operation, mg/l
Volatile fraction of MLSS
Yield coefficient $a$, mgVSS/mgCOD
Oxygen coefficient $a'$, mgO$_2$/mgCOD
Endogenous coefficient $b$ @ 20°C, d$^{-1}$
Temperature coefficient $\theta$ for BOD or COD removal
Average influent NH$_3$-N or TKN converted to NH$_3$-N, mg/l
Average influent P, mg/l
Average effluent soluble BOD, mg/l
Maximum effluent soluble BOD, mg/l
Average effluent suspended solids, mg/l
Maximum effluent suspended solids, mg/l

### CALCULATION OF THE REQUIRED AERATION VOLUME

Winter operation will usually control the design except in the case of a two-tiered permit

- Correct $K$ at 20°C ($K_{20}$) to the winter operating temperature

$$K_T = K_{20} \cdot \theta^{(T-20)}$$

- For average conditions compute the required detention time using the winter operation MLVSS

$$t = \frac{S_{o_a}(S_{o_a} - S_a)}{KX_vS_a}$$

in which $S_{o_a}$ and $S_a$ are the average influent and effluent soluble BOD, respectively.

- For maximum conditions, recompute the required detention time:

$$t = \frac{S_{o_m}(S_{o_m} - S_m)}{KX_vS_m}$$

The detention time employed will be the larger of the two calculated. If a large difference exists between the two values consideration should be given to increased equalization or in-plant process modifications.

- The average $F/M$ can now be computed

$$F/M = \frac{S_{o_a}}{X_vt}$$

- Calculation of required MLVSS during summer operation: Correct $K_{20}$ to summer conditions. Compute $X_v$ required to meet both average and maximum conditions

$$X_v = \frac{S_o(S_o - S)}{KSt}$$

## CALCULATION OF THE EFFLUENT COD

- The nondegradable influent COD ($COD_{ND}$) is

$$COD_{ND} = COD_{INF} - \frac{BOD_5}{f} \cdot \frac{1}{0.92}$$

in which $f$ = the ratio between $BOD_5$ and $BOD_{ULT}$ and 0.92 is assumed ratio of $BOD_{ULT}$ to COD.

- The degradable effluent COD is

$$COD_D = \frac{BOD_5}{0.65} \cdot \frac{1}{0.92}$$

The COD due to $SMP_{ND}$ is assumed to be 8 percent of the degradable COD removed.

- The total effluent COD is therefore

$$COD_{Eff} = COD_D + COD_{ND} + 0.08COD_{INF-D}$$

## SLUDGE PRODUCTION

Sludge production is calculated for average conditions. In order to compute sludge production, the degradable fraction of the VSS under average operating conditions must be known. $X_d$ is computed from the relationship

$$X_d = \frac{0.8}{1 + 0.2b\theta_c}$$

At this point, both $X_d$ and $\theta_c$ are unknown. A trial and error procedure can be employed in which $\theta_c$ is assumed and $X_d$ is calculated.

The average sludge production based on COD removal is

$$\Delta X_v = aS_r - bX_dX_vt$$

The SRT, $\theta_c$, is then checked

$$\theta_c = \frac{X_vt}{\Delta X_v}$$

If the original assumption is wrong, a second solution is made. Alternatively $\theta_c$ can be calculated

$$\theta_c = \frac{-(aS_r - bX_vt) \pm \sqrt{(aS_r - bX_vt)^2 + 4(abX_vS_r)(X_vt)}}{2abX_vSr}$$

## OXYGEN REQUIREMENTS

Oxygen requirements will usually be controlled by summer operation. Correct the endogenous coefficient $b$ to summer conditions

$$b_T = b_{20} \; 1.04^{(T-20)}$$

Assume $b' = 1.4b$. Recompute $X_d$ for summer operation and calculate the oxygen required.

$$O_2 = a'S_r + b'X_dX_vt$$

## NUTRIENT REQUIREMENTS

The nitrogen lost in the WAS is computed

$$N = 0.123 \frac{X_d}{0.8} \Delta X_v + 0.07 \frac{0.8 - X_d}{0.8} \Delta X_v$$

The ammonia-nitrogen in the influent wastewater is

$$NH_3 - N_{INF} \cdot Q_{INF} \cdot 8.34$$

The nitrogen to be added is

$$NH_3 - N_{ADDED} = N_{WAS} - NH_3 - N_{INF}$$

Phosphorus requirements are calculated in a similar manner:

$$P = 0.026 \frac{X_d}{0.8} \Delta X_v + 0.01 \frac{0.8 - X_d}{0.8} \Delta X_v$$

## SECONDARY CLARIFIER DESIGN

- Select the maximum SVI for the process—usually 150 ml/g unless available data indicates some other value.
- For a RAS concentration of 10 g/l select the solids flux, $G$, from the design diagram.
- For the maximum MLSS, $X_a$, compute the sludge recycle rate

$$R/Q_{INF} = r = \frac{X_a}{X_r - X_a}$$

- Compute the required clarifier area, A:

$$A = \frac{Q_{INF}(1 + r)(X_a)}{G}$$

- Compute the overflow rate, OR

$$OR = Q_{INF}/A$$

*Example 4.12*

An activated sludge plant is to be designed for an organic chemicals wastewater. Since filamentous bulking is not a problem, a complete mix design will be employed. The following conditions apply.

Influent flow rate = 3.0 mgd
Average influent $BOD_5$ = 420 mg/l
Maximum influent $BOD_5$ = 505 mg/l
Average influent COD = 630 mg/l
Maximum influent COD = 750 mg/l
Average $K_{20}$ (50%) = 6.5 $d^{-1}$
Minimum $K_{20}$ (5%) = 3.4 $d^{-1}$
Average influent $NH_3 - N$ = 5 mg/l
Average influent $PO_4$–P = 10 mg/l
$X_v$ = 2,500 mg/l
$a$, mg VSS/mg COD = 0.45
$a'$, mg $O_2$/mg COD = 0.37
$b$ @ 20°C = 0.10
$\theta$ = 1.06

The effluent requirements are an average soluble BOD of 10 mg/l and a maximum day soluble BOD of 25 mg/l.
Compute:

(1) Required aeration basin volume, the *F/M* and SRT
(2) Effluent COD
(3) Sludge production at 20°C
(4) Oxygen requirements for maximum loading conditions at 35°C
(5) Nutrient requirements
(6) $X_v$ at 10°C

*a.* Aeration basin detention time

Average conditions (20°C)

$$t = \frac{S_o(S_o - S)}{KX_vS}$$
$$= \frac{420(420 - 10)}{65 \cdot 10 \cdot 2,500} = 1.05 \text{ days}$$

Maximum conditions (20°C)

$$t = \frac{505(505 - 25)}{3.4 \cdot 25 \cdot 2{,}500} = 1.14 \text{ days}$$

The aeration basin volume will be designed for 1.14 days and the average F/M is

$$F/M = \frac{420}{2{,}500 \cdot 1.14} = 0.15/d$$

b. Average effluent total COD

The nondegradable influent COD ($COD_{ND}$) is

$$COD_{ND} = COD_{INF} - \frac{BOD_5}{0.8} \cdot \frac{1}{0.92}$$

$$= 630 - \frac{420}{0.8} \cdot \frac{1}{0.92} = 60 \text{ mg/l}$$

The degradable effluent COD ($COD_D$) is

$$COD_D = \frac{BOD_5}{0.65} \cdot \frac{1}{0.92} = \frac{10}{0.65} \cdot \frac{1}{0.92} = 17 \text{ mg/l}$$

The COD due to $SMP_{ND}$ is

$$SMP_{ND} = 0.08 \cdot COD = 0.08 \cdot (630 - 60) = 46 \text{ mg/l}$$

Total effluent COD = 60 + 17 + 46 = 123 mg/l
By similar calculations under maximum influent loading conditions, the total effluent COD is 166 mg/l at 20°C and 134 mg/l at 35°C.

c. Sludge production

The degradable fraction at 20°C is determined by

$$X_d = \frac{0.8}{1 + 0.2b\theta_c}$$

Assume $\theta_c = 32$ days

$$X_d = \frac{0.8}{1 + 0.2 \cdot 0.1 \cdot 32}$$
$$= 0.48$$

The average sludge production at 20°C based on COD removal is

$$\Delta X_v = aS_r - bX_dX_vt$$
$$= 0.45 (630 - 123) - 0.1 \cdot 0.48 \cdot 2,500 \cdot 1.14$$
$$= 89 \text{ mg/l}$$
$$= 89 \cdot 3 \cdot 8.34 = 2,227 \text{ lbs/day}$$

Check actual $\theta_c$

$$\theta_c = \frac{X_vt}{\Delta X_v}$$
$$= \frac{2,500 \cdot 1.14}{89} = 32 \text{ days}$$

Since the actual $\theta_c$ equals the assumed value, no further iterations are required. The sludge production will be slightly larger during cold weather operating conditions ($T = 10$°C).

d. Oxygen requirements

The SRT and $X_d$ at 35°C are calculated as follows

$$b \text{ @ } 35°C = 0.1 \cdot 1.04^{(35-20)} = 0.18$$

Assume $X_d = 0.37$
$$\Delta X_v = 0.45(750 - 134) - 0.18 \cdot 0.37 \cdot 2,500 \cdot 1.14$$
$$= 87 \text{ mg/l}$$

$$\theta_c = \frac{2,500 \cdot 1.14}{87} = 33 \text{ days}$$

$$b' \text{ @ } 35°C = 0.18 \cdot 1.4 = 0.25$$

$$O_2 = a'S_r + b'X_dX_vt$$
$$= 0.37 (750 - 134) + 0.25 \cdot 0.37 \cdot 2,500 \cdot 1.14$$
$$= 491 \text{ mg/l}$$
$$= 491 \cdot 3 \cdot 8.34 = 12,285 \text{ lbs/d}$$

e. Nutrient requirements at 20°C

$$N = 0.123 \frac{X_d}{0.8} \Delta X_v + 0.07 \frac{0.8 - X_d}{0.8} \Delta X_v$$

$$= 0.123 \frac{0.48}{0.8} \cdot 2{,}227 + 0.07 \frac{0.8 - 0.48}{0.8} \cdot 2{,}227$$

$$= 226 \text{ lbs/day}$$

The influent $NH_3$–N is

$$5 \text{ mg/l} \cdot 3 \cdot 8.34 = 125 \text{ lbs/day}$$

The $NH_3$–N to be added is

$$226 - 125 = 101 \text{ lbs/day}$$

$$P = 0.026 \frac{X_d}{0.8} \Delta X_v + 0.01 \frac{0.8 - X_d}{0.8} \Delta X_v$$

$$= 0.026 \frac{0.48}{0.8} \cdot 2{,}227 + 0.01 \frac{0.8 - 0.48}{0.8} \cdot 2{,}227$$

$$= 44 \text{ lbs/day}$$

The influent P is

$$0.5 \text{ mg/l} \cdot 3 \cdot 8.34 = 12.5 \text{ lbs/d}$$

The $P$ to be added is 31.5 lbs/d.

f. Required $X_v$ at 10°C and average influent loading condition at 10°C

$$K = 6.5 \cdot \theta^{(10-20)}$$
$$= 6.5 \cdot 1.06^{(10-20)}$$
$$= 3.6/d$$

The effluent soluble BOD will be

$$S = \frac{S_o^2}{S_o + KX_v t}$$

$$= \frac{420^2}{420 + 3.6 \cdot 2{,}500 \cdot 1.14}$$

$$= 16.5 \text{ mg/l}$$

In order to maintain an effluent BOD of 10 mg/l the MLVSS must be increased to

$$X_v = \frac{S_o(S_o - S)}{KSt}$$
$$= \frac{420 \cdot (420 - 10)}{3.6 \cdot 10 \cdot 1.14}$$
$$= 4,200 \text{ mg/l}$$

## DESIGN PROCEDURE FOR A WASTEWATER CONTAINING DEGRADABLE INFLUENT VOLATILE SUSPENDED SOLIDS

The following additional (to those of Example 4.12) data are required:

- influent VSS, $X_i$, mg/l
- fraction of influent VSS which are degradable, $f_x$
- degradation rate coefficient of influent VSS, $K_p$
- fraction of degradable influent VSS degraded, $f_d$

### PROCEDURE

- Compute the required detention time as in the soluble BOD case assuming all the VSS are biological.
- Compute $\Delta X_v$ assuming only bio-solids, as in the soluble case.
- Estimate $X_d$ and compute $\theta_c$.
- For this $\theta_c$ compute the influent VSS which will be degraded

$$(1 - f_d) = e^{-K_p\theta_c}$$

- Compute the new $X_v t_{bio}$ assuming that 1 mg/l VSS solubilized generates 1 mg/l COD

$$X_v t = \frac{\theta_c a S_r}{1 + \theta_c b X_d}$$

- Compute the new $\Delta X_v$ biological

$$\Delta X_{vb} = a[(S_o - S) + f_d f_x X_i] - b X_d X_{vb} t$$

- Compute the fraction of the MLVSS which is biological

$$f_b = \frac{\Delta X_{vb}}{\Delta X_{vb} + (1 - f_d)f_x X + (1 - f_x)X}$$

- Compute the total $\Delta X_v$

$$\Delta X_v = \Delta X_{vb} + (1 - f_d) f_x X_i + (1 - f_x)X_i$$

- The total $X_v t$ is $\dfrac{X_v t_{bio}}{f_b}$

Select an appropriate $X_v$ and $t$ for the new operation.

*Example 4.13*

The wastewater in Example 4.12 contains 200 mg/l of VSS, 60 percent of which is degradable. The VSS degradation rate coefficient, $K_p$, has a value of 0.15 days at 20°C. Compute the new MLVSS required to maintain the same process performance.

From Example 4.12, $\Delta X_v = 89$ mg/l (2,227 lbs/day) and $\theta c = 32$ days = 0.48. The influent degradable VSS remaining is

$$(1 - f_d) = e^{-0.15 \cdot 32} \cong 0$$

If it is assumed that 1 mgVSS generates 1 mg COD, then an additional 120 mg/l of COD will be produced. For a $\theta_c$ of 32 days the new $X_v t$ can be calculated

$$X_v t = \frac{\theta_c a S_r}{1 + \theta_c b X_d}$$
$$= \frac{32 \cdot 0.45 \cdot 627}{1 + 32 \cdot 0.1 \cdot 0.48}$$
$$= 3,650$$

if $t = 1.14$ days, $X_v = 3,202$ mg/l

$$\Delta X_v = 0.45 \cdot 627 - 0.1 \cdot 0.48 \cdot 3,650$$
$$= 107 \text{ mg/l}$$
$$f_b = \frac{107}{107 + 80} = 0.57$$

The total VSS will be

$$X_v = \frac{3,202}{0.57} = 5,617 \text{ mg/l}$$

The total waste sludge will be

$$\Delta X_v = 107 + 80 = 187 \text{ mg/l}$$

In order to maintain a reasonable MLSS level, it will be necessary to increase $t$ to 2.0 days and carry 3,000 mg/l MLVSS.

## DESIGN FOR PRIORITY POLLUTANT REMOVAL

In a CMAS the effluent concentration of a specific pollutant is a function of SRT in accordance with the Monod relationship

$$\theta_c = \frac{K_s + S}{aq_m S - bX_d(K_s + S)}$$

The maximum degradation rate $q_m$ and the coefficient $K_s$ are specific for the pollutant, the wastewater, and the operating characteristics of the system. They must therefore be determined using the modified FBR technique.

*Example 4.14*

A wastewater from a chemical plant has the following characteristics:

BOD = 425 mg/l
$\quad K = 5.0$/day @ 20°C
DCP = 10 mg/l
$\quad K_s = 0.2$ mg/l
$\quad q_m = 0.8$ g/g · d
$\quad a = 0.6$ mg VSS/mg BOD
$\quad b = 0.1$ @ 20°C

Permit requirements call for an effluent soluble BOD of 15 mg/l and DCP of 20 µg/l. Develop a process design to meet these requirements at 20°C.

*a.* Requirements for BOD removal

$$\frac{S_o\ S}{X_v t} = K\frac{S}{S_o}$$

$$\frac{425 - 15}{X_v t} = 5.0\frac{15}{425}$$

$$X_v t = 2{,}342 \text{ mg-d/L}$$

$$F/M = \frac{425}{2342} = 0.18\text{d}^{-1}$$

Selecting $X_v = 3{,}000$ mg/l, then $t = 0.67$ days
Since $\Delta X_v = aS_R - bX_d X_v t$

$$\Delta X_v = 0.6(415) - 0.058(2{,}342)$$
$$= 113 \text{ mg/l}$$

$$\theta_c = \frac{X_v t}{\Delta X_v} = \frac{2{,}342}{113} = 21 \text{ days}$$

$$X_d = \frac{0.8}{1 + 0.2b\theta_c}$$

$$= \frac{0.8}{1 + 21} = 0.57$$

The BOD requirement will be met with an SRT of 21 days.

b. Requirements for DCP removal

$$\theta_c = \frac{K_s + S}{aq_m S - bX_d(K_s + S)}$$

Assume $X_d = 0.31$

$$\theta_c = \frac{0.2 + 0.02}{0.6 \cdot 0.8 \cdot 0.02 - 0.03 \cdot 0.22}$$
$$= 79 \text{ days}$$

Check $X_d$

$$X_d = \frac{0.8}{1 + 0.2(0.1)(79)}$$
$$= 0.31$$

For DCP removal

$$\theta_c = \frac{X_v t}{\Delta X_v}$$
$$79 = \frac{3,000t}{113}$$
$$t = 2.98 \text{ days}$$

c. Two-stage design

Reduce the BOD to 75 mg/l in Stage 1

$$\frac{425 - 75}{X_v t} = 5 \cdot \frac{75}{425}$$

at 3,000 MLVSS

$$t_1 = 0.13 \text{ days}$$

DCP will be removed in Stage 2

$$X_v t_2 = \frac{\theta_c a S_r}{1 + \theta_c b X_d}$$

Assume the soluble BOD is reduced to 5 mg/l

$$X_v t = \frac{79 \cdot 0.6 \cdot 70}{1 + 79 \cdot 0.031}$$

if $X_v = 3,000$ mg/l, $t_2 = 0.32$ days
The total $t$ for a two-stage process is 0.45 days.

## REFERENCES

1  Lester, J. C. 1987. *Heavy Metals in Wastewater and Sludge Processes—Volume II, Treatment and Disposal.* CRC Press.

2   Patterson, J. W. and Kodukula, P. S. 1984. Metals Distribution in Activated Sludge Systems, *Journal WPCF,* Volume 56, Number 5.

3   Lankford, P. and W. Eckenfelder. 1990. *Toxicity Reduction in Industrial Effluents.* New York: Van Nostrand Reinhold.

4   Ford, Davis. Private Communication.

5   Matter, Muter et al. 1980. Non-biological elimination mechanisms in a biological sewage treatment plant. *Prop. Wat. Tech.* 12:299.

6   NamKung & Rittman, 1987. Estimating Volatile Organic Compound Emissions from Pubically Owned Treatment Works, *Journal WPCF* 59(7):670.

7   Broughman and Panss, 1981. Microbial bioconcentration of Organic Pollutants from Aquatic Systems—a critical review. *Critical Review in Microbiology,* 8:205.

8   Dobbs et al. 1989. Sorption of Toxic Organic Compounds on Wastewater Solids. *Enviro. Sci. and Tech.* 23(9):1092.

9   Eckenfelder, W. 1980. *Principles of Water Quality Management.* Boston: CBI Publishing Company.

10  Pitter, P. and J. Chudoba. 1990. *Biodegradability of Organic Substances in the Aquatic Environment.* Boca Raton: CRD Press.

11  Tabak, H. H., S. A. Quave, C. I. Mashni and E. F. Barth. 1981. ''Biodegradability Studies with Organic Priority Pollutant Compounds,'' *J. Water Pollution Control Federation,* 53:1503.

12  Tabak, H. H. and E. F. Barth. 1978. ''Biodegradability of Benzidine in Aerobic Suspended Growth Reactors,'' *J. Water Pollution Control Federation,* 50:552.

13  Watkin, A. 1986. ''Evaluation of Biological Rate Parameters and Inhibitory Effects in Activated Sludge,'' Ph.D. dissertation, Vanderbilt University.

14  Blum, J. W. and R. E. Speece. 1990. A Database of Chemical Toxicity to Environmental Bacteria and Its Use in Interspecies Comparisons and Correlations. Vanderbilt University.

15  Templeton, L. L. and C. P. Grady. 1988. ''Effect of Culture History on the Determination of Biodegradation Kinetics by Batch and Fed Batch Techniques,'' *J. Water Pollution Control Federation,* 60:5, 651.

16  Philbrook, D. M. and C. P. Grady. 1985. ''Evaluation of Biodegradation Kinetics for Priority Pollutants,'' *Proc. 40th Industrial Waste Conference,* Purdue University.

17  Hoover, P. 1989. M.S. dissertation, Vanderbilt University.

18  Chudoba, J. 1985. ''Inhibitory Effect of Refractory Organic Compounds Produced by Activated Sludge Microorganisms on Microbial Activity and Flocculation,'' *Water Research,* 19:2, 197.

19  Volskay, V. T. and P. L. Grady, 1988. ''Toxicity of Selected RCRA Compounds to Activated Sludge Microorganisms,'' *J. Water Pollution Control Federation,* 60:10, 1850.

20  Watkin, A. and W. W. Eckenfelder. 1988. ''A Technique to Determine Un-steady State Inhibition in the Activated Sludge Process,'' *Water Science Technology,* 21:593–602.

21  Larson, R. J. and S. L. Schaeffer. 1982. ''A Rapid Method for Determining the Toxicity of Chemicals to Activated Sludge,'' *Water Research,* 16:675.

22  Williamson, K. J. and P. L. McCarty. 1975. ''Rapid Measurement of Monod Half-Velocity Coefficients for Bacterial Kinetics,'' *Biotechnology and Bioengineering,* 17:915.

23  Jenkins, D., M. Richard and G. T. Daigger. 1986. *Manual on the Causes and Control of Activated Sludge Bulking and Foaming.* Lafayette, CA: Ridgeline Press.

**24** Richard, M. G. Filaments in Industrial Wastewater Activated Sludge Treatment, Activated Sludge for Industrial Wastes Workshop 1997 WEF Industrial Wastes Tech. Conference, New Orleans, LA 1997.

**25** Palm, J. C., D. Jenkins and P. S. Parker. 1980. "Relationship Between Organic Loading, Dissolved Oxygen Concentration, and Sludge Settleability in the Complete Mix Activated Sludge Process," *J. Water Pollution Control Federation, 52:2, 484.*

**26** Kincannon, D. F. and A. Fazel. 1986. "Volatilization of Organics in Activated Sludge Reactors," *Proc. 41st Industrial Waste Conference,* Purdue University.

**27** Weber, W. J. and B. Jones. 1983. "Toxic Substances Removal in Activated Sludge and PAC Treatment Systems," U.S.EPA NTIS No. PB-86-18242J/AS.

**28** Koczwara, M. K. 1987. "Activated Sludge Treatment of Selected Aqueous Organic Hazardous Waste Compounds," *Proc. 42nd Industrial Waste Conference,* Purdue University.

**29** Sutton, P. M., T. R. Bridle, W. K. Bedford and J. Arnold. 1979. "Nitrification and Denitrification of an Industrial Wastewater," First Workshop, Canadian-German Cooperation Water Pollution Control for the 80's, Wastewater Technology Center, Burlington, Ontario, Canada.

**30** Anthonisen, A. C. 1976. "Inhibition of Nitrification by Ammonia and Nitrous Acid," *J. Water Pollution Control Federation, 48:835.*

**31** Prakasam, T. B. S., W. E. Robinson, and C. Lue-Hing. 1977. "Nitrogen Removal from Digested Sludge Supernatant Liquor Using Attached and Suspended Growth Systems," *Proceedings of 32nd Industrial Waste Conference,* Purdue University.

**32** Franta, J. R. and Wilderer, P. A., 1997. "Biological Treatment of Papermill Wastewater by Sequencing Batch Reactor Technology to Reduce Residual Organics," *Water Science Technology, 35;1:67.*

**33** Dietrick, M. J., W. M. Copa, A. K. Choudhury and T. L. Randall. 1988. "Removal of Pollutants from Dilute Wastewater by the PACT Treatment," *Process Environmental Progress, 7:143.*

**34** Briddle, T. R., D. C. Coimenhage and J. Stelzig. 1979. "Operation of a Full-Scale Nitrification-Denitrification Industrial Waste Treatment Plant," *J. Water Pollution Control Federation, 51:1,127.*

**35** Norbaitz, R. M. et al. 1997. "Pact Process for Treatment of Kraft Mill Effluent," *Water Science Technology, 35(23):283.*

**36** Daigger, G. T. and R. E. Roper. 1985. "The Relationship Between SVI and Activated Sludge Settling Characteristics," *J. Water Pollution Control Federation, 57:859.*

**37** Argaman, Y. and W. J. Kaufman. 1970. "Turbulence and Flocculation," *ASCE Sanitary Engineering Journal,* April.

**38** Paduska, R. A. 1979. "Operation, Control, and Dynamic Modeling of the Tennessee Eastman Company Industrial Wastewater Treatment System," *Proc. 34th Industrial Waste Conference,* Purdue University, p. 167.

**39** Grady, C. L. P., et al. 1996. Use of Monod Kinetics for Predicting Exposure Concentrations of Organic Compounds in Wastewater Effluents SETAC Biodegradation Kinetics Int. Conf. Port Sunlight, U.K.

**40** Sun, P. T. and Cano, M., 1997. The Fate of Xenobiotic Compounds in Industrial Activated Sludge Systems Activated Sludge for Industrial Wastes Workshop, 1997 WEF Ind. Wastes Tech. Conf. New Orleans.

# Index

**333**

For Product Safety Concerns and Information please contact our EU
representative GPSR@taylorandfrancis.com
Taylor & Francis Verlag GmbH, Kaufingerstraße 24, 80331 München, Germany